Best practices for

Wildlife Control Operators

Best Practices for Wildlife Control Operators

Paul D. Curtis
Jill Shultz

This book was based upon *Best Practices for Nuisance Wildlife Control Operators: A Training Manual*, produced by the New York State Department of Environmental Conservation, Cornell Cooperative Extension, and the New York State Integrated Pest Management Program.

THOMSON
DELMAR LEARNING™

Australia Canada Mexico Singapore Spain United Kingdom United States

THOMSON
DELMAR LEARNING

Best Practices for Wildlife Control Operators
Jill Shultz, Paul D. Curtis PhD

Vice President, Career Education Strategic Business Unit:
Dawn Gerrain

Director of Learning Solutions:
John Fedor

Acquisitions Editor:
David Rosenbaum

Managing Editor:
Robert L. Serenka, Jr.

Product Manager:
Christina Gifford

Editorial Assistant:
Scott Royael

Director of Production:
Wendy A. Troeger

Production Manager:
Mark Bernard

Content Project Manager:
Jeff Varecka

Technology Project Manager:
Sandy Charette

Director of Marketing:
Wendy Mapstone

Channel Manager:
Gerard McAvey

Marketing Coordinator:
Jonathan Sheehan

Art Director:
Joy Kocsis

Interior Design:
Judi Orozco

Cover Design:
Studio Montage

Printed in United States.
1 2 3 4 5 XXX 11 10 09 08 07

For more information contact Delmar Learning,
5 Maxwell Drive, PO Box 8007, Clifton Park, NY 12065-2919.

Or you can visit our Internet site at
http://www.delmarlearning.com.

ISBN-13: 978-1-4180-4094-9
ISBN-10: 1-4180-4094-0

Library of Congress Cataloging-in-Publication Data

Best practices for wildlife control operators / Paul D. Curtis and Jill Shultz, eds.
p. cm.
Includes bibliographical references.
ISBN 1-4180-4094-0
1. Wildlife pest—Control. I. Curtis, Paul D. II. Shultz, Jill.
SF84.3.B47 2008
639.9—dc22

2007018921

NOTICE TO THE READER

Publisher does not warrant or guarantee any of the products described herein or perform any independent analysis in connection with any of the product information contained herein. Publisher does not assume, and expressly disclaims, any obligation to obtain and include information other than that provided to it by the manufacturer.

The reader is expressly warned to consider and adopt all safety precautions that might be indicated by the activities herein and to avoid all potential hazards. By following the instructions contained herein, the reader willingly assumes all risks in connection with such instructions.

The Publisher makes no representation or warranties of any kind, including but not limited to, the warranties of fitness for particular purpose or merchantability, nor are any such representations implied with respect to the material set forth herein, and the publisher takes no responsibility with respect to such material. The Publisher shall not be liable for any special, consequential, or exemplary damages resulting, in whole or part, from the readers' use of, or reliance upon, this material.

CONTENTS

PREFACE

A Practical "How-To" Book for Wildlife Control Operators

This book focuses on the best practices for wildlife control and the prevention of wildlife damage problems. It offers a step-by-step approach to implementing integrated wildlife damage management, which is broadly accepted as state-of-the-art pest control by wildlife biologists and managers. Comprehensive and scientifically-based, this book describes a broad array of tools, techniques, and resources; an easily understood decision-making model; the basic biology and natural history of the species most likely to cause nuisance problems, with specific management recommendations for each species; and the safety, ethical, and legal issues related to wildlife control activities.

Who's It For?

This book was written specifically for wildlife control operators (WCOs), but it was designed with the needs of state wildlife agencies in mind, to serve as the core book in educational programs. Wildlife biologists and extension educators may find the book useful as a reference, especially for training staff who answer public hotlines. It's a good introduction for biologists who specialize in other areas. The species accounts in the appendices can serve as workshop handouts. Vocational programs might choose to build a course based upon this book to meet the educational needs of this rapidly expanding industry. It's also a useful reference for other professionals involved with wildlife, land, or building management, such as pest control professionals, parks department personnel, general contractors, veterinarians, and wildlife rehabilitators.

Why We Created This Book

In 2002, the New York State Department of Environmental Conservation asked Dr. Paul Curtis, an associate professor and Extension Wildlife Specialist in the Department of Natural Resources at Cornell University, to help create training materials and a licensing exam for wildlife control operators. A team of 11 professionals worked on the project. We received extraordinary support throughout the process from New York State Department of Environmental Conservation staff, the National Wildlife Control Operators Association, the New York State Wildlife Management Association, and the Zoonoses Program of the New York State Department of Health.

The New York State manual was reviewed by nearly 100 professionals from across the nation, including wildlife biologists, wildlife control operators, trappers, animal rights and animal welfare organizations, veterinarians, wildlife rehabilitators, and Cooperative Extension educators. Both the manual and its complementary test bank were validated by an educational psychologist. It's received national praise and awards from The Wildlife Society and the Association of Natural Resources Professionals, and a commendation from the New York State Wildlife Management Association.

As that project proceeded, twelve state wildlife agencies and two national trade associations inquired about the possibility of adopting this curriculum. At the Northeastern Fish and Wildlife Conference in 2002, the project team was strongly encouraged by wildlife agency administrators to create a regional or national version. Many states face increasing public pressure to strengthen their oversight of this industry but lack adequate educational materials and the resources to develop them.

The NYS Department of Environmental Conservation generously agreed to grant permission for a revised edition. With additional support from the Northeastern IPM Center, two members of the Cornell team, Dr. Curtis and Jill Shultz, a science writer at Cornell's New York State Integrated Pest Management Program, revised the work to meet the needs of a general audience. Many people graciously shared their expertise and time to help us create both editions.

How To Use This Book

This book is meant to help WCOs learn the best practices approach to wildlife damage control and to guide readers to useful information sources. Its first six chapters describe the best practices approach and many of the tools and techniques used by wildlife control operators. The appendices include wildlife fact sheets that summarize basic biological facts and control techniques for each species. They were written specifically with WCOs in mind, to answer questions that other sources don't address. A list of equipment suppliers, professional organizations, government agencies, and other resources can also be found in the appendices.

Please note: This book covers a broad array of methods used in wildlife control. The use of some of the tools or techniques may be restricted or illegal in your state. Check your state and local laws and regulations.

The fact sheets focus on the wildlife species that are most likely to come into conflict with people in the eastern United States. This covers the area from Maine through North Carolina, west up to (but not including) the Plains States. Many of these species will be handled by WCOs in other regions of North America also, but they may need to address other locally important species and wildlife diseases. (Plague, which is a concern in the western U.S., is not prevalent enough in the East to be included in this book, for example.) But the strategy—best practices—will work anywhere.

You need to understand the best practices approach, which is also called "integrated wildlife damage management," to make good use of the wildlife fact sheets. They're presented in a streamlined fashion to make it easier for you to quickly find information. So, for example, the chipmunks fact sheet advises the use of 10 × 3 × 3 inch Sherman traps but doesn't explain why you might choose that tool or how the use of a live trap would fit into your overall strategy. That's covered in the first six chapters.

The fact sheets include information about wildlife activity patterns that has been generalized for the region. You may want to tweak them to better match your local situation because even within a state, the timing of certain activities may vary between far northern and far southern locales. Climate, weather, and the availability of food are among the factors that influence the behavior of wildlife. Those elements will vary from place to place, and from year to year. Local field guides and wildlife biologists can provide you with more precise information about wildlife activity patterns in your area.

The format we've chosen for this book works for many people, but if it doesn't match your learning style, ignore it and follow your own methods. How you learn the material doesn't matter much—what counts is how well you master the subject.

Each chapter follows this format

- Learning objectives
- Chapter text
- Optional activities
- Review questions

Chapter Two focuses on the decision-making model. Chapters Three and Four describe some of the important aspects of your job that influence your choices. Then we get down to the details in Chapter Five, which gives you an overview of the tools and techniques of wildlife control. Finally, in Chapter Six we discuss professionalism, the attitude that will help you continue to use and improve your best practices. Chapters Four and Five cover a lot of material, so we've broken them into sections. You may wish to tackle them one section at a time.

Before you launch into the chapter, read the learning objectives. They may not make sense yet but even so, they'll give you a sense of what's coming and what matters most. If there's a term you don't know, check the glossary. When you're done reading the chapter, answer the review questions, which will help you decide if you have mastered the material.

Our Vision for Wildlife Control

We hope that in the future, more WCOs will have the opportunity to use their special combination of skills and knowledge to help people learn how to prevent wildlife damage problems from happening in the first place. Stronger marketing of inspection services and prevention packages is a natural development for the industry.

Many people now believe that it's sensible to have their furnaces inspected every year. They don't wait for a problem—they rely on that inspection to help prevent one. This same approach works for wildlife control. Some WCOs now routinely inspect their customers' buildings and grounds and advise them about situations that might lead to wildlife problems. In many cases, a simple but timely repair, such as replacing a torn screen, can prevent a problem from developing. Currently, the majority of customers seeking this service manage large properties, such as apartment complexes. They see the financial and social value of the service. In time, perhaps even individual homeowners will hire WCOs to conduct a yearly inspection of their properties.

Many good reasons exist to develop this part of a WCO business. Inspection and "animal-proofing" services could be offered during slower periods in the year, addressing cash-flow concerns while also developing the WCO's client base. Because these jobs are not emergencies, they can be scheduled with flexibility. With a steadier year-round income, some full-time WCOs might choose to focus on a smaller geographic area, which could make their jobs easier. These services can also be lucrative, and they provide opportunities to generate positive publicity about wildlife control. It's a great business opportunity that promotes wildlife conservation and helps to prevent property damage and wildlife conflicts.

A well-trained WCO is the right person for the job. Wildlife biologists often lack the construction knowledge needed to properly animal-proof a structure. Building contractors usually don't know enough about wildlife or exclusion devices. Other pest management professionals may include only a few select wildlife species in their business services, but probably can't match the knowledge base of an experienced WCO, so during an inspection, they might miss problems that would be evident to the WCO.

The authors readily admit that as a society, we're not there yet. Currently, in most cases a WCO will be hired to solve a problem that already exists. The best practices approach described in this book works equally well to help solve or prevent wildlife conflicts. It's a flexible guide that can be used by part-timers and full-timers, those who work alone or with a team, specialists and generalists. However you define the wildlife services you offer, we hope you'll make best practices a part of them.

About the Authors

Paul D. Curtis, Ph.D., is an Associate Professor and Extension Wildlife Specialist in the Department of Natural Resources, Cornell University. Dr. Curtis has coordinated the Wildlife Damage Management Program for Cornell Cooperative Extension during the past 17 years. He received a Ph.D. in Zoology from North Carolina State University in 1990 and a M.S. in Wildlife Biology from Colorado State University in 1981. His research interests include wildlife damage management in urban and agricultural landscapes, wildlife fertility control, and resolving community-based wildlife issues. Extension programming has included public policy education and a variety of wildlife-related publications and videotapes. He's won awards from The Wildlife Society, the Berryman Institute, the Association of Natural Resources Extension Professionals, Epsilon Sigma Phi, Gamma Sigma Delta Honor Society, and the American Society for Horticulture Science, among others.

Jill Shultz is a science writer, and received a M.S. in Environmental Science from Antioch University in 1988 and a B.S. in Biology from Cornell University in 1985. For the New York State Integrated Pest Management Program (part of Cornell Cooperative Extension), Ms. Shultz developed a broad range of educational materials on wildlife, agricultural, and home pests. She's also managed wildlife sanctuaries, taught environmental science, served as program director for a statewide literary organization, and worked as a zookeeper. Ms. Shultz has won awards from The Wildlife Society, the Association of Natural Resources Extension Professionals, and the U.S. Environmental Protection Agency, among others.

Northeastern IPM Center

Acknowledgments

This publication is supported, in part, with funding from the Northeastern Integrated Pest Management Center. The Northeastern IPM Center fosters the development and adoption of IPM, a science-based approach to managing pests in ways that generate economic, environmental, and human health benefits. We work in partnership with stakeholders from agricultural, urban, and rural settings to identify and address regional priorities for research, education, and outreach.

The New York edition was funded by the New York State Department of Environmental Conservation, Bureau of Wildlife, and was developed in cooperation with the Department of Natural Resources, Cornell Cooperative Extension, and the New York State Integrated Pest Management Program.

Project Team for the New York Edition

New York State Department of Environmental Conservation: Gordon Batcheller, project leader; Louis Berchielli, project leader; Steve Heerkens; Mark Kandel; Ron Newell; Michael Putnam; and Scott Smith.

Cornell University: Paul D. Curtis, project leader, Department of Natural Resources; Lynn A. Braband and Jill Shultz, New York State Integrated Pest Management Program.

With additional support from: Bill Altman, Psychology and Human Services Department at Broome Community College (curriculum and test design, validation of manual and test bank); Raj Smith and Siwing Tsoi, Cornell University (website development); and helpful review comments and suggestions from Diana Bryant; Rich Chipman; Rich Daniotti; Anthony DeNicola; Susan Greene; Tim Julien; Carrie Koplinka-Loehr; Don LaFountain; Al LaFrance; Kirk La Pierre; Lois Levitan; Frank Phillips; Jerry Pickel; Chris Ryan; Jack Ryan; Dale Stockton; Paul Stringer; Dr. Amy Willsey, DVM; Dr. Gwendolyn Wollney, DVM; and Greg Yarrow.

We offer our deepest gratitude to the reviewers of both editions, who donated countless hours of their time—in one case, with retirement only a few weeks away. Their perspectives, expertise, and attentiveness greatly improved this book. In addition, 83 volunteers from around New York State participated in the validation of the manual and the test bank. We are also deeply grateful to the WCOs who allowed us to accompany them in the field, answered questions, shared tips and photographs, and provided encouragement.

Special appreciation is granted to the following reviewers who provided their input on the revision: Thomas Barnes, University of Kentucky; Becky McPeake, University of Arkansas; and Paul DeBow, DeBow Wildlife Service.

Illustrations were provided by: Donna Curtin and Susan MacKay, freelancers; Jim Engel and Karen English, New York State Integrated Pest Management Program; Joe Goodman, Michael Stickney, and Wayne Trimm, NYS Department of Environmental Conservation.

Photographs were provided by: Most of the photographs are used with the permission of the Connecticut Nuisance Wildlife Control Operators

Association, Inc. Additional photographs provided by Bird Barrier, Inc., Paul D. Curtis, Lynn A. Braband, Scott Smith, and Angie Berchielli.

The following graciously allowed us to excerpt from their work:
Centers for Disease Control and Prevention (in Chapter Four)
Dr. Robert H. Schmidt, Utah State University (in Chapter Six)

This book is a synthesis of information, some of which was gleaned from these fine sources:

- NYS Department of Environmental Conservation, Region 1 NWCO training course materials, 2002.
- State of Connecticut, Department of Environmental Protection, Wildlife Division, *Nuisance Wildlife Control Operators Training Outline and Policies and Procedures Booklet*, 1999.
- *A Guide for Certifying Wildlife Control Technicians* (2002 unpublished draft). T. Julien and K. Clark, eds. National Wildlife Control Operators Association.
- *Prevention and Control of Wildlife Damage.* S. E. Hygnstrom, R. M. Timm, and G. E. Larson, eds. 1994. University of Nebraska Cooperative Extension.
- *North Carolina Wildlife Damage Control Agent Certification Notebook*, 2002. NC Wildlife Resources Commission and the NC Cooperative Extension Service.
- University of Kentucky, Department of Forestry, Cooperative Extension Service. Eastern Nuisance Wildlife Control Operators Shortcourse.

The authors welcome comments

Please contact Dr. Paul D. Curtis with suggestions.

Department of Natural Resources
Cornell University
114 Fernow Hall
Ithaca, NY 14853
E-mail: pdc1@cornell.edu

Paul D. Curtis
Jill Shultz

Chapter 1
Introduction

Please note: This manual covers a broad array of methods used in wildlife control. The use of some of the tools or techniques may be restricted or illegal in your state. Check your state and local laws and regulations.

LEARNING OBJECTIVES

1.1 Describe one way that you will have to refine the wildlife fact sheets at the end of this book for your use. List two sources for this additional information.

1.2 Name one way in which hunters, trappers, and wildlife control operators contribute to statewide wildlife management efforts.

1.3 List three public attitudes about wildlife.

1.4 Define "best practice," "culprit," and "nuisance species."

Speaking the Same Language

The authors are fascinated by wildlife, as are most wildlife control operators (WCOs). Yet this manual, indeed, the very field of wildlife damage control, focuses on conflicts between wildlife and people. Some may consider that a negative way of viewing wildlife. Is it?

Not when it's considered in its proper context. Wildlife control (or "wildlife damage management," as some prefer to call it) is part of the larger field of wildlife biology. In fact, this work supports wildlife conservation and wildlife population management programs. Unfortunately, like many branches of science, wildlife biology has become highly specialized. And with specialization comes the risk of fragmented communication. That can be a problem for WCOs and biologists, because the public may not make the connections between this work and wildlife conservation efforts.

Add in the natural tendency of professionals to use jargon, and it's easy to see how misunderstandings can arise. Jargon can be a great way to communicate complex ideas with just a word or two. It can also be a barrier to communication. Sometimes jargon is just incomprehensible. (What's a "thin space" to anyone but a typographer?) But when a common English word becomes industry jargon, people may think they know what you mean when they really don't. We've tried to minimize our use of jargon, but it's hard to avoid entirely. We also use one phrase to mean two things (but you won't be confused, honest). This is what we mean when we say:

Best practices:

1. The strategy (also known as a "decision-making model") that helps WCOs develop a customized solution for their customers' wildlife damage problems. This is also called the "integrated wildlife damage management theory."
2. The individual techniques used by a WCO that are considered the best options for accomplishing a particular goal in that situation.

For example, to live trap an animal, a WCO might select from among the many designs of cage traps, multiple capture traps, foothold traps, nets, and snares. The choice should be suitable for the species and the situation. A snap-trap sized for a mouse is not suitable for an adult rat. Using a live trap that was specially designed for raccoons to catch

Canada geese would not be a best practice. Because skunks are more likely to enter a covered trap, some WCOs will use a plastic box trap—but only when the weather is cool, because box traps can overheat, killing the animal inside.

Please remember that the best practices approach endorses the use of a broad array of tools and techniques, including lethal controls. Check your state and local laws. You may need a special permit or license or may not be able to legally use some of the best practices described in this book.

Culprit: The specific animal that is responsible for the problem. This is one of the most important ideas in wildlife damage control: you must deal with the right animal, the one that actually caused the problem, not any member of that species that just happens to live in the same neighborhood. (Please note that some problems may be caused by a group of animals, not just an individual.)

The word "culprit," as it is commonly used, directly links a particular individual to a specific event. Evidence establishes an individual as a culprit. Some people prefer the phrase "target animal" or "nuisance animal."

Nuisance: An annoying situation that doesn't involve property damage. So, for example, an animal that makes a lot of noise may be causing a nuisance, according to some people. Others might not mind. Check your state laws regarding the legal status of each species that you handle.

Nuisance species: These are the wildlife species most likely to contain individuals that will run afoul of people. Not all species of wildlife are equally likely to come into conflict with people. The ruby-throated hummingbird rarely troubles anyone; European starlings often do. We'll explain this idea in Chapter Two.

Bear with us, there's some more important background information. As we said before, to understand wildlife control, you need to consider it in its proper context.

The Big Picture: Wildlife Management in the East

In the past, government agencies and Cooperative Extension staff handled most wildlife damage complaints. But the demand for these services grew at a phenomenal rate, far outstripping their ability to respond. WCOs met the expanding need. For example, in New York State, the industry grew over 300% from the mid-eighties to the mid-nineties, according to Cornell researchers. The Pennsylvania Game Commission reports a doubling of the number of license holders from 1992 to 2004.

Now, many wildlife damage complaints are handled by wildlife control operators. In some places WCOs are agents of the state, while elsewhere they are private businesspeople. Regardless of the details, your partnership with the state wildlife agency matters.

Most state wildlife agencies juggle several goals for their environmental programs. They want to promote habitat protection, wildlife conservation, and the sustainable enjoyment of nature through a variety of activities—while minimizing conflicts between people and wildlife.

Your job presents many opportunities to contribute to statewide efforts to manage wildlife. First, there's the obvious one: you solve some of those conflicts between people and wildlife. But there's much more at stake. The success of wildlife management depends on the public believing that it is possible to balance the needs of people and wildlife. When you solve your customers' problems in a responsible, scientifically-based manner using the best practices approach, you can prove the case. And you can do that right in their homes, when they're really paying attention.

So when you can, take the time to promote the benefits of sound wildlife management to your customers and the public. Share not only your knowledge of wildlife, but also your appreciation for it. You won't get paid extra for that. Why bother? Because you may be the first—or only—"wildlife

WCOs solve mysteries

Sometimes the customer's problem is obvious, such as the raccoon in the bedroom (Figure 1–1a). The nest in the attic fan poses several potential problems, including one your customers might overlook: it's a fire hazard (Figure 1–1b). The nest stops the fan's blades from moving, which could cause the fan to overheat and ignite the nest. See the mothballs in the damaged gutter in Figure 1–1c? That's a clue to a failed control attempt. Investigation is the first step in the best practices approach.

person" your customer ever meets, especially in urban areas. After you've solved a problem that's been hurting someone's quality of life, you have a lot of credibility. And one-on-one conversations can be very persuasive.

Wildlife control operators use some of the same tools and techniques used by fur-trappers, hunters, wildlife biologists, and wildlife rehabilitators. Some people assume that anyone holding a trap is a trapper, or that everyone who uses a firearm is a hunter. So whether or not you intend it, your actions may reflect on many others, including the wildlife control industry as a whole and your state wildlife agency. All of these groups play an important role in wildlife management in your state. It's something to think about.

After suffering property damage or a nuisance situation created by wildlife, some people may become hostile to the presence of wildlife. Maybe "hostile" is a mild term. Perhaps a customer has been unsuccessfully trying to rid his attic of flying squirrels. With every day he becomes more upset. By the second week, it's not enough to remove the den from his attic—he wants to destroy every flying squirrel on his block. Then he backs his car into the mailbox because he's so sleep-deprived from all their nocturnal shenanigans. He decides it would really be better to wipe out every last squirrel in the neighborhood. By the time he calls you, he's not embarrassed to say that he thinks the state ought to be doing something about those animals because they're no good and it's just a matter of time before they cause trouble. Really, the state ought to get rid of all of them.

He turns on the TV and listens to a report about the state wildlife agency's plan for protecting piping plovers. Sections of his favorite beach will be off-limits just when he'll be on vacation. "They can't take care of the squirrels that are ruining my home but they want to use my tax dollars to take care of some stupid little bird? Why don't you protect *my* habitat," he mutters.

So you can see that science, its application, and public attitudes connect wildlife control and wildlife management.

Public Attitudes about Wildlife and Its Management

"Half your business is wildlife—the other half is people," advises one wildlife control operator. That's true on many levels. Your business success will probably depend as much on the way you talk to your customers as it does on the clever way you assess situations and modify cage traps. Public opinion about wildlife control influences the state legislature, which sets most of the laws that regulate your industry. Ideally, those laws reflect good science and good sense.

There are three things you need to know about the public's attitude toward wildlife and wildlife management. Many surveys have shown that:

1. The public cares deeply about the treatment of wildlife.
2. The public does not take the killing of wildlife lightly.
3. The public doesn't really understand trapping, hunting, wildlife management, or the science of wildlife biology.

FIGURE 1–1a

(*Connecticut Nuisance Wildlife Control Operators Association, Inc.*)

FIGURE 1–1b

FIGURE 1–1c

The public cares about wildlife on two levels. They care about the way individual animals are treated, and they care about the health and success of wildlife populations.

Because the best practices approach will guide you to humane methods that are effective, you can reassure your customers that you can solve their problems and treat the individual wild animals in the most humane way possible for the circumstances. What if your suggested strategy includes lethal techniques and they're worried about how this will affect the local wildlife populations? All of the species described in this manual are abundant. Killing an individual animal will not unduly harm the overall population. It may sound coldhearted, but it is reality.

Some people may have moral concerns about killing any animals. They may understand and accept the science but they choose to use a different standard. If you do not feel you can honor a particular request and provide an effective solution, don't take that job. (In cases where public health or safety is threatened, you may be obligated to act, following the directions of the health department.) However, if your customers have taken a particular stance because they think something is "good" or "bad" for wildlife, and their belief runs counter to the best available science, please try to provide them with resources to make an informed decision.

Like you, fur-trappers, hunters, wildlife rehabilitators, and wildlife biologists all contribute to the maintenance of healthy, sustainable wildlife populations. It is in everyone's best interest to promote sound wildlife management and to help the public appreciate the important role that each group plays in the state's wildlife management efforts.

One common misconception is that regulated harvests are bad for wildlife populations. Yet annual, regulated harvests are one of the most practical ways to manage game species; this may also help prevent conflicts with people. For that to make sense, a person needs to understand a little bit about the lifestyle and biology of that animal, and the interactions between wildlife and their habitats.

There's only enough food, water, and shelter to support a certain number of individuals of each species within an area. That number is called the "**carrying capacity**." It changes with the seasons and over time. A 10-acre swamp may be able to support 5 pairs of nesting wood ducks during the summer when food is abundant, but none during the winter after ice covers the open water. If it's a mild winter, the carrying capacity may not drop as much because there's more food or cover available than usual.

How does this relate to hunting and trapping and wildlife control? Many species produce far more young than could possibly survive in that habitat—way beyond the limit of its carrying capacity, to put it in wildlife biology terms. To a large degree, predators and other natural forces take care of the excess production each year.

Unfortunately, in many places, particularly suburban areas, there are too few predators to provide a check on the abundance of some prey species. Another problem is that we have so altered the landscape that some of the other natural forces that used to limit population growth are no longer as effective in stabilizing the numbers of certain species. For example, there are limits to the amount of natural food that's available to wildlife. Yet if people provide a feast for wildlife at bird feeders or in gardens with fertilized plants, animal densities in these subsidized habitats may far exceed those found in more natural areas.

We could, and should, do a better job of managing garbage and keeping buildings in good repair so they're not vulnerable to invasion, which would reduce the amount of artificial sources of food and shelter that's allowing some wildlife populations to grow. That's the kind of advice and assistance that WCOs offer. In addition to solving the damage problem, WCOs help in the maintenance of healthy, sustainable wildlife population levels.

We also rely on the managed removal of a certain number of animals through hunting and trapping to help keep certain wildlife populations within a healthy limit. The same scientific principle (carrying capacity) applies to these groups.

Carrying capacity is one of those common-sense ideas that many people will understand easily. WCOs work to reduce the artificial sources of food and shelter that raise the carrying capacity of an area, while hunters and trappers assume the role of predators.

Reassure people that only abundant species can be hunted or trapped—not endangered or threatened species. Like wildlife control and wildlife rehabilitation, hunting and trapping are highly regulated activities. And those regulations are based on science and socially acceptable practices.

Trapping is an important part of wildlife control, management, and research. Many people don't realize that wildlife biologists trap animals to conduct research that is often vital to the protection of wildlife populations, the environment, and sometimes even our health. (For instance, scientists used carcasses donated by trappers to collect data about the levels of the dangerous chemical, PCB, in the environment.) Wildlife restoration projects often

Does setting a trap make you a "trapper"?

WCOs and fur-trappers use some of the same equipment but there are important distinctions between these groups. Both are forms of regulated trapping but different laws apply. Wildlife control work isn't restricted by trapping seasons.

Traps are *a* tool for WCOs, not *the* tool. Just look in the truck! Sure, there are traps, but you could also find catchpoles, hardware cloth, one-way doors, expanding foam sealant, and chimney caps. That's because WCOs are there to fix a problem. To do this, they often trap animals, but they might repel or exclude the animals instead.

WCOs use cage traps much of the time, while fur-trappers most often use foothold or body-gripping traps. What WCOs hope to capture differs, too. They usually target different species. Fur-trappers don't deal with birds or snakes at all. There are only a few species that both groups commonly handle, such as raccoons, skunks, beaver, and muskrat. (Some WCOs may handle complaints about other furbearers, such as mink, weasels, coyotes, and foxes.)

One of the easiest differences to spot is location. Most WCOs work in urban or suburban areas and often set traps inside a building or on the roof. Fur-trappers usually trap in rural areas and more natural habitats.

rely on trapping. River otters, for example, were once nearly impossible to find in most of the Midwest, where they had once thrived. Otters were trapped in states that had many, such as Maryland, and relocated to 18 states that had lost their otter populations, such as Ohio. Now, they're making a comeback in the Midwest.

Those positive aspects of trapping all relate to wildlife populations. But the public also cares deeply about what happens to individual animals, which may raise concerns about hunting and trapping. Because fur-trappers and WCOs use some of the same tools and techniques, public misconceptions about trapping are of great concern to many WCOs.

Gunshot is considered a humane method for killing wildlife when it's done correctly. What about traps? Right now, the international Association of Fish and Wildlife Agencies is conducting research to evaluate traps and trapping methods. They're following a protocol developed by the agency that sets product standards for many industries worldwide (the international Organization for Standardization, known for its "ISO" certifications). This best practices research is providing data about the selectivity and efficiency of the traps and the welfare of animals caught in them. They're also looking at whether those traps are practical and safe for the trappers. The effort has already resulted in new recommendations to improve traps and trapping methods.

In Chapter Five, we will address the advantages and disadvantages of particular traps and trapping methods. Know that in most cases, your skills will determine whether a trapped animal is treated humanely. Regardless of the advertising, there are no traps that can guarantee the humane treatment of wildlife. The tools must be chosen wisely and used correctly.

For scientific reasons, state wildlife agencies and wildlife biologists focus on the fate of wildlife populations, not that of individual animals. Hunters and trappers play a crucial role in helping state agencies to achieve their management goals for wildlife populations.

Wildlife control operators and wildlife rehabilitators deal with the fate of individual animals. Wildlife rehabilitators rescue and care for orphaned, sick, or injured animals, hoping to release them back into the wild. Both WCOs and wildlife rehabilitators provide services in response to public concerns about wildlife.

Although the public cares very deeply about wildlife, as the interest in rescuing even an individual animal shows, the increased demand for wildlife control services indicates that they also want to be able to stop wildlife damage. That's where WCOs fit in. Sometimes they're hired by communities or work with federal or state agency staff to assist in large wildlife control efforts, but usually they're hired by a homeowner to solve a problem that affects just one property. Because WCOs may be involved in wildlife control efforts at the level of populations or individual animals, it helps to understand both the biology and the social implications of the strategies and methods.

There are important ways in which WCOs differ from hunters, trappers, and rehabilitators. First of all, most WCOs are running a business. Some trappers may try to earn a living from trapping but that's increasingly rare. While WCOs are paid for their services, hunters and trappers pay to pursue those activities—and those fees help support other wildlife management programs. Wildlife rehabilitators must rely on donations; they are not allowed to charge for their services.

As we suggested in the preface, WCOs can offer a service that helps promote wildlife conservation and prevent wildlife damage at the same time. They can inspect homes and properties for signs of vulnerability to wildlife invasion and advise their customers about the risks and options for addressing the situation. Better still, they can "animal-proof" the site. And frankly, no one is better suited for this task than an experienced, well-trained WCO.

All of these groups—wildlife biologists, wildlife control operators, hunters, trappers, and wildlife rehabilitators—contribute to efforts that help people coexist with wildlife. Without any one of these groups, state agencies would find it harder to achieve their environmental goals. There are natural connections between these groups, especially because some people belong to more than one of them. Many members of these different groups care deeply about wildlife and work for its conservation.

Other People to Consider

Many seasoned WCOs advise new professionals to develop networks right from the start. Earn the trust and respect of your regional state wildlife agency and law enforcement staff, local police, animal control officers, health department staff, wildlife rehabilitators, veterinarians, the staff at local animal shelters, town clerks, and your fellow WCOs. Some of these professionals may assist you during an emergency, such as the local police and animal control officers. There are situations that require you to follow the health department's directives, so it's helpful to have a cordial relationship with them.

You're probably aware of the practical reasons to maintain friendly relations, but this is also an excellent opportunity to market your business. People often don't know whom to contact for help with a nuisance animal, so they may turn to any professional who works with animals or enforces laws. If you earn the trust and respect of these professionals, they may refer customers to you and enhance your business. You may also get invitations to participate in community events and demonstrate some of the specialized equipment available.

Wildlife control is not a nine-to-five job, especially during the busy season. Some people refer jobs to other WCOs when they're too busy to handle them. Or maybe you're suddenly confronted with a nine-foot-long snake in someone's apartment, and you're not quite sure what it is. You may want to call another professional who's more experienced with exotic species. Some WCOs partner with construction professionals, too. Say you had to cut through an interior wall to recover a trapped raccoon or squirrel. Instead of repairing the hole in the drywall yourself, which would require multiple trips to the house over a period of days, you might suggest the homeowners hire someone else to do the repair work.

These examples point to one aspect of professionalism: recognizing your level of technical expertise and making sure that you have the skills and knowledge needed to do a job well. There's a bit more to it, of course, and we'll highlight some of those issues in Chapter Six.

Some people, including customers, will consider another factor as they weigh your professionalism: how well do you get along with the state wildlife agency personnel? Do your business practices help promote wildlife conservation? Now, that question might make some WCOs' blood pressures rise because they believe that it's impossible to promote wildlife conservation and control wildlife damage. That's a myth.

In fact, you can even accomplish both goals on the same property. For example, if your customer suffered property damage caused by moles, she could still plant flowers that would provide food for hummingbirds or install a bat box. She could leave a fallen log or standing dead tree in place; both provide valuable shelter for many wildlife species. Using their knowledge and skills, WCOs can help people achieve both goals.

This book promotes the best practices approach to wildlife control. It's a step-by-step method that will help you investigate the situation and evaluate the possible solutions in the context of the "big picture."

To learn more...

These are some optional activities that may help you explore other perspectives about the topic:

- Check a few websites that relate to wildlife damage management, such as the Internet Center for Wildlife Damage Management (see Appendix E).
- Develop a presentation for a local school class to explain what a WCO does at work. (You don't have to actually do the presentation, just consider how you'd describe your job to a six-year-old.)
- Review the websites of a wildlife rehabilitation group, a trapping organization, and an animal welfare organization. How do their perspectives about the control of nuisance wildlife differ from your own? Consider joining these groups and participating in their activities.
- Get to know other WCOs. Join your state wildlife control association, sign up for a listserve, or attend meetings and conferences.

Review Questions

Optional. Answer these questions to see if you've learned the material.

1. What's a best practice?

a. It's a method that's cheap and fast.

b. It's a way to solve a nuisance problem that doesn't hurt the wildlife.

c. It's the simple, single solution for all the problems you'll encounter.

d. It's an effective way to solve a nuisance problem that minimizes the risks to people and the environment.

2. You don't have to worry about your customers' opinions because most people don't care about wildlife. (Circle the correct answer.)
True False

3. How do the state's wildlife management efforts benefit from trapping?
   a. Trapping is a good way to capture animals that will be relocated to reestablish a wildlife population in a different area.
   b. Trappers provide furs that are used in school programs and as door prizes.
   c. Trapping solves all of the wildlife complaints, allowing biologists to spend more time on other programs.
   d. There is no connection.

4. You may partner with another professional to finish a job because:
   a. You don't have the proper equipment, but your roofer friend does.
   b. You're not as comfortable handling that species.
   c. It's your busiest season, and you don't have the time to spare for the building repair or clean-up services your customer needs.
   d. Any of the above.

5. To use the wildlife fact sheets most effectively, you have to:
   a. Pick one control technique from each category.
   b. Attend a wildlife biology course.
   c. Consider them in terms of the best practices strategy.
   d. Follow them exactly.

Chapter 2
Balancing the Needs of People and Wildlife

LEARNING OBJECTIVES

2.1 Name the six major questions that WCOs should think about as they decide how to handle a job.

2.2 Name at least two reasons why an animal would be considered "nuisance wildlife."

2.3 Explain what attracts wildlife to a particular site. Can you sum it up in two words?

2.4 Define "humane."

2.5 Describe one ecological concern related to a WCO activity.

"Life is Robbery"—Alfred North Whitehead

Prey eat plants. Predators eat prey. Individuals of the same species fight to secure shelter and prime territories. The web of life is defined by the flow of resources from one individual to another, a form of "robbery" that is fundamental to survival. It is also the key to understanding why wildlife damage problems occur.

Many wildlife conflicts can be explained with two words: food and shelter. Provide these crucial resources intentionally or accidentally, and some wild animal will probably accept the invitation. Surprised? As human development spreads, wildlife conflicts happen more frequently. As the competition for natural sources of food and shelter increases, some animals seek their living in our world. At the same time, some species adapt to these new opportunities so well that their populations rise dramatically, further increasing the chance of a conflict with people.

Some wildlife species are more likely than others to come into conflict with people. Adaptable species that tolerate the presence of humans are far more likely to take advantage of opportunities to find food and shelter in our neighborhoods than are reclusive wilderness species. Finicky animals that rely on a highly specialized diet may not find what they need near us. Yet to a species that can thrive on whatever's available, garbage collection day may turn a neighborhood street into a free all-you-can-eat buffet. In addition to its diet and home range requirements, the type of shelter preferred by a wildlife species may help determine whether or not it will thrive in human environments. Most wild animals just mind their own business and never cause a conflict with people. In fact, many people deeply enjoy their interactions with wildlife.

Some wildlife can find what they need to survive near us, while others cannot. The ones that can adapt to human situations aren't maliciously trying to annoy people. So it's not that some species are "bad" and others are "good." To a large degree, we've created this problem.

Or are nuisance animals just "entrepreneurs"? If a businessperson recognized and then aggressively pursued a market niche, making a lot of money as a result, would you applaud this entrepreneur's success? How does that exploitation of opportunity in the pursuit of valued resources differ from the actions of a squirrel raiding a bird feeder, which is easier than seeking wild sources of food, or a raccoon choosing to den in an attic, which is less accessible to its predators than a tree cavity?

Often, species are not viewed as pests in their normal habitats but may clash with people when they enter our world. How many people worry about a skunk in the woods? Yet a skunk under the porch is an entirely different matter.

So when does a wild animal become a nuisance? Are certain species always nuisances while others never qualify? There's no universal answer because people react differently to wildlife. Consider a huge population of Canada geese in a city park, whose droppings foul lawns and ball fields. Some people are concerned about catching diseases from the birds; others feel they've been deprived of the use of the park because they find the bird droppings disgusting. One group may seek a strategy for controlling the goose population but another may feed the birds during lunch breaks. These people enjoy their daily contact with nature and may react angrily to any attempt to remove the geese. Who decides what to do?

In some states, landowners have the right to decide whether a particular animal is a nuisance. But that doesn't mean that if someone has overreacted, you can't try to help them see the situation in a different way. We hope you will.

State regulations may also set standards, which may vary by species, so ask your state wildlife agency for the details. (They may have already provided that information as a supplement to this book.) If public health or safety is threatened, a public official may have the right to declare an individual animal a nuisance and take appropriate action.

Property damage is usually readily observable, but how do you decide if a wild animal's behavior qualifies as a nuisance? State agencies do not intend for wildlife to be controlled simply because someone is intolerant of their presence. A red fox crossing your yard does not constitute a menace. A squirrel sitting on the railing of a deck that borders a woodlot is not destructive. Those are normal behaviors. Wildlife control activities that require federal permits (e.g., taking of migratory birds) must meet the guidelines set out by the federal agency that issues the permit.

Typical Wildlife Damage Problems

For most people, an animal is considered "nuisance wildlife" if it is destructive or menacing. It may damage property such as buildings, crops, pets, livestock, gardens, or public parks. Wildlife may threaten human health and safety by spreading diseases, through direct attacks, or accidentally, through collisions with cars, airplanes, or trains.

According to a 2001 report by the U.S. GAO (the federal General Accountability Office), collisions between deer and cars caused over $1 billion in damage each year. Worldwide, another billion is lost each year due to birds colliding with airplanes. Beavers, woodchucks, squirrels, and other species cause additional millions of dollars of damage to roads, bridges, dams, and electrical utilities. Estimates for the damage to U.S. agriculture range from $600 million to over $1 billion annually, with more than half of all farmers and ranchers experiencing some type of wildlife-related damage each year. (From "Wildlife Services Program, Information on Activities to Manage Wildlife Damage," GAO-02–138.)

Some nuisance species may even threaten the survival of other wildlife or destroy their habitats. That's a significant concern in areas with endangered or threatened species. In several Northeastern states, researchers worry about raccoons and gulls preying on the eggs of the federally endangered piping plover. They also believe that excessive browsing by white-tailed deer may play a significant role in the disappearance of some rare wildflowers. So to ensure greater success for wildlife and plant restoration programs, federal or state agencies may control certain nuisance species—sometimes with the help of WCOs—a direct example of how wildlife control efforts can help promote wildlife conservation.

To protect ...	The agency might control these nuisance species:
Piping plover	Red fox, raccoon, striped skunk
Least tern	Red fox, raccoon, coyote, opossum
Sea turtles	Raccoon, coyote
Rails	Red fox
Nesting waterfowl	Red fox, raccoon, striped skunk
Karner blue butterfly	Woodchuck

Hunting and trapping may help reduce the populations of some nuisance species and alleviate potential conflicts. However, if the damage is caused by species that aren't normally harvested, such as red squirrels, or if the problem happens in an urban area, or outside of the regular harvest season, other control methods will be necessary. That's where WCOs fit in. This book focuses on the wildlife species that are most likely to come into conflict with people in the eastern United States—only about three dozen species out of hundreds of mammal, bird, and reptile species.

Best Practices: The Theory and Its Techniques

There is no magic pill for nuisance wildlife control, no single, simple remedy that can be relied upon to solve all of the problems you'll encounter. At times, you may even need a different solution for the "same" problem. Your favorite method for removing a raccoon from a chimney may not work at a particular house, for example, because the roof is rotted and unsafe to walk upon. Also, effective long-term solutions often require a combination approach that includes several techniques.

Best practices are effective methods for solving a nuisance wildlife problem that also minimize risks to the environment and our health and well-being. They are based on the best available information, which includes scientific knowledge and the field experience of wildlife biologists and wildlife control operators. To use the best practices approach, WCOs must balance a variety of factors when deciding how to resolve a nuisance wildlife problem, such as the need to:

- reduce risks to human safety;
- ensure the humane treatment of wildlife;
- choose practical techniques that will be effective in this case;
- respect landowners' rights to manage their own properties;
- work in such a way that your actions will not harm (or better yet, will promote) the state's efforts to protect wildlife populations and habitats;
- consider a variety of ethical issues related to the treatment of wildlife and the possible consequences of certain control activities;
- pick methods that are legal for that locale;
- keep the costs appropriate for the customer; and
- maintain the look (or aesthetics) of a property.

The solution for a particular problem is customized to fit the situation. Which factors matter most will vary, based on both the circumstances and the customer's needs and values. For example, some customers may balk at the suggestion of a particular repair method because they think the result would be ugly and "ruin the house." Others might not care about that factor at all. In most cases, you'll be able to tailor your services to meet your customers' needs, but there are two occasions when you can't: if doing so would violate a law, and when public health or safety is threatened.

Here's a hypothetical scenario that illustrates the interplay of these different factors in a WCO's decision-making approach. Say a customer asks you to remove a fox from her property because it's crazy. She's had foxes under her deck before, but this one nipped her for no good reason. She doesn't want the fox hurt but she's afraid it will attack her grandchild, so she wants it removed. When you arrive, you notice her hand is indeed bandaged. The fox is cornered under the deck, near the door.

Foxes can infect people with rabies, a deadly disease. Although you can't be sure just from its behavior, the animal did attack without reason, which is one of the warning signs of rabies. And your customer was bitten, which is the most common way that people are infected with rabies.

As required by state law, you call the health department to report the possible exposure. They tell you to kill the fox so a sample can be collected and tested for rabies. You explain to your customer that you are legally required to do what the health department says. Your customer is not happy, but she understands. She asks that you use the most humane method to kill the fox. Of the methods available to WCOs, two of the humane choices for killing foxes are shooting the animal, or capturing it and killing it with carbon dioxide in a chamber. Is either a best practice in this situation?

How would you safely transfer the fox to the carbon dioxide chamber? Your top concerns are the need to minimize your risk of being exposed to rabies and the possibility that the fox could escape, because if it did, your customer would have to get the preventive rabies shots.

A cage trap would offer you the most protection while handling the fox, but it's hard to catch adult foxes in cage traps. You don't want to risk a "miss" that might make this animal trap-shy because you really need to capture this particular fox. Catch poles work well for restraining foxes, and you'd be happy to use yours, except there's so much junk under this deck that you can't reach the fox. If you pulled out some of the debris, you might scare the fox and prompt an escape attempt, which could turn ugly. So you consider shooting it, then clearing a path for its removal.

But you can't. Local ordinances prohibit the use of firearms in this town. Also, you prefer to target the brain, but that could interfere with the collection of the sample needed for the rabies test. The heart/lungs shot, which doesn't interfere with the rabies test, is a humane target and would be a good option—if not for that local ordinance. Then you remember that the ordinance doesn't apply to animal control officers or the police. In the end, you decide to call animal control.

This example illustrates a few of the factors that you must weigh in your decisions. In this case, your customer asked you to provide the most humane death possible, but legal restrictions and concerns about human safety and the relative effectiveness of different tools also influenced your actions. This situation could easily have been more complicated.

Simplifying the approach

How can you determine the best course of action to solve the problem for your customer? The best practices approach can be boiled down to six questions that will help you evaluate your options when choosing capture, handling, transportation, and disposal methods for wildlife:

1. Is it safe?
2. What are the likely ecological consequences of this action?
3. Is it practical?
4. Is it humane?
5. Is it legal?
6. How would your actions play on the evening news?

Your understanding of these issues will deepen over time, as you gain experience working with the public and with wildlife. Here are some basic points related to each of these big questions.

Is It Safe?

Many WCOs are most concerned about safeguarding people: themselves, their customers, and the public. There are health and safety risks posed by the presence of nuisance wildlife, and by wildlife control techniques. Wild animals might expose people to diseases or parasites, for example, or cause fires or road accidents (see Chapter Four).

Those are the routine risks. Some WCOs face unusual risks because they handle exotic wildlife species, such as pet snakes left behind in dorms when the college term ends. It's not always easy to identify the species. Could it be venomous? Obviously, if you're tangling with an animal that could hurt nearby people, extra caution is needed.

The way you go about your business could also affect the safety of other species and the environment. A careful WCO can help prevent the spread of a wildlife disease into a new area or to a different species. Likewise, by using traps and pesticides carefully, that WCO can reduce the chances of catching or killing the wrong animal, or contaminating the environment.

How safe is the situation, and the method you want to use? Consider such things as:

- Is the nuisance animal sick, aggressive, or otherwise dangerous?
- Do you have the right gear for the job?
- Is the building safe? (See Figure 2–1.)
- How well can you control access to any pest control devices, such as traps or pesticides?
- Could children, pets, or other animals gain access to the animal or to your devices and injure themselves?
- Do weather conditions make the job too dangerous? (For example, icy, wet, or windy conditions can make it unsafe to walk on a roof.)

FIGURE 2–1 Roof work is one of the more common WCO job hazards. Falls caused 11% of the deaths associated with job-related injuries in upstate New York (1993 statistics from the NYS Dept. of Health). How do other WCOs stay safe and get the job done? *(Connecticut Nuisance Wildlife Control Operators Association, Inc.)*

What Are the Likely Ecological Consequences of This Action?

Think beyond your client's property. Quite a few of the techniques mentioned in this manual could affect other wildlife. Any modification of the landscape, even something as simple as mowing, will be good for some species and bad for others.

For this reason, when you're planning your strategy, try to keep your control activities contained within the smallest possible area. To do this, imagine drawing two circles around the area you're trying to protect.

The inner circle defines the area that requires the most protection. In most cases, this includes the customer's house and a barrier zone around the perimeter. In a public park, the inner circle could be a playground or the area right next to a pond. Make that barrier impenetrable. Beyond that barrier lies the rest of the property in the outer circle. Here, you can relax your defenses. If the barrier around the vulnerable area is strong, wildlife can live in that outer circle and not cause a problem.

The size of the barrier zone depends on the abilities of the species you're trying to keep out. For example, if you're trying to stop squirrels from getting into an attic, you need to trim overhanging branches to leave about a ten-foot gap between the end of the branch and the roof. Anything less and the squirrels will make the leap from the branch onto the roof. Once there, they may chew their way in. So in this case, the inner circle includes the house and a ten-foot-wide perimeter. With squirrels, it's very important to focus up high, at roof level.

If you were trying to keep mice out of the same house, a twelve-inch perimeter around the foundation would do the trick. For extra protection, you might make that a gravel perimeter instead of bare soil, because the gravel would inhibit plant growth, which would reduce the amount of shelter for the mice. So you can see that the habits of nuisance wildlife also define the type of barrier you select.

This idea is easier to picture on larger properties but it can be applied in many urban situations. In some cases, the space is too tight and neighboring properties may present problems that you can't fix, which will provide access for wildlife onto your customer's property. But property owners who have some elbow room have more options. They might want to draw a map of their property and define different goals for different sections, just as park managers do with their land. For example, the customers may want to keep all wildlife out of the house and garage, and woodchucks and deer out of the garden. But they'd actually love to see all sorts of wildlife everywhere else, from turtles and fish to bats and swallows and even deer.

The two-circle approach works because you don't need to get rid of all wildlife, everywhere—you just need to remove the opportunity for them to cause problems for your customers. Let's return to our squirrel example. If the house is well maintained and it's nearly impossible for them to get into the attic, then they don't pose much of a risk if they're nesting in a tree that's fifty feet away from the house in the outer circle.

As a WCO, you have many tools to solve nuisance wildlife problems. Good judgment is the most important one. You may well be able to prevent wildlife damage and promote wildlife conservation—on the same property, at the same time.

The location of your control activities and the size of the "managed" area are not the only factors that determine the ecological consequences of your actions. Extra care and thoughtfulness is needed with certain techniques because of their greater potential to harm other species or the environment. This is especially true when using lethal tools such as body-gripping traps or pesticides. You could accidentally capture or injure someone's pet or a protected species.

You don't want to catch any member of the wrong species, or individuals of the target species that aren't causing the conflict. You're after the culprit, remember? If you catch the wrong individuals, you haven't solved the problem for your customer.

Is it possible to catch too many animals? As we explained in Chapter One, many of the wildlife species that most commonly cause conflicts with people are abundant, so capturing and killing an individual isn't going to threaten the species' survival. But what if you're asked to remove hundreds or thousands of animals?

This could happen with starlings, crows, Canada geese, or bats. Some conservationists would be happy if you killed European starlings because they're an exotic species that competes with our native wildlife. Crows and Canada geese are abundant in this area, so wildlife biologists aren't overly concerned about the effects of flock reduction on the larger population. Bats are a different story.

The two species of bats that are most likely to cause nuisances in the region are the little brown bat and the big brown bat. In both species, females gather in large colonies to raise their young. Each female has only one or two pups per year. Typically, you'll see a few hundred brown bats in a maternity colony, but in a very successful site you might find as many as several thousand females and their pups (see Figure 2–2). Their low breeding rate, coupled with their colonial habit, makes them vulnerable. Think of the phrase, "putting all your eggs in one basket."

For this reason, we recommend a nonlethal approach to bat removal unless a situation involves a possible rabies exposure. The problem is that if there

FIGURE 2–2 The most important tool for a WCO is good judgment. A solution that works when you're dealing with 70 bats may not be appropriate when you're dealing with 7,000 bats. *(Connecticut Nuisance Wildlife Control Operators Association, Inc.)*

are bats in a house or garage, there may be bats in living areas, where they cannot be tolerated. So the solution should focus on getting the bats out of the building. Because bats can't chew holes in buildings, careful bat-proofing will do the trick. That solves your customer's problem in a way that's sensitive to the big ecological picture. Lethal control wouldn't make your solution more effective. Also, removing a large bat colony might harm local bat populations, especially if many people in the area were destroying colonies.

Sometimes, just moving an animal can cause problems. We'll talk about that technique, called "trap and transfer" or "translocation," in Chapter Five.

Is It Practical?

Some methods sound great until you actually try to make them work. And some methods are great, but customers balk at the cost. Many factors will affect whether you can actually use a certain approach, including your skill levels, the available gear, weather conditions, timing, and the cost (see Figure 2–3).

Of course, your solution has to fix the problem. After you've determined which species to target and

FIGURE 2–3 Even if you could afford everything on the market, how would you fit it into your truck? Here's a simple answer to a situation that called for a covered trap. Solutions don't have to be fancy to work. *(Connecticut Nuisance Wildlife Control Operators Association, Inc.)*

placed your devices so they'll most likely capture the offending animals and no others, you still need to figure out how many animals are probably there. If there are six and you've caught three, chances are that's not good enough.

And have you provided a long-term solution or just temporary relief? You may want to avoid quick-fix jobs because no matter what the customers say when they hire you, they might not be happy later, and might blame you. (Well-written contracts may help, especially if the paperwork includes descriptions of the options offered, the recommended actions, and the likely consequences.)

Sites sometimes present unexpected challenges. The area you really want to get to may be inaccessible. Or maybe the neighbors are feeding the squirrels that are overrunning your customer's yard. Without their cooperation, your options are limited. Your traps might be vandalized or stolen. Some sites, because of their location or use, suffer from a curse—people think the area is good for releasing unwanted animals, and don't consider the landowners' needs. Many people abandon animals at farms or on public lands, for example, and over time this can become overwhelming.

Timing can present practical problems, too. What if no one knows who owns the potential release site? Will you tell your customer to wait while someone does the research? (Who?) Or what if the owner refuses to give you permission to release animals onto that land, and it would take hours to drive to another suitable site? What if the customer wants you to deal with the problem right now, but certain techniques require federal and state permits?

Then there's the dilemma of the mysterious "someone." Your customer asks, "isn't there someone you can just take the animal to?" In nearly all cases, the answer is "no." Wildlife rehabilitators accept injured, sick, and orphaned wildlife, not healthy adult animals. Don't use them as a dumping ground for unwanted wildlife. If on-site release or relocation are not viable options, then lethal techniques may be a more appropriate solution for that situation.

Also, make sure that you don't accidentally create wildlife orphans. That's a matter of professionalism. You're there to solve problems, not create them. It's also a matter of practicality, because a good relationship with your local rehabber will serve you well. If they trust the quality of your work, then they'll be more likely to accept animals from you. Even if you work very carefully, given the sheer number of animals that you're likely to handle over your career, it's likely that a few will be injured. Perhaps you'd even be called in specifically to rescue an injured animal and transport it to a rehabilitation facility, because the rehabber lacks some of the tools that would facilitate the capture. Another thing to keep in mind is that not all wildlife rehabilitators handle the same species. In some states, they must have special training and additional permits to handle rabies vector species.

One last practical point to consider as you choose a method for capturing, handling, evicting, transporting, or disposing of wild animals: Could you make things worse? If a female who has young is removed from the den, she might cause further damage to the building in her attempts to return to her young. It takes more skill to capture certain species, such as coyotes. Careless attempts might train an animal to avoid traps, which would make it harder to capture later. Skunks are attracted to sardines and cat food, but those are lousy baits to use in a cage trap because they'll also attract house cats. A commercial sweet bait is a far better choice.

Your customers may accidentally make your job harder with a failed control attempt. Perhaps they thought the animal had left, so they closed up the hole to keep it from getting back in, only to trap the animal inside. Or maybe they had a raccoon family in the attic. They used a loud radio to harass the female so she'd move her young. It worked—but not the way they hoped. The female did feel threatened enough to seek a more secure place for her young, but instead of moving them outside, she placed them in a wall cavity.

Is It Humane?

"**Humane**" does not mean that you only use nonlethal techniques to solve a wildlife conflict. There are circumstances in which a nonlethal technique may be considered less humane than killing the animal. When we use the word "humane," we mean that you cause no unnecessary pain or stress to the animal. Through each stage of your work, whether capturing, handling, excluding, transporting, or disposing of wildlife, you act respectfully.

Most WCOs enter the industry because of their love of wildlife. If you want to work in a way that reflects that value, you'll need skills, knowledge, compassion, and a mature way of thinking. Those abilities may be sorely tested at times, because some of your customers may focus on a particular method and believe that it's always the right answer.

We'll discuss the details of handling and disposal methods in Chapter Five, but here's the big picture. If you want to treat wildlife in a humane fashion, you have to think about what this means through every stage of the job. How is the animal affected by the strategy and techniques you've chosen? What's likely to happen after you leave?

Questions about the humane treatment of wildlife usually focus on the choice between non-lethal and lethal techniques, but it's more complicated than that. Although it seems odd, a nonlethal technique will not guarantee humane treatment for the animal in all situations. For example, relocating squirrels during the summer may be appropriate, but harsh winter conditions may not give a relocated squirrel enough time to find shelter before it dies of exposure. If it escapes predators and finds shelter, it may starve to death if you've moved it too far away from its food cache. Another example: if you release an injured, sick, or highly traumatized animal in a new place, it may not be strong enough to ward off the attacks of animals that are already well established in that area. You must also consider the possibility of introducing a disease to a new wildlife population. Does humane treatment only apply to the animal that's been captured?

To make this discussion of humane treatment more complicated, there are nonlethal techniques that are meant to stress the animal. Aversive conditioning is used to train the animal to change its behavior. For example, WCOs may use border collies to chase Canada geese away from a park, or rubber buckshot to scare bears. This is a legitimate approach that may solve the wildlife conflict. It does make the nuisance animal feel some pain or stress—but only as much as is necessary to reduce a problem. You'll see this general approach described by such terms as "frightening techniques," "scare devices," "hazing," and "aversive conditioning." These techniques support other wildlife management efforts, too.

Here are a few issues to consider when you're trying to decide how to humanely capture an animal. Many types of traps are selective, effective, and humane when they're used properly. It all depends on your skill and carefulness. A cage trap can be misused and cause a cruel death, while a foothold trap could be used well and aid in humane capture. Some simple modifications to your traps and capture methods may provide more comfort to the restrained animal and increase its chance of survival. Try to protect the trapped animal from exposure to bad weather. In some cases, you'll also need to protect the captured animals from people (especially children), pets, and predators, or their own behavior.

Rain, or extreme heat or cold, can kill some animals. You can prevent many of these accidental deaths by checking your traps as soon as you can (at least once a day); by using a plastic box trap (except in the summer, when it can heat more quickly than other designs); by covering a standard metal cage trap with plywood, cardboard, burlap, towels, blankets, tarps, or aluminum; by setting the trap in a less exposed place; or by waiting for better weather, if possible.

Trap placement is critical if you need to keep children, pets, or predators away. Some people might intentionally harm an animal or release it. Kids might be too young to understand the effects of their actions. Even if all they do is watch the animal, their presence will increase its stress. If a person is scratched or bitten, that could translate into a death sentence for a mammal, should rabies testing be necessary. Pets and predators may also investigate, causing even more stress for the animal.

One trapper cautions, "a live trap is not a pillow trap." When captured in a live trap, some species will settle down while others, like squirrels and raccoons, will paw furiously at the trap or rub their bodies against the metal. Given enough time, they could rub off fur and bloody themselves. The shape and size of some new trap designs helps to minimize this problem. Checking traps frequently helps, too.

Those concerned with the humane treatment of wildlife also take special care to avoid turning young animals into orphans. This means you may need to alter your methods when there are young in the den or nest that are not independent and can't leave on their own. Some of the ways to accomplish this are species-specific. The most humane approach is to wait until the young are capable of taking care of themselves. At that point, they'll leave on their own. Sometimes when that's not an option, a nonlethal approach is still preferred. In that case, try to remove the family as a unit. If the mother is separated from the young, some WCOs report success at reuniting them by placing the young in a special "release box," from which the mother can retrieve them and move them to another shelter. (This technique will be addressed in Chapter Five.)

Some customers may request bat-proofing during the winter, when bats are hibernating. If the attic is an active hibernacula, you need to plan for the possibility that bats might be hibernating inside it. The species of bats that overwinter in buildings are small and can hide in very narrow cracks, so even if you searched carefully, you know it's possible to overlook a bat or two. If that happened and you'd bat-proofed so successfully that they couldn't find a way out of the attic when they emerged from their hibernation in the spring, the bats might fly into the living spaces to seek another escape route. This might bring them into contact with people, which is risky. Or the bats might starve to death in the attic. That's not humane, and their stinky, rotting carcasses might also attract other pests. Either scenario could create new problems for your customer.

You suggest limited bat-proofing and the use of a "checkvalve," or "one-way door." Right now, you'll seal all of the holes except for the main exit, over which you'll install the checkvalve. This device allows animals to leave but keeps them from re-entering the building. So if there were any bats hibernating in the attic, they'd have a safe escape route.

In the spring, after the bats have emerged from hibernation, you return to the house. At dusk, you stage a "bat watch" to see if any bats emerge from the attic. When you're confident that there are no bats inside the building, you remove the checkvalve and seal the hole, completing the bat-proofing. (One caution about checkvalves: don't use them during the pup-rearing season, when young, immobile bats may be left behind.)

Finally, when you have to kill an animal, choose one of the best practices described in Chapter Five whenever you can. These techniques are more likely to provide as painless a death as possible. You may achieve this using one, or perhaps a combination, of techniques. Ideally, you either kill the animal quickly, or first quickly make it unconscious, then kill it quickly.

Is It Legal?

Of course, all of your wildlife control activities must be legal under local, state, and federal laws, and must follow the state wildlife agency's regulations. There are times when the state health department also has jurisdiction over your actions, and you're legally required to follow their directions. And you must obey federal regulations. You may not take, possess, or transport most migratory birds without a federal permit, for example.

Are you, as a WCO, allowed to handle the species? Do you need other permits or additional licenses for the methods you'd like to use, or because of the location of the site? To solve a beaver complaint, you may also need wetlands permits, for example. A pesticide applicator license may be required if you'd like to use chemical repellents on your customer's property (in addition to the WCO permit).

Have you secured all of the necessary permissions? If you want to release an animal on someone else's property, do you need permission? In writing? Some people mistakenly believe they can release animals onto public land, such as parks or wildlife management areas, without permission. Nope! Talk to the property manager first.

Local restrictions might affect which techniques you use to capture or kill an animal. Shooting is frequently regulated by local ordinances, but some areas will even control the type of fences you can install.

How Would This Play on the Evening News?

"Wildlife belongs to the public. Regional and national surveys demonstrate that the public has concerns about how their wildlife is treated. Because of this, [WCOs] should manage wildlife as the public asks, whenever possible and practical. In all cases, [WCOs] will use [their] professional judgment to determine the best course of action, understanding that every damage prevention job is a customized operation."

—excerpted from Dr. Robert H. Schmidt's column, "The Professional Touch: A Professional Code of Ethics for Wildlife Damage Management" in *Animal Damage Control* 1, no. 5 (October/November, 1993): 8

Many WCOs are deeply committed to improving the industry and achieving professional recognition for their work. Communication skills are critical to this effort and to the success of your business. The greatest wildlife expert in the world won't win customers or the public's trust if that person is dismissive of other people's feelings and beliefs, arrogant, or just plain confusing.

Nuisance wildlife problems can be highly emotional for some clients. You're entering someone's home, perhaps at a time when they're not even close to being at their best. Be at yours. Let your professionalism shine. Be respectful and honest with your customers. Listen. Try to understand what they want. One thing that's probably true about most successful business people is they make their customers happy. That's a balancing act, of course. It's okay to refuse a job if someone insists on something that you think is wrong, or if you know that "solution" won't work. You also have to maintain your reputation. You'll probably have some difficult customers. It can be challenging to listen sometimes, but do your best. Remember, listening quietly to someone's point of view doesn't mean you support it.

If there are legal or practical reasons why you can't do what your customers ask, explain this to them. Describe the available options in the amount of detail that your customers request. Some customers won't care how you do the job, while others will have strong opinions, especially about the disposal of animals. They may want to hear every detail.

Unfortunately, you may find yourself in a delicate position. Your customer may approve of the strategy that you've chosen together, but what if the neighbors disagree? Even if they don't have the legal right to intervene, you don't want to cause friction between neighbors if it can be avoided. If you're using best practices—the approach and techniques that most WCOs and wildlife biologists prefer—chances are, this won't become an issue. But keep it in mind. During the busy season, when you're rushing from one call to another, don't let your careful approach slip. Would your customer's neighbor be upset if he saw a dead animal in a body-gripping trap hanging over the side of the roof? Well, wouldn't you? Perhaps a blunder like that would even cause him to reconsider his opinion of trapping in general.

Such mistakes can polarize communities and lead to a media feeding frenzy. Whether you like it or not, the public cares about how you do your job. So imagine the cameras are always rolling. What reaction would you expect if someone videotaped your work and then showed it on the evening news?

Think about the average person, who probably feels either neutral about wild animals, or likes the idea that they're "out there." There's a good chance that this person believes that people have the right to protect their families and their property, and that it's perfectly reasonable to trap and remove wild animals from a home. But maybe he doesn't want to see it. In fact, even customers who want lethal techniques integrated into the control strategy may feel squeamish about the possibility of seeing a dead animal. They might prefer that you cover the trap and remove the animals when they're not looking. Whenever you have to work outdoors, be as discreet as possible. Consider how your work might be interpreted by a passerby if you didn't have the opportunity to explain what you are doing.

Do neighbors and other community members ever have a right to influence the wildlife control work you do for a particular customer? With the techniques of on-site release and relocation, this raises some tricky issues. Although those methods might solve the problem for your customer, they might create one for the neighbors. There's no way to predict this with certainty. There are some cases in which the on-site release or relocation of a wild animal is highly unlikely to cause any problems for other people. Unfortunately, in some situations—especially with certain species—this is a real cause for concern.

Many people believe that once an animal has learned about the "good pickings"—the food and shelter homes can provide—it will seek the same situation wherever it lives. Remove a raccoon from your customer's attic and it may move into a neighbor's attic. Now imagine that the problem occurred in a housing complex in which all the buildings are the same. That means they share the same design vulnerability that allowed the animal to invade. Maintenance is probably provided by the complex's owner, so if one building needs attention, it's likely that others do, too.

In the best-case scenario, this problem would persuade the owner to fortify the other buildings and improve the maintenance to help prevent wildlife damage problems from developing in other vulnerable buildings. Wouldn't that be a great job! (Consider taking photographs of the damage and developing an estimate of the costs of the repairs vs. the costs of future control services to aid in your pitch.)

But what do you do when that's not a possibility? Is it fair to do something that might transfer your customer's problem to someone else? Some people might say that you're planting a problem in order to generate more business. Yet, the animal was there in the first place. (Unless you relocated an animal to a site that was close to vulnerable properties, you didn't introduce the problem.) It's a thorny issue. You'll hear different opinions, even among WCOs, and will have to develop your own standards.

As we learn more and refine the tools and techniques of the trade, best practices will get better and better. That's why we urge you to keep learning throughout your career. For example, you may wish to join a professional organization that puts you in touch with other WCOs, or read magazines and listserves.

Public attention can be a good thing. Many WCOs volunteer to help educate the public about wildlife and wildlife damage management. They participate in public events, such as community fairs, write columns for local newspapers, appear on local TV or radio shows, or maintain websites to help others find credible information.

Such actions help people get to know and respect you. It's a chance to help others understand the facts and complexities of various wildlife situations. A quick letter to the editor that says "the bats will be returning soon, so this is a good time to make sure they can't get into your attic" is good for the community, your business, and the wildlife control industry overall. A postcard mailing to your customers reminding them that skunk season is just around the corner, with a tip about how to tell if skunks might get under the house, would probably be appreciated, especially if you offer a prevention package.

To learn more...

- How do you define the word "humane"? Try explaining this to a few friends.
- Find out if there are any legislative bills that would affect the wildlife laws of your state. You can check the state government website, or ask a librarian for help. What's your position?
- Start a file that lists opportunities to talk to the public about wildlife or animals. Write down your ideas about how you might participate.

Review Questions

1. To qualify as a nuisance, an animal must:

a) Take up residence in someone's attic.

b) Damage a crop (agricultural, fish stocks, forestry, nursery).

c) Cause an accident.

d) Attack a child.

e) It depends on the species and the situation.

2. Animals often enter human environments to fulfill basic survival needs. What do they seek?

a) Food and shelter
b) Mates
c) Cable TV
d) Protection from diseases

3. When providing wildlife control services, you are always required to do exactly what your customer requests. (Circle the correct answer.)

True False

4. WCOs who use the best practices approach to solving nuisance wildlife problems think about:

a) Safety, wildlife rehabilitation, finding a quick-fix for the problem, and the wildlife agency's goals for its wildlife programs.
b) The humane treatment of wildlife, whether the method is practical, safety issues, and landowner rights.
c) Legal and ethical concerns, and how much this method will help increase the deer herd.
d) Costs, fur trapping, human safety, and getting rid of as many of the animals as you can to make sure the problem doesn't happen to the neighbors.

5. Trap and transfer (a.k.a. "relocation" or "translocation") can be bad for local wildlife because:

a) It may overload a habitat, forcing all of the members of that species to struggle more to find food and shelter.
b) The public doesn't like this technique.
c) It can spread a wildlife disease into an area that wasn't affected.
d) Answers "a" and "c" are both correct.
e) Answers "a" and "b" are both correct.

6. A best practice takes the following factors into account (check all that apply):

____ Practical limitations
____ Landowner rights
____ Public beliefs about how wildlife should be treated
____ Human safety
____ Protection of the environment
____ Public health
____ Legal restrictions
____ "Management by objective" theory

7. To ensure a humane death for a wild animal, you should:

a) Kill it as quickly as possible, using a best practice for that species.
b) Use anesthesia.
c) Quickly make the animal unconscious, then kill it quickly.
d) All three answers are correct.
e) Answers "a" and "c" are correct.

8. Which questions help WCOs choose the best ways to solve their customers' problems? (Check all that apply.)
   ____ Are these methods legal?
   ____ Is this practical?
   ____ How will this affect the local wildlife population and the environment?
   ____ Does this minimize the animal's suffering?
   ____ What's the public relations angle?

Key Points to Get You Started with the Best Practices Strategy

- Many wildlife conflicts can be explained with two words: food and shelter.
- A best practice is an effective solution for a wildlife damage problem that also minimizes risks to the environment and to our health and well-being.
- The best practices decision-making strategy balances the needs of people and wildlife. Many issues must be taken into account. To address them, ask these six big questions when considering a technique:
  1) Is it safe?
  2) What are the likely ecological consequences of this action?
  3) Is it practical?
  4) Is it humane?
  5) Is it legal?
  6) How would your actions play on the evening news?
- Customize your approach to the species and to the situation. There is no magic pill for nuisance wildlife control.
- Right tools. Right place. Right animals.
- Humane treatment causes no unnecessary pain or stress to the animal.
- It is often possible to prevent wildlife damage and promote wildlife conservation at the same time, even in the same place. Use the two-circle approach. If the barrier around the vulnerable inner circle is secure, the presence of wildlife in the outer circle should not lead to problems.
- Effective long-term solutions to wildlife damage problems usually involve the combination of several best practices.
- Whenever possible, once the problem is solved, try to prevent it from happening again by using exclusion techniques.

Chapter 3
Legal Issues

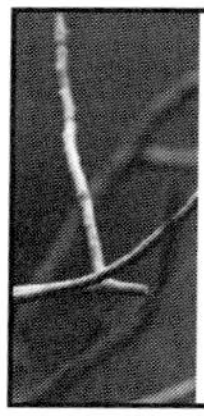

SECTION ONE: Federal Laws Related to Wildlife Control

LEARNING OBJECTIVES

3.1 List three federal agencies that regulate wildlife control.

3.2 Identify one way in which the Endangered Species Act might affect your business.

3.3 List three groups of birds that aren't protected by the Migratory Bird Treaty Act.

3.4 FIFRA (Federal Insecticide, Fungicide, and Rodenticide Act) defines pesticides. Give three examples of pesticides used in wildlife control.

3.5 Identify two subject areas important to WCOs that would be covered in depth on the Occupational Safety and Health Administration (OSHA) website.

A number of local, state, and federal laws are designed to protect wildlife or to safeguard the public and the environment from the improper use of pesticides. You need to be aware of the current status of these laws at all levels because state and local laws are sometimes much more restrictive than federal regulations.

Different laws apply to WCOs, pesticide applicators, hunters, trappers, wildlife rehabilitators, and to those who control the populations of domestic animals, such as dogs and cats. For example, if you want to use pesticides such as repellents, rodenticides, Avitrol bait, or fumigants in your WCO business, you will need a pesticide applicator license from your state.

This chapter introduces the relevant federal laws and describes the types of state and local laws that may affect your business. If you have any questions, contact the regulatory agency or seek the assistance of a qualified legal advisor. Laws and regulations change, so stay up to date.

The major agencies involved in regulating the WCO industry at the federal level are the U.S. Fish and Wildlife Service (USFWS), the U.S. Environmental Protection Agency (USEPA), and the Occupational Safety and Health Administration in the Department of Labor (OSHA). In special circumstances other federal agencies may have jurisdiction, such as the Federal Aviation Administration, which supervises wildlife control at airports.

The U.S. Department of Agriculture, Animal and Plant Health Inspection Service, Division of Wildlife Services (USDA APHIS-WS) provides federal leadership in addressing wildlife damage issues. Their staff manages wildlife to reduce damage to agriculture and natural resources, to minimize risks to human health and safety, and to help protect endangered and threatened species. This agency may become involved with the permit processing for migratory birds in collaboration with the USFWS, or with pesticide registrations in cooperation with the USEPA and the state wildlife agency.

Here are brief descriptions of the most pertinent federal laws affecting the WCO industry. These laws (which are part of the U.S. Code) may be found online and at most public libraries. We've included the website addresses and the citation for the relevant section of the U.S. Code. Reference librarians can help you find the law in the books, especially if you present the citation, such as "16 U.S.C. 1531–1544, 87 Stat. 884."

Endangered Species Act: It's Not Your Job

Regulatory agencies: USFWS, and each state's lead wildlife agency

Applicable to: All plants and animals on the federal endangered or threatened species lists

Read the law:
Online—http://www.fws.gov/laws/laws_digest//esact.html
In print—16 U.S.C. 1531–1544, 87 Stat. 884

The federal government maintains lists of federally threatened and endangered plant and animal species. Each state tracks those species that are rare only within its borders. This law protects endangered or threatened plant and animal species. What does it mean for you? It's simple: it means an endangered or threatened species should not be injured by your wildlife control activities. You can't harass them or disturb their habitats. These species cannot be killed, harmed, or collected except under some carefully described circumstances, and then only with federal and state permits.

You're not likely to encounter threatened or endangered wildlife while on the job, especially not while you're inside a building. (Although a biologist did find an Indiana bat in a structure with little and big brown bats, so it's good to be able to recognize unusual species, even if you don't expect to find them.) If there are endangered or threatened species living in the areas where you work, you must take special precautions. For example, this might affect how you set traps or bird exclusion devices, or apply pesticides. To exclude geese from certain areas, a grid of wires may be installed over a pond or field. If threatened or endangered birds use that area, you'd need to ensure that the wires wouldn't affect those species.

In a rare situation, you might receive a special state permit that would allow specific nonlethal controls for a wildlife species that is threatened or endangered within your state, but is not on the federal lists. Your chances of securing a permit for a species on the federal list are even slimmer.

One measure of a pest manager's professionalism is how diligently you try to protect other species from control activities, whether or not those "non-target species" are endangered. Chances are you'll only get calls about a few federally protected species. Most likely, these would include Canada geese, woodpeckers, gulls, cormorants, turkey vultures (a.k.a. "buzzards"), and some birds of prey (usually hawks and owls). Most of these species are protected by the Migratory Bird Treaty Act, but not by the Endangered Species Act, so you may have some more options. In some circumstances, the landowners may receive permission to capture or otherwise harm a federally protected species. This is handled on a case-by-case basis. Special depredation permits from both the federal and the state government are required.

Migratory Bird Treaty Act: Special Jobs Require Special Permits

Regulatory agencies: USFWS, and each state's lead wildlife agency

Applicable to: all migratory birds (such as ducks, geese, songbirds, gulls, shorebirds, wading birds, birds of prey) with these exceptions:

- Three nonnative birds: the pigeon (now known officially as the "rock pigeon"); the house sparrow (a.k.a. "English sparrow"); and the European starling
- Upland game birds that don't migrate, and are managed by the state wildlife agency
- Certain blackbirds in certain agricultural situations (see page 31)

Read the law:
Online—http://www.fws.gov/laws/laws_digest//migtrea.html
In print—16 U.S.C. 703–712; Ch. 128; July 13 1918; 40 Stat. 755

This law protects all migratory birds, including their feathers, nests, and eggs (with the few exceptions listed above). You may not take, possess, or transport a migratory bird without a special federal permit. Before you attempt to control a migratory bird, the

Avoid the "oops"

FIFRA is probably the law that is most misunderstood by WCOs. Why? Because in many cases, they don't even realize they're dealing with a pesticide. Any substance that's meant to prevent, destroy, or repel pests, or reduce their damage, is legally classified as a pesticide. It doesn't matter if it's a commercial product or a home remedy.

Chemical repellents are pesticides. So are fumigants, such as phosphine gas tablets or a carbon monoxide gas cartridge used to control rodents in their burrows. Poison baits such as rodenticides are pesticides.

There's an easy way to figure this all out. Read the product label. If it's a legal pesticide, the label will include all the information you need. And—the label is the law.

What about "home remedies"? That term covers concoctions such as Auntie Sue's chipotle-garlic deer repellent and the novel use of a household product as a pesticide (for example, household ammonia used as a repellent to drive raccoons out of a chimney). When a product hasn't gone through the regulatory process, you lack vital information. What's the recommended dose? How exactly are you supposed to use and dispose of the product? That's really important information. Ammonia is a legal product. It's also toxic to people. With poor instructions, someone trying to use ammonia as a repellent could get hurt.

In certain cases, the restrictions may seem ludicrous. A raccoon is living in the home, urinating wherever it pleases, but you can't apply a little bit of a commercial repellent that only contains urine? What's the problem?

Do you trust everyone? Not all manufacturers are honest or competent. They might accidentally include a dangerous substance in the product. That's why we rely on scientific studies and government regulatory processes. That's how you know that the list of ingredients is complete and accurate. That's how you can be certain that a 2% solution is just that, not a 0.2% solution in this bottle, and a 20% solution in that one.

Remember, too, that just because a pesticide is sold through a trade magazine or the Internet, that doesn't mean you can legally use it in your state. (Legal pesticides are registered by the EPA and your state agency with an EPA registration number, or are in a special category, "FIFRA 25-B minimum risk exemptions.") In some states, commercial applicators cannot legally use home remedies or FIFRA 25-B compounds. To be safe, use only EPA- and state-registered pesticides.

landowners must obtain the 50 CFR Depredation Permit from the U.S. Fish and Wildlife Service. This permit allows the taking of migratory birds that have become a nuisance, are destructive to public or private property, or are a threat to public health or welfare. The permit spells out the conditions under which the birds may be controlled and the methods that may be used. Permit holders may control migratory birds that are clearly shown to cause, or are about to cause, serious damage to crops, nursery stock, or fish in hatcheries. (USDA APHIS-WS staff can help you apply for this permit. It's usually assigned to a specific landowner or municipality, and a WCO can be designated as a control agent.)

That said, the Migratory Bird Treaty Act has a special provision about blackbirds: "A federal permit shall not be required to control yellow-headed, red-winged, rusty, and Brewer's blackbirds, cowbirds, all grackles, crows, and magpies when found committing or about to commit depredations upon ornamental or shade trees, agricultural crops, livestock, or wildlife, or when concentrated in such numbers and manner as to constitute a health hazard or other nuisance...."

State and local ordinances may further define control activities. For example, in some states, crows are listed as game birds and may receive protection. Noise ordinances may rule out the use of pyrotechnics in certain locales. Check local and state laws before attempting to control any bird species.

In most states, nonlethal wildlife control actions do not require permits. You may be able to persuade the birds to leave by hazing them with dogs, or frightening them with devices such as noisemakers and scarecrows. Alter the conditions that make the site attractive to the birds, for example, by removing unnatural food sources such as garbage. Exclude them from the site by erecting fences. Just remember that you can't destroy eggs or disturb young or a nest if there are eggs or young in it, because those actions are classified as "taking." If that approach doesn't work, call your state wildlife agency to discuss the situation and seek the proper federal and state permits. That delay may not be appealing to your customers, so you may opt to try a combination of nonlethal control methods first.

Federal Insecticide, Fungicide, and Rodenticide Act (FIFRA): Training and Special Permits Required

Regulatory agencies: Responsibilities are split between the USEPA (federal level) and the designated state agency

Applicable to: All pesticides

Read the law:
Online—http://www4.law.cornell.edu/uscode/html/uscode07/usc_sup_01_7_10_6.html
In print—Title 7, U.S.C. Ch. 6

Sit up straight, because this one is complicated. FIFRA is the federal law that regulates pesticides. However, in most states, certain provisions of FIFRA are enforced at the state level through state pesticide regulations. A pesticide is any substance designed to prevent, destroy, repel, or mitigate any pest.

Because of FIFRA:

- Most pesticides, as well as each use of that pesticide, must be registered by the USEPA. The EPA must also approve the product label.
- The EPA also classifies pesticides as either "general use" or "restricted use." (A "restricted use" pesticide is one that, even if used as directed, might possibly harm people or the environment.)
- A restricted-use pesticide may only be used by, or under the direct supervision of, a certified applicator.
- The amount of pesticide residue allowed on food is described.
- There are regulations for storing and disposing of pesticides and their containers.
- The EPA provides standards for worker safety and re-entry into a pesticide-treated area.

- There are criminal or civil penalties for violations of FIFRA. A commercial applicator could be fined up to $25,000, serve one year in prison, or both. WCOs using pesticides on the job qualify as commercial applicators.
- States may set stricter standards but they can't establish a more permissive one. That means that no state can allow the sale or use of a pesticide that is not permitted by federal law.

Each state has laws governing the sale, use, disposal, storage, and transportation of pesticides. Nearly every state controls the certification of pesticide applicators within its borders (in Colorado and Nebraska, federal programs handle this duty).

Occupational Safety and Health Act (OSHA): Federal Worker Safety Regulations May Apply

Regulatory agency: OSHA division of the U.S. Department of Labor

Applicable to: all employers who have employees (unless all of the workers are members of the employer's family)

Read the law:
Online—http://www.osha.gov/comp-links.html
In print—Public Law 91–596, 91st Congress, S. 2193, Dec. 29, 1970

This law sets standards to promote worker safety. If you have *any* employees, even one person, you are covered by OSHA unless all of the workers are members of your family.

OSHA requires the employer to provide a workplace that is free from recognized hazards that are causing, or are likely to cause, death or serious physical harm to the employees. It also calls for the investigation of employee safety or health complaints including those that may be related to the use of pesticides.

Employers must inform their workers about possible job hazards, such as the kinds of injuries that might result from the use of ladders or the risk of exposure to wildlife diseases from contact with potentially contaminated environments. They must always display the OSHA poster describing the law where workers and job applicants can see it. The poster is available at every OSHA office, and can be downloaded or ordered from the website at www.osha.gov.

If you require workers to use protective gear (such as respirators) or equipment that could be dangerous if misused (such as ladders), you must train them in the proper use of that equipment. The OSHA website provides safety tips and training recommendations.

Record keeping requirements under OSHA depend on the size of the business.

- *All businesses covered by OSHA* must report job-related deaths and catastrophes (anything that causes overnight hospitalization of three or more workers). They must contact OSHA within eight hours of the incident.
- *Larger businesses (11 or more workers)* must keep records of all work-related injuries and illnesses using the OSHA 300 Forms.
- *Small businesses (10 or fewer workers)* and those in certain "low hazard" industries (retail, service, finance, insurance, real estate) may be partially exempt from keeping injury and illness records unless the Bureau of Labor Statistics or OSHA sends a written request to do so.

There are penalties for OSHA violations. Several factors are considered when setting the penalty, including the size of the company; a firm with 300 workers may be charged a larger fine than a business with 15 workers for the same violation.

SECTION TWO: What You Need to Know About Your State and Local Laws

LEARNING OBJECTIVES

- The most common state and local laws that affect a WCO business include laws about wildlife, the use of pesticides, firearms ordinances, and building codes.
- Which situations a WCO can handle—and which are beyond your authority.
- When it comes to handling wildlife, what can any citizen of your state legally do, and what can only be done by a WCO?
- The requirements and procedures for receiving and keeping a state wildlife control operator permit or license, if such is required.
- The details for handling any species that may receive different consideration, such as game species, rabies vector species, and domestic species (feral or stray dogs and cats).

It's Probably Not Your Job, But...

In most states, dogs and cats are legally defined as domestic species, even if they are feral and behave like wild animals. Customers may turn to you for help with feral cats but there are thorny liability issues that make this a complicated and risky business. Some of these issues are addressed in Appendix C.

It *Is* Your Job, But Say Hello to a New Boss

Special care is given to the handling of mammals that are most likely to be carrying rabies. Known as "rabies vector species," the ones to watch will differ from state to state, but generally include bats, raccoons, skunks, foxes, and coyotes.

You are required by law to follow your state health department's directions concerning the disposal of rabies vector species, no matter what the landowner requests. Of course, you still have to follow your state wildlife agency's regulations, too. You are also legally required to follow any of the health department's other directions related to the control of rabies.

Talk to Your State Wildlife Agency Staff

State and federal agencies manage game species primarily by controlling hunting and trapping. Some game species, such as muskrat, beaver, white-tailed deer, turkey, and black bear may become a nuisance on private property. What should you do if you're asked to take care of a problem with one of these species? First, read any materials that your state wildlife agency may have provided. If in doubt, call them.

Talk to the Local Governments Where You Work

Some local regulations address wildlife control techniques such as trapping or the use of firearms or pesticides. Health codes or other regulations may restrict the transportation of wildlife. Building codes and fire codes dictate the design and construction of exclusion devices, such as chimney caps. There may even be specific regulations about electric fences. Certain species may receive added protection in some local areas.

The first page in the resource section includes a form for your local contacts. Fill it in, so the information is handy when you need it.

To learn more...

- Contact your state wildlife agency to learn which state and local laws affect your wildlife control business.
- Contact your state wildlife agency for the most current lists of federally endangered and threatened species. While you're at it, ask for the state lists, too. Learn how to identify those species that are found in your area, and know their habitats and habits.
- Learn what you can legally do to deal with problems caused by Canada geese, gulls, and woodpeckers without the federal permit (see Appendix C).
- If you want to use pesticides in your wildlife control business, seek the proper training and license. Learn how to choose the least toxic materials that provide effective results.
- A directory of state agencies for structural pest control can be found at: http://www.aspcro.org/htbin/asdirs.com.
- Browse the OSH. A website at www.osha.gov for safety tips.

Review Questions

1. Some birds are not protected under the Migratory Bird Treaty Act. Check all of the ones that are *not* covered by this law.

___ Canada geese
___ House (English) sparrow
___ Upland game birds
___ European starling
___ Pigeon
___ Herring gull
___ Woodpecker

2. A PCO told you about a very effective deer repellent. You don't have a commercial pesticide applicator license. Can you use it at work?

a) Yes.
b) No.
c) Maybe, but only under the supervision of a veterinarian.

3. Your business requires the use of respirators. If you employ your son and eight non-related workers, you must train them all in the proper use of respirators. (Circle the correct answer.)
True False

4. Which federal agencies have regulations that affect the WCO industry? Check all that apply.
___ Occupational Safety and Health Administration (OSHA)
___ Federal Aviation Agency (FAA)
___ U.S. Fish and Wildlife Service (USFWS)
___ U.S. Department of Education (USDOE)
___ U.S. Environmental Protection Agency (USEPA)

5. The same restrictions apply to homeowners and commercial pesticide applicators. (Circle the correct answer.)
True False

6. If there's a threatened plant species near your work site, you must not:
a) Tell anyone.
b) Set any traps in that area.
c) Apply any pesticides in that area.
d) Injure it in any way.

7. It is illegal to apply a pesticide in a way that's not described on the label. (Circle the correct answer.)
True False

Chapter 4
A Common Sense Approach to Job Safety

Overall learning objectives for this chapter

- Identify risks associated with the presence of nuisance wildlife and with wildlife control techniques.
- Understand how to protect yourself from falls, bites, heat stress, and wildlife diseases.
- Know which protective gear and safety equipment belongs in the WCO "wardrobe."
- Be familiar with the wildlife diseases that you're most likely to encounter.
- Feel confident you can answer the common questions customers ask about wildlife diseases.
- Know whom to contact if you're dealing with a sick animal.

Relax, but don't fall asleep at the wheel

"Had this squirrel job in an old house. I stepped off my ladder onto the metal roof and pow! Those squirrels must have chewed through some wires that were touching the roof. The entire thing was electrified. I was lucky I didn't fall off the roof."

—Eric, WCO in Connecticut

Throughout this chapter, we'll discuss risks related to WCO work or the presence of wild animals inside homes or businesses. In most cases, you'll be called in after an animal has caused damage. The risk is easy to understand because the results are right there. What about those times when you see evidence of possible problems?

What's the worst that could happen? How likely is it? Can the situation be prevented? These are the kinds of questions that will help you put things into perspective. For example, wildlife could potentially expose a person to about 200 different diseases. We'll only discuss nine of them because the chances of being exposed to the others are very low.

WCOs are more likely to be hurt by a car accident or a fall from a ladder than from a wildlife-related disease, but it's West Nile virus that makes the news, isn't it? We hope this chapter will help you better understand the risks posed by your work so you can make good choices.

SECTION ONE: Risks Faced by Your Customers

LEARNING OBJECTIVES

4.1 List four risks that nuisance wildlife pose to your customers' health and safety.

4.2 Identify an example of wildlife damage that might concern a: business owner, apartment dweller, wildlife biologist, homeowner, farmer, or government official.

4.3 Name two ways that a method used to remove wildlife pests can be dangerous to the environment and people.

Risks That May Come With the Pests

When some wild species move in, they can put your customers and their property at risk. Some people don't understand the possibilities until it's too late. Others overreact (Figure 4-1). By offering credible information in a professional manner, you can help your customers make sensible decisions.

There are safety risks. Rodents, raccoons, and birds can cause fires by chewing through wires or blocking vents or fans with their nests (fan motors might overheat and ignite the highly flammable nest materials). If a nest blocks a chimney, dangerous fumes could be trapped inside. Chewed wires may also cause electronic systems to fail—imagine the consequences in a jail or hospital. As previously mentioned, wildlife may collide with airplanes and cars.

And health risks. You, your customers, and their pets and livestock might be bitten, scratched, or exposed to a wildlife disease, such as rabies. WCOs are more likely to encounter a wildlife disease than the average person, because they often handle wild animals and spend a lot of time in disease hot spots such as attics and crawl spaces. The close presence of wild animals (and their fur, dander, droppings, or parasites) can also trigger allergies in some people. Wild animals are often noisy at night, which might deprive your customers of sleep. That doesn't sound too bad until it's happened for several nights.

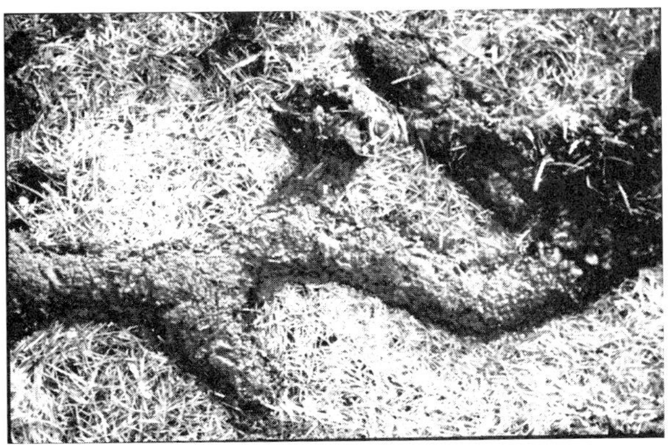

FIGURE 4–1 Has a giant alien worm invaded the lawn, ready to strike your customer's child? Of course not. These are mole tunnels, made by a creature that poses few health or safety risks to people. If this happened to the greens of a golf course, the manager might consider it a financial risk. Whether or not your customers call this a nuisance depends on their perspectives. *(Paul D. Curtis)*

Nuisance wildlife pose financial risks. To gain entrance to a building, some animals might destroy parts of the exterior. Once inside, they might chew or soil woodwork and many other materials; items stored in attics are particularly vulnerable. Raccoons and mice often ruin insulation, causing heating and cooling bills to rise. Chewed wires, of course, might need replacement, which can be expensive.

They may threaten other wildlife or change habitats. A nuisance animal may introduce a disease to another species. In large numbers, the nuisance species might kill and eat many of that other species, such as gulls preying on plovers and terns, or it might destroy their habitat.

Some of the removal methods present their own dangers to people and the environment. An improperly set trap may capture or injure the wrong species, and could even be hazardous to people. If misused, pesticides can contaminate water, soil, and air. They can kill other species too, including beneficial organisms that help control pest populations. Certain pesticides and euthanasia products can also be dangerous to people. Even exclusion, one of the favored methods, has risks associated with it, because a highly motivated animal may damage the building to get back in, especially if it has young inside.

To learn more...

- Walk around your neighborhood and look for signs of wildlife damage.
- Start a file of newspaper clippings of stories about nuisance wildlife problems.
- Ask your friends, neighbors, or local business people if they've experienced any losses caused by wildlife.

SECTION TWO: Risks That Come With Your Job

LEARNING OBJECTIVES

4.4 Draw a diagram showing the recommended way to position a ladder against a building.

4.5 List three tips for the safe use of ladders.

4.6 Define "zoonotic disease" and "zoonoses."

4.7 Describe the type of clothing that will help you avoid being stung by insects.

4.8 Name six warning signs of heat stroke and one way to prevent it.

4.9 List three ways to quickly cool down someone who's suffering from heat stress.

4.10 Identify the best way to protect yourself from a tetanus infection.

Some of the equipment used to remove wildlife can be dangerous, so it's sensible to stay alert to the latest information about the safe and effective use of traps, firearms, and ladders.

This section briefly discusses some of the safety issues you'll confront on the job and describes tips for avoiding accidents. For more detailed information, check the U.S. OSHA (Occupational Health and Safety Administration) website at www.osha.gov or the other resources listed in Appendix E.

The Most Dangerous Thing You Handle Isn't Furry

"I was working on this bird job in an equipment shed. We were nearly done; I just needed to treat one last area. I was using a 20-foot straight ladder, but the roof was about 15 feet in that spot. Instead of getting a shorter ladder, I set the long one against the rafter and started up. Just as I reached the rafter, the ladder slipped, and down I went. End result: cracked rotator cup in my elbow, dislocated toe, multiple fractures in my feet, a large gash across my knee."

—Wayne Langman, WCO in Indiana

Like contractors, WCOs spend a lot of time on ladders and roofs, but, unlike roofers, WCOs also contend with another hazard: the unpredictable actions of wild animals. Carrying a trap containing a scared or aggressive animal down a ladder is a bit more exciting than toting a bucket of nails. So WCOs have a few more items to add to their list of safety issues.

Safety precautions can be a pain: they slow you down and inhibit your mobility. It's hard to justify taking the time during the busy season, especially if you just need to quickly check a trap, and you have many other jobs waiting. Very few people die from diseases they catch from wildlife, but accidents associated with ladders are fairly frequent and often serious. In 2000, for example, falls were the second-leading cause of all work-related deaths in the U.S., averaging close to two deaths each day (these statistics are from NIOSH's National Traumatic Occupational Fatalities Surveillance System; "NIOSH" stands for "National Institute for Occupational Safety and Health" and is part of the CDC).

Accidents usually happen when someone is hurried or distracted and not concentrating on safety. Sometimes the condition of the ladder is at fault. Sometimes it's your shoes, or wet or icy surfaces. Wind might overcome the stability of the ladder and tip you over.

If you run your own business, it's up to you to decide how much risk you're comfortable with; however, if you have even one employee who is not a member of your family, you're covered by OSHA regulations. Even if your business is exempt, you may still want to check out the OSHA website, www.osha.gov/SLTC/fallprotection/index.html. Here's a basic summary of the OSHA recommendations for the safe use of ladders.

FIGURE 4–2 This WCO wears a safety harness attached to a retractable lanyard system that would arrest a fall. This protection still provides enough mobility to install the bird netting on the bridge. *(Bird Barrier, Inc.)*

Setting Up the Job

- First, make sure that the ladder is the right design and strength for the job. Its load rating must be high enough to ensure the ladder can support the weight of the user, materials, and tools. Don't load a ladder beyond its rated capacity.
- Step ladders should not exceed 20 feet in length. A single-section ladder shouldn't be more than 30 feet long. Two-section ladders shouldn't be more than 48 feet long (if metal) or 60 feet long (if wooden or fiberglass). Ladders with more than two sections shouldn't be longer than 60 feet.
- Follow the manufacturer's instructors for correct use and maintenance. Don't improvise with ladders (for example, don't use a folded step ladder in place of a straight ladder).
- Inspect ladders before use for signs of damage (to the rails, rungs, steps, or cleats), missing or damaged safety devices (such as rung locks, lock spreaders, or safety shoes), and grease, dirt, or anything else that could cause a slip.
- If using fall protection equipment, such as lifelines, shock-absorbing lanyards, and harnesses, seek proper training in the use and selection of those materials (Figure 4-2).
- Make sure you're using a climbing rope that is rated for people.
- Keep the area around the top and bottom of the ladder clear. If you must place the ladder in front of a door, make sure the door is barricaded so no one can open it and knock over your ladder.
- Place the ladder at a safe angle, according to the manufacturer's recommendations. There should be no more than a 4:1 ratio between the height to the roof's edge and the distance from the eave (Figure 4-3). In this illustration, the roof edge is 20 feet up, which means that the base of the ladder needs to be 5 feet out from the eave to achieve the 4:1 ratio (not drawn to scale).
- The top of the ladder should extend 3 feet above the roof.
- Secure ladders properly to a strong structure, especially if you're going to step off the ladder onto the roof, or when you'll be working on the ladder for any length of time. Keep the bottom of the ladder from slipping by using braces or an anchor board. Tie the "fly" section (the moveable part) and base together at their overlap.

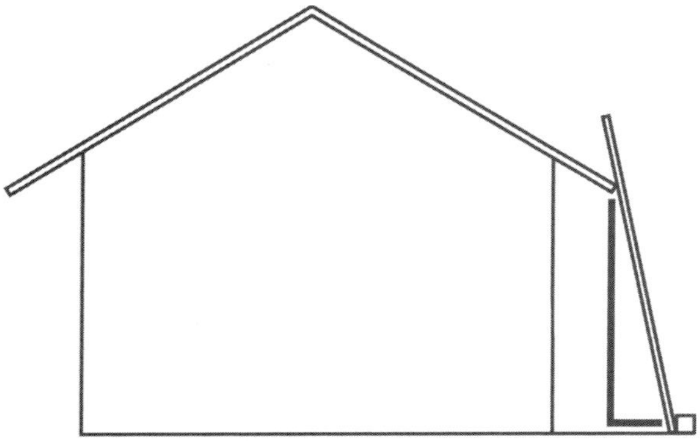

FIGURE 4–3 When using a ladder, maintain a 4:1 ratio between the height to the roof's edge and the distance out from the eave. *(Karen English)*

When fully extended, each section of an extension ladder must overlap the adjacent section by a specified number of feet (see manufacturer's label). Secure the top of the ladder using ladder rungs or "ropes" (the types of ropes that are rated for this use are more properly called "lifelines" and "lanyards").

- The ladder must be as firm as possible. Both rails must maintain equal contact with the supporting structure. The base must be flat, level, and secure. Use equipment to stabilize the base if the ladder's on uneven ground. If you can't secure the base, choose another spot for your ladder. Some ladders have adjustable legs for use along walls above steps.
- Inspect fall-arrest equipment frequently and don't use damaged pieces. Replace lanyards and harnesses after a fall.
- Pad ropes so they don't chafe against the roof's edge.
- If using an aluminum ladder, watch out for electrical services to avoid electrocution.

Climbing and Dismounting

- The likelihood of falling is not directly related to your weight or size.
- Climb slowly and surely and always face the ladder. Avoid the temptation to lean off the side of the ladder because that may make you lose your balance. Follow the belt-buckle rule: always keep your belt buckle inside the side rails of the ladder.
- Keep three parts of your body in contact with the ladder (either both hands and a foot, or both feet and one hand).
- Don't carry heavy or bulky items as you climb. Pull them up and lower them down on a towline, attach them to your tool belt, or have them handed up to you.
- If you have a live animal in a cage trap, lower it to the ground using a towline.
- Towlines, electrical cords, and airlines for air-powered equipment should not be used above the roof eaves.
- Wear shoes with strong soles, and keep them clean for maximum traction.
- Dismount by stepping sideways onto the roof; don't step over the ladder.

Tips from WCOs and Safety Experts

- If you don't feel safe, don't do it. Consider renting a bucket lift (Figure 4-4). The rental costs may even make sense financially, because you could lose a lot more money if you had an accident that keeps you away from work for a long time.
- Watch out during the spring and fall, when temperatures fluctuate a lot. That can lead

FIGURE 4–4 Consider renting a bucket lift if you need to install bird control devices on tall buildings and bridges. *(Bird Barrier, Inc.)*

to condensation on the roof and very slippery conditions.

- Carry a cell phone with you, especially if you're working solo. That way, if you have an accident and need help, you can call someone.
- To check whether there's an animal in a trap without climbing onto the roof, try this simple device: Fasten a car's side-view mirror onto a long pole. Raise the pole to see the trap from the ground.
- For more details, see the OSHA website at www.osha.gov. Another great source of advice is the building industry. Roofers have a lot of relevant experience and, like you, they're up there trying to get their work done.

Wildlife Diseases

Zoonotic diseases or **zoonoses** are illnesses that people can catch from animals or from contact with their habitats. There are about 200 zoonotic diseases. (Details for the most relevant zoonoses will be discussed later.)

As previously mentioned, risks associated with ladders are far more significant for WCOs, but some wildlife diseases can also be fatal to people. Even if you're comfortable with your personal risk, you owe it to your customers to be cautious. You have no way of knowing how healthy your customers are; some may have compromised immune systems because they suffer from cancer, diabetes, AIDS, or other illnesses. A disease you could shrug off might be no laughing matter for them.

Diseases can also spread to other wildlife species and devastate their populations, a major worry if the affected species is endangered.

As a professional, behave in ways that minimize the risk of exposing other people to diseases. Help prevent the spread of diseases to other areas or species. This concern influences your animal handling and disposal procedures, your choice of gear, customer education, and clean-up strategies for the site and your equipment.

Want to read about a disease right away?

Rabies:	p. 55
Raccoon roundworm:	p. 62
Histoplasmosis:	p. 65
Toxoplasmosis:	p. 68
Hantavirus:	p. 71
Mange and distemper:	p. 74
West Nile virus:	p. 77
Lyme disease:	p. 81

Rabies management provides an example of the important role WCOs play in protecting public health. Rabies is widespread in wildlife, so state health departments often recommend treating any skunk, raccoon, bat, or fox as "rabid until proven innocent." Rabies is almost always fatal once symptoms appear. (Four people who had been given some vaccination—but not in the recommended way—did develop the disease and still survived, although they suffered severely.) Yet there are only about one or two human deaths caused by rabies each year in the U.S., according to the CDC (Centers for Disease Control and Prevention). One reason for this remarkably low number is the vigilance of health department staff, WCOs, veterinarians, and many others.

Bites and Stings

When practical, avoid situations in which you might get bitten or stung, especially if you're allergic to wasps or bees. There's gear that can help protect you from bites, such as animal handling gloves, catchpoles, and traps with protective plates around the carrying handle (gear will be discussed in more detail later). In some cases, a strategic retreat may be in order. If holding onto the animal means you're probably going to be bitten, just let it go and try to catch it again later. If you are bitten by a mammal, call the Department of Health for advice.

To reduce your chance of being stung by an insect, wear light-colored but not colorful clothing. When you approach a wasp or bee nest, be careful not to vibrate it or shine a light directly on it. "Bee suits" may be warranted if you're dealing with a large nest.

What about sprays? Check with the state agency that regulates the commercial use of pesticides. Some make reasonable allowances for workers in dangerous situations, such as WCOs who are up on ladders. You may be allowed to use an over-the-counter spray to protect yourself from stinging insects, but you can't provide stinging insect control as a service without a commercial pesticide applicator license.

Heat-related Illnesses

"I was removing a large starling nest from an attic. The nest was 6–8 feet tall and almost as big around. Because of the dust, I was wearing a respirator. I wasn't moving around much, just bagging up the nest. After an hour, I noticed I was getting light-headed. As soon as I moved, the dizziness really hit me. It was all I could do to get to the ladder and get down in one piece. If I hadn't recognized the symptoms, I could have collapsed up there and maybe died from the heat."

— Wayne Langman, WCO in Indiana

WCOs have to go where the animals are. Often, that takes you into an enclosed space that's hot and stuffy. To make things worse, there's a good chance you'll be wearing protective gear that will make you even hotter. This can lead to a variety of heat-related conditions, such as heat rash, heat cramps, heat exhaustion, and heat stroke.

Heat stroke is the most serious condition—it's a life-threatening emergency. Twenty percent of heat stroke victims die, regardless of health or age, according to a study by Ohio State University. Heat stroke can kill quickly or cause permanent brain damage. Your body temperature can rise to 106°F or higher within 10 to 15 minutes.

Milder forms of heat-related illnesses, such as heat exhaustion, can develop into heat stroke if untreated. Even milder conditions can be serious for WCOs because they might lead to accidents, especially falls. Dizziness, fogged safety glasses, slippery, sweaty palms, compromised balance, and outright fainting could make you fall off a ladder or beam.

Under hot conditions, some workers can lose as much as 2–3 gallons of water a day through sweat. You need to drink about as much water as you lose to sweat to avoid dehydration. Don't count on thirstiness to signal when you need to drink because it's not a reliable indicator. Just plan on drinking 5–7 ounces of water every 15 minutes, which is equal to 1 quart every hour.

Warning Signs of Heat Stroke

- Hot skin, usually dry, red, or spotted
- Extremely high body temperature, 105°F or higher
- No sweating
- Confusion or delirium
- Unconsciousness
- Convulsions
- Vomiting

Warning Signs of Heat Exhaustion

- Moist, clammy skin
- Pale or flushed complexion
- Normal or slightly elevated body temperature
- Heavy sweating
- Giddiness, dizziness, or fainting
- Headache, perhaps really throbbing
- Queasiness
- Extreme weakness or tiredness
- In more serious cases, vomiting or unconsciousness

What to Do

Call 911 and seek immediate medical attention if: the symptoms are severe; you have heart problems or high blood pressure; the symptoms worsen; or the symptoms last more than an hour.

Cool down as fast as you can. Seek immediate medical care—heat stroke victims need to go to the hospital. Remove heavy clothing. Douse yourself with cool water from a shower or garden hose. Go to an air-conditioned room. Place ice packs or wet towels or sheets around your neck, under your armpits, and near your groin (these are areas with higher blood supply). Drink cool, nonalcoholic beverages; ideally, about 5–7 ounces of water every 15 minutes. Do whatever it takes to get your body temperature to drop to 101–102°F. **If the person is vomiting,** turn him on his side to keep his airway open. **If he's experiencing convulsions,** make sure he doesn't hurt himself. Do not place any objects in his mouth, and do not give him fluids.

Tetanus ("Lockjaw")

Tetanus is an infection caused when the toxin of the bacterium, *Clostridium tetani*, is released into

Best ways to prevent heat-related problems

- Give yourself a few days (ideally, four to seven days) to get used to the heat by gradually increasing the amount of time you spend in hot areas. If you're out of the heat for more than a week, you need to do this again.
- Take a break every hour while you're working in a hot spot (80°F or hotter), especially if the humidity is high. Move to a cool spot.
- During your cool break, drink at least 1 quart of water.
- Eat light, cold meals, preferably those that are lower in fat, because fat is harder to digest in hot weather.
- Do not drink alcoholic beverages.
- If you take medication, read the label to know if your medicine affects your ability to tolerate heat or sun. If you have any questions, talk to your doctor or pharmacist.
- Do the job early in the morning whenever possible, especially if it requires a lot of time in an attic or crawlspace.
- Work with another person if you can, so you can check each other for signs of heat stress.
- Special gear, such as a cooling vest, is available.

For more information, see:

- "Working in hot environments," a CDC report, at www.cdc.gov/niosh/hotenvt.html#prepare.
- "Protecting workers in hot environments," OSHA report 95-16, at www.osha.gov/pls/oshaweb/owadisp.show_document?p_table=FACT_SHEETS&p_id=167.
- "Heat stress," the Ohio State University Extension Agricultural Tailgate Safety Training Series, at www.cdc.gov/nasd/docs/d001701-d001800/d001702/d001702.html.

a wound, usually through a deep puncture. It's often fatal. This disease is also known as "lockjaw" because the muscles of the jaw and neck contract spasmodically.

WCOs are most likely to suffer puncture wounds in two ways: either through an animal's bite or by accidentally impaling themselves on a nail, which is a common hazard in attics and barns. Be especially alert for nails in horse barns because many horses are infected with the tetanus bacteria.

The right gloves provide excellent protection from bites but they're not foolproof. Some larger rodents and carnivores, such as woodchucks and coyotes, can bite through even heavy-duty gloves. To protect yourself from a tetanus infection, get a tetanus immunization every nine years (or sooner, if recommend by your doctor). If you receive a puncture wound and it's been more than five years since your last tetanus shot, your doctor may recommend a booster dose.

Electrocution and other Safety Risks

Many buildings contain old wiring systems, such as "knob and tube." If there was any insulation on the wires, it may have deteriorated. You could be working around bare, "live" wires. Touch one and you could be electrocuted. (This also happens to rodents and birds.) Anything that's made of metal, such as a roof, can become electrified. Be especially alert for dangers when you're working in an old or poorly maintained building. Will the attic floor bear your weight? Are the joists solid? Slate roofs can be very slippery, especially when wet. To learn how to recognize and avoid hazards associated with particular building styles, talk to building contractors.

To learn more...

- Check the OSHA website for more information about fall protection. Write the best tips on an index card, laminate it, and attach it to the visor of your truck.
- Check MedLine Plus, a government site that includes information from the National Institutes of Health and the National Library of Medicine, for information about heat-related illnesses at www.nlm.nih.gov/medlineplus/heatillness.html.
- Find out which types of wasps most frequently sting people. Are there any harmless insects that could be mistaken for wasps?
- Seek training in first aid and CPR (cardiopulmonary resuscitation).

Review Questions

1. Diseases that people can catch from animals or from contact with their habitats are called:
   a) Epinotic diseases.
   b) Zooposies.
   c) Zoonotic diseases.
   d) Mammalian diseases.

2. If a ladder is placed 20 feet high, the base of the ladder should be:
   a) 20 feet from the side of the building.
   b) 10 feet from the side of the building.
   c) 5 feet from the side of the building.
   d) Firmly attached to the building.

3. Wear colorful clothing to avoid being stung by insects. (Circle the correct answer.)
   True   False

4. The odds of falling are related to:
   a) Your weight and height.
   b) The condition of the ladder.
   c) Weather conditions.
   d) The type of shoes you're wearing.
   e) More than one answer is correct.

5. What's the best way to prevent suffering from heat stroke or heat exhaustion if you have to work in a hot attic?
   a) Take a break every hour and drink one quart of water.
   b) Finish up the job quickly so you can go get a drink.
   c) Wear protective gear.
   d) Schedule jobs for late in the afternoon.

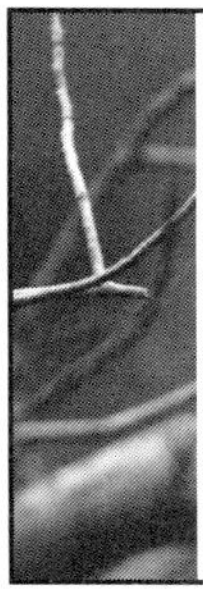

SECTION THREE: Common Sense Precautions Require Some Uncommon Gear

LEARNING OBJECTIVES

4.11 List the safety equipment you're likely to use when working in an attic.

4.12 Identify the kinds of gloves you should wear to handle large mammals, small mammals, birds, and snakes.

4.13 Know which government agency rates respirators.

If you're running your own show, you can make personal decisions about whether or not to take certain precautions. But if you have even one employee who is not a member of your family, the government insists on certain standards that are enforced by OSHA regulations.

Gloves are an indispensable safety tool. The type needed depends on the situation, but disposable vinyl or latex gloves are the most versatile choice (some people are allergic to latex and must use vinyl). When applying certain chemicals, such as some pesticides, you may need to wear a specific type of gloves. You'll probably want to keep a variety of gloves handy.

Lightweight leather work gloves are usually adequate for handling small birds and snakes.

The smart, stylish WCO wardrobe includes:

- Animal handling gloves (Kevlar or heavy leather gauntlets)
- Disposable rubber or plastic gloves
- Protective eyewear, such as goggles
- Respirators and dust masks
- Disposable coveralls with hoods
- Disposable shoe covers
- Rubber boots
- Knee pads
- Helmet, safety ropes, harnesses (to connect to all the devices used to secure ladders)

With larger birds and mammals, wear thick leather gloves or gloves made with Kevlar (the material in bulletproof vests), or perhaps even welder's gloves. This is especially important if you think the mammal might be rabid or if it's agitated. Unfortunately, some carnivores can bite through the strongest gloves, and no glove can protect you from a crushing injury.

Although safety will probably be the most important factor when you choose gloves, there are other things to think about, too. For comfort, mobility, and out of concern for the animal, select the lightest pair that you can. Thick, heavy gloves make it harder to feel how the animal is responding to being handled, and may cause you to grasp it too tightly.

Protective clothing can range from long-sleeved cotton shirts that help prevent scratches to disposable coveralls used in areas that could be contaminated with diseases, or when you're working with pesticides. (If you use pesticides, you'll have to pay extra attention to your clothing, including the way you wash it.)

Knee pads protect you while you're crawling around in attics and crawlspaces and on roofs.

Goggles or similar eye protection is important in many circumstances.

A specialized respirator will help protect you from inhaling such disease agents as fungal spores, viruses, and bacteria, but only if it fits properly. Wearing a respirator is a good precaution if you're in a disease hot spot (such as an attic, crawlspace, or an area near a large bat or bird roost) or if you're likely to disturb droppings or encounter contaminated soil or rodent nests.

There are many different kinds of respirators. Your choice will depend on many factors, such as which disease you're trying to avoid, whether or not you have a beard or mustache, how much weight you can carry, and how much mobility you need. One of the most versatile designs is the half-face respirator with a filter. Look for a NIOSH-approved respirator. For the OSHA standards about respirators, check http://www.osha.gov/publications/OSHA3079/osha3079.html.

Safety helmets range from "bump caps" to helmets with visors. Some helmet designs attach snugly to the head to give added protection in case of a fall.

When you've finished with the job, disinfect your gloves with a household or commercial disinfectant, or a dilute bleach solution, before you remove them. If the situation requires a respirator, keep it on until you've safely dealt with the rest of your clothing and gear. You may want to wear a respirator and goggles when you clean your truck, especially if you use a power washer, which could splash contaminants around.

Disposable clothing can be placed with other contaminated materials and double-bagged for removal to a landfill or other approved disposal facility. Then wash your hands thoroughly with warm water and soap.

Another good general precaution is to tell your doctor about what you do for a living. This will alert your doctor to consider some of the wildlife-related diseases, which are of little concern to other patients.

To learn more...

- Imagine you are about to hire an employee. List the safety equipment you'll need to buy for that worker, and price it out.
- Find a few options for additional training in the safe use of ladders and respirators. Pursue whichever educational opportunity you like best.
- Attend an OSHA training program. They are sometimes offered by community colleges.

SECTION FOUR: What You Need to Know About Wildlife Diseases

LEARNING OBJECTIVES

4.14 Know which zoonotic diseases are relatively common, how they're spread, and which symptoms are seen in wildlife and people.

4.15 Recognize the signs of a disease hot spot.

4.16 Understand how to reduce your chances of becoming infected with a wildlife disease.

4.17 Know which vaccinations you should receive because you're a WCO.

4.18 Understand what you are required to do by law when handling an animal that's likely to be rabid.

How Do You Catch Them?

"I was scratched on the wrist while banding a Canada goose. About a week later, another goose scratched off the scab. It took me a few hours to notice that the open wound was completely covered in goose poop. Our field station was a day's helicopter flight away from the nearest hospital, and I'd forgotten a first-aid kit. So I washed the wound as best I could with snow. The next morning, the entire area was swollen and tender and I had difficulty bending the wrist." [He recovered fully.]

—Arthur Smith, wildlife biologist, SD

Wildlife diseases can be caused by bacteria, viruses, fungi, or parasites. The ones that can be transmitted to people are called "zoonotic diseases" or "zoonoses." There are several different ways you can become infected. Most often, this happens after a bite or scratch from an infected animal (Figure 4-5). Disease agents may enter your body through wounds, or through your eyes, nose, or mouth.

You can also pick up diseases indirectly when you're bitten by a mosquito, tick, or flea that fed on an infected animal. Certain mosquitoes spread West Nile virus, black-legged ticks spread Lyme disease, and fleas may carry plague and typhus.

FIGURE 4–5 When you see something like this in your customers' home, what crosses your mind? (After "good thing they called me.") This exclusion attempt could signify that the nuisance animals entered the living spaces. Ask if any person or pet came into contact with the wild animal. If that's true, consult with the health department. You may need to submit a specimen for a rabies test. That need might change your approach to capturing and removing the animal. *(Connecticut Nuisance Wildlife Control Operators Association, Inc.)*

Some diseases are transmitted through the air, such as hantavirus and histoplasmosis. You can breathe them in, especially while stirring up dust in a confined space. Touching your mouth after you've touched something that's contaminated, or eating infected meat that hasn't been properly cooked, may also cause an infection. This is a significant problem for young children, especially when they're playing outdoors. Their sandboxes, play areas, or toys may become contaminated by the droppings or urine of wildlife. Kids may put soil, wood chips, or droppings into their mouths. Raccoon roundworm is spread this way; the parasite's eggs are found in contaminated soil.

Hopeless? Helpless? No Way.

So how can you protect yourself, and make sure that you don't bring diseases or parasites into your home? Practice good personal hygiene, wear protective gear such as disposable gloves, disinfect your equipment, maintain your rabies and tetanus vaccinations, and use safe animal capturing and handling techniques. Good hygiene and sanitation will also reduce the chance of developing allergies to animals.

Probably the single most important thing you can do to reduce your risk of catching a zoonotic disease is to wash your hands. Ordinary soap and water will do. Wash your hands the way your parents taught you to, thoroughly and often. And always wash your hands before you eat, drink, or smoke.

Another healthy habit is to avoid contact between your hands and your face, eyes, or mouth. Pay attention when you're eating, drinking, smoking, adjusting your glasses, applying cosmetics, taking medication, and when you sneeze.

Keep your gear clean, too. Many of the objects you handle are often fouled by blood, feces, urine, saliva, or body tissues. If your hands or gloves are dirty, it's easy to contaminate doorknobs, car doors, clipboards, telephones, computer keyboards, faucet handles, and many other objects. The nuisance animal may also have made quite a mess; if you don't offer clean-up services, you may want to tell your customers how to deal with it safely.

At the end of the day, clean and disinfect all of the equipment you used with dilute bleach and water (a 10% chlorine bleach solution, which is one part bleach to nine parts water), or a household or commercial disinfectant. Wipe down your truck's seat, steering wheel, and door handles. Some WCOs keep a quart spray bottle of disinfectant in their truck's cab because bleach solutions don't keep long, so it's better to use it in small batches. Just don't mix bleach and ammonia, or use bleach to clean up droppings, which contain ammonia. Use a household or commercial disinfectant instead. Antibacterial wipes may seem even more convenient but they weren't designed to kill parasites, fungi, or viruses. Those agents cause all but one of the diseases discussed later.

Consider the time you spend cleaning as a marketing effort—some customers will interpret cleanliness as a sign of professionalism and competence.

Review Questions

6. Which simple activity may be the most effective way to protect yourself from catching zoonoses?
   a) Use antibacterial wipes.
   b) Become an accountant.
   c) Wash your hands throughout the day.
   d) Replace wooden traps with metal ones.
   e) Exercise.

7. Wildlife diseases are caused by
   a) bacteria only
   b) bacteria and viruses
   c) germs and dirt
   d) bacteria, viruses, parasites, and fungi

8. How do people catch "wildlife" diseases?
   a) They handle an infected animal.
   b) They breathe in the disease organism.
   c) An infected animal bites or scratches them.
   d) They rub their eyes, nose or mouth when their hands are dirty.
   e) They eat unwashed vegetables.
   f) All these answers are correct.

A few wildlife-related safety tips from the CDC that you may want to share with customers

These CDC tips focus on disease prevention. As we mentioned earlier, you can take measures to prevent nuisance situations while leaving part of your landscape for wildlife. There are many things to consider when managing land for wildlife, including the risk of exposure to wildlife diseases.

- Teach children never to handle unfamiliar animals, wild or domestic, even if they appear friendly. Tell them, "Love your own, leave other animals alone."
- Enjoy wild animals from afar. Never adopt wild animals or bring them into your home. Do not handle, feed, or unintentionally attract wild animals to your home or yard (see below).
- Don't try to nurse sick animals to health. Do not "rescue" baby birds or other baby animals. They usually don't need rescuing. (Direct questions to a wildlife rehabilitator or to state wildlife agency staff.)
- Before traveling abroad, consult with a healthcare provider, travel clinic, or your health department about the risk of exposure to rabies. They can advise you about whether pre-exposure rabies vaccinations are sensible, and how you should handle an exposure in that country, should it arise.

Medically Important or Just Media Hype?

Consider this an introduction to some health concerns that affect WCOs. For example, you'll often be cautioned to "wear a proper respirator." That's shorthand to alert you to the need to protect yourself from breathing in microscopic disease agents.

How much do you need to know? You're not a doctor, after all. You don't have to be able to throw around words like "sapro-zoonoses," but you do need to know enough to protect yourself and others, and to answer your customers' questions. In this manual, we've focused on those diseases that are most likely to be encountered by WCOs while they're working. So, for example, we haven't discussed chronic wasting disease, which is associated with deer, because most WCOs don't handle deer. Chronic wasting disease is spreading eastward. If you have a contract with the highway department to pick up deer carcasses from the road, for example, we recommend that you learn about this disease and safe handling methods for deer.

Sometimes our fears about these wildlife diseases are much greater than our actual risks of catching them, or the likely results of an infection. Even for an ethical WCO who's not trying to sell a job by frightening customers with an overblown assessment of the risk of catching a wildlife disease, it can still be tricky to share the necessary information in the right context. It's also important to resist jumping to conclusions. For example, distemper can cause symptoms that look like rabies. The only way to be sure is to test the animal.

Some of these diseases are potentially fatal. That's something your customer will probably want to know—what's the worst-case scenario? But the chance of catching most of these diseases is low, and even then, many of them are treatable.

The trick is to have good, complete, and credible information from a trusted source. Two extremely valuable sources for current and accurate information are the national Centers for Disease Control (CDC) at www.cdc.gov and the Southeastern Cooperative Wildlife Disease Study (SCWDS) at www.uga.edu/scwds/index2.htm. Other good sources for information about wildlife diseases include physicians; veterinarians; medical entomologists; the state's health, wildlife, and agriculture departments; trade magazines; professional organizations; wildlife conferences; books; fact sheets; videos; listserves; and websites. (See Appendix E.)

One last medically related point: When you remove wildlife from people's homes, it's also important to plan for the parasites that may be left behind. Birds and mammals are hosts for a variety of parasites including fleas, ticks, mites, and lice. Although these parasites generally prefer their original host species, if you remove those animals, the hungry parasites may enter the home looking for a meal. Many of these parasites will bite people and they can be extremely annoying. (And itchy customers are generally not happy, which isn't good for business.)

These pests may fly or crawl into the home through windows, ventilators, cracks, and crevices. Droppings, feathers, fur, food, and carcasses can also attract other pests such as flies and carpet beetles. This can be a significant problem if animals have died in inaccessible locations (like the walls). That can happen when an animal is poisoned, and it's one of the risks of using pesticides such as toxicants and rodenticides as a management method.

Effective clean-up should remove any parasites present in the home. If a site is badly infested, you should wear protective clothing before entering. Even if you don't offer clean-up services, know how to properly advise your clients. In addition to commercial services, most states have an "integrated pest management" or "IPM" program that's usually associated with the land grant university. These programs offer information about how to deal with many pests, including the ones that are shed by wildlife. IPM stresses a combination of practical, economical pest management methods that minimize the risks to people and the environment.

How do you protect yourself from wildlife diseases?

How people contract the disease	Precautions for WCOs
Bites or scratches: • **Rabies** (mammal bites or scratches) • **West Nile virus** (mosquito bites) • **Lyme disease** (tick bites) • **Hantavirus** (possible from rodent bite but uncommon)	Mammal bites or scratches: • Get rabies pre-exposure vaccine and keep it current • Wear animal handling gloves • Use a restraining device such as a catchpole • Capture animal in trap • Avoid contact with animal's mouth and saliva • Shower soon after work, every day Mosquito or tick bites: • Wear loose-fitting, light-colored clothing (it's harder to bite through and makes it easier to see small ticks) • Use repellent • Tuck your pants into your socks (keeps ticks from crawling onto legs) • Check yourself for ticks at lunchtime and remove any that you find
Breathe it in: • **Histoplasmosis** • **Hantavirus** • **Canine distemper?** (not definitively established)	• Wear a proper respirator, disposable clothing, goggles, gloves, and hood • Ventilate the area, if possible • Dampen contaminated materials, wipe with wet sponge • Spray contaminated area or dead animals with disinfectant • If possible, schedule the job for cool, damp weather
Dirty* hands touch mouth, eyes, or nose: • **Raccoon roundworm** (mouth) • **Toxoplasmosis** (mouth) • **Rabies** (infected saliva gets into mouth, eyes, or nose) • **Hantavirus** (mouth, eyes, or nose)	• Wear a proper respirator, disposable clothing, rubber gloves • Wash your hands thoroughly with soap and water, especially before eating, drinking, or smoking • Avoid contact between your hands and your face. Pay attention when you sneeze, eat, drink, smoke, adjust eyeglasses or put in contacts, or take medication.
Disease gets into wound: • **Rabies** • **Hantavirus**	• Protect wounds with bandages, if practical • Wear gloves or clothing that covers wounds • Check wounds and keep them clean
Eat contaminated food or put dirty object into mouth: • **Raccoon roundworm** (dirty object) • **Toxoplasmosis** (contaminated meat or dirty object) • **Hantavirus?** (may be possible via contaminated food or water)	• Wash your hands thoroughly after outdoor activities and especially before eating, drinking, or smoking • See advice for customers on pages 63, 64, 69, and 70.
Handle an infected animal or contaminated equipment: • **Mange**	• Wear gloves • Minimize contact with mangy animal by using restraining devices • Minimize contact with contaminated clothing, equipment • Dry clothing at high heat to kill any mites on it

Note: Boldface type indicates an individual disease.
* "Dirty" means that the hand, glove, or object is contaminated with the disease-causing agent (such as a virus, bacterium, or a parasite's eggs). These agents are often microscopic.

Now, on to the zoonoses. The zoonotic diseases that are potentially fatal for people are listed first. So, for example, we've listed hantavirus before mange, which a WCO is far more likely to encounter, because hantavirus can become a much more serious health problem. At the end of the chapter we include two wildlife diseases that people generally encounter from contact with a "middleman," either a mosquito or tick.

There's a lot of information in this section. The chart on page 53 should help you remember the answers to three critical questions: Who? What? How? Who is likely to transmit the disease to people? What disease? How do people catch it?

RABIES

Caused by: A virus

Most common way people catch it: Bite from an infected mammal

Worst-case scenario: Death

How common in our region? The virus is widespread and may infect any mammal, including people. Specific strains of the virus cycle through certain wildlife populations, so at times, it may be common in some species, especially skunks, raccoons, and bats. On average, rabies causes only one to two human deaths in the U.S. each year. It's not commonly encountered from small rodents (such as squirrels, mice, and rats), opossums, rabbits, and hares.

Most vulnerable groups: People with unvaccinated pets; animal handlers, especially those working with wildlife.

Rabies attacks the nervous system and is effectively always fatal once symptoms occur. One person is documented to have survived rabies without vaccinations (this happened in 2004). But the good news is that people and pets can be protected beforehand with pre-exposure vaccinations. If unvaccinated people have been exposed, the disease can still be prevented—swift medical attention is critical.

How Do You Catch It?

Rabies virus is found primarily in saliva and in the tissues of the central nervous system, especially the brain. It's usually spread through the bite of an infected animal.

It can also be transmitted if the animal's saliva or nervous tissue gets into your eyes, nose, mouth, or an open wound or scratch. Airborne transmission is possible but rare—it's more of a concern for laboratory workers who handle animals, or in moist caves with little ventilation.

You cannot catch rabies from contact with blood, feces, urine, or scent glands. The rabies virus infects the nerves, traveling directly from the bite wound to the brain. Later on, it may travel from the nerves to other organs, but it never enters the blood. That said, if the animal's head has been damaged, there could be spinal tissue or fluid mixed in with splattered blood.

Rabies might be passed from mother to offspring in the womb. However, when people encounter very young animals that are rabid, it's more likely that they were infected after they were born, either from contact with their mother or another rabid animal.

Although deadly, the rabies virus is actually fragile. It can be destroyed by exposure to sunlight (UV light). This is relevant for the clean-up of your gear and of sites. Although we recommend that WCOs mist their gloves with a 10% bleach solution after finishing a job in which rabies exposure was a concern, simply allowing the gloves to dry outdoors is generally sufficient.

Symptoms

An animal may be infected with rabies for a long time before it shows symptoms, anywhere from two weeks to many months. (The incubation period is usually two to three weeks.)

Different species show different signs of the disease. Expect variations even within the same species, because few animals show all of the signs of rabies. Some signs are subtle and easily missed. Unfortunately, you can't tell whether an animal is rabid just by its behavior. Other diseases, such as distemper or toxoplasmosis, can also cause similar symptoms. An animal that's been poisoned by lead, mercury, or antifreeze may also act "rabid." The only way to prove that an animal is rabid is to test its brain tissue in a laboratory. That's why it's smart to take precautions.

Here are the rabies symptoms you may see in wild animals:

- Unprovoked aggression ("furious" rabies). Some animals may attack anything that moves, or even inanimate objects.
- Unusual friendliness ("dumb" rabies).
- An animal may stumble, fall, appear disoriented or uncoordinated, or wander aimlessly.
- Paralysis, often beginning in the hind legs or throat. Paralysis of the throat muscles can cause the animal to bark, whine, drool, choke, or froth at the mouth.
- Vocalizations ranging from chattering to shrill screams.
- Nocturnal animals may become unusually active during the day (Remember, some daytime activity is normal, especially when nocturnal animals are feeding their young. Daytime activity may also be caused by unusual disturbances to their habitat, such as construction or tree-cutting).
- Raccoons walk as if they're on very hot pavement.

Skunks, raccoons, foxes, and dogs usually display furious rabies. Bats often display dumb rabies, and may be found on the ground, unable to fly. This can be very risky for children, who are more likely to handle wild animals than adults. In domestic animals, you should suspect rabies if you see that the animal has a sudden change in disposition, fails to eat or drink, or becomes paralyzed or runs into objects.

You may become aware of another similarity between rabies and distemper. Both are "density dependent" diseases. That means they spread more easily when wildlife populations are higher, because there's more contact between individual animals. When enough animals die from rabies or distemper, these viruses can't spread as easily. That's why the number of cases spikes and then drops off. This rollercoaster pattern repeats over time.

Protection on the Job (and Legally Required Actions)

Situations involving potentially rabid animals should take priority over all other work.

Contact the local health department immediately. Follow their directions. You must report bites or other exposures. (Even such details as how to package the specimen matter, because you don't want to endanger anyone who may handle the package.)

Contact doctors and veterinarians as needed. Local police and dog control officers may also help, especially in emergencies.

Follow these guidelines with all species:

- Use caution when approaching the suspect animal, because many are aggressive and can bite, even if paralyzed. Wear animal handling gloves and use restraining devices to minimize contact with the animal. Avoid bites, scratches, and direct skin contact. Handle dead animals with care, too, especially when your hands are near their mouths. You could scratch yourself on the animal's teeth and be exposed to the virus.
- Restrain and isolate the suspect animal.
- The animal may be choking. Beware the impulse to clear the "obstruction" from its throat! Do not put your hand in or near the animal's mouth.
- If rabies testing is required, you must kill the animal without damaging its head. Brain tissue is needed for the rabies test—that's why the test can't be done on a live animal. Even with this restriction, it's still possible to humanely dispatch animals in most circumstances. For example, raccoons and skunks can be captured in cage traps. The trap can then be placed in a CO_2 chamber and the animal can be euthanized. If the animal is aggressive, and you're in an area where you can legally discharge a firearm, you may prefer to restrain the animal with a catchpole and then shoot it in the heart and lung area, using a low-caliber rifle or pistol. That method reduces contact with the animal and may be safer for the operator.
- If you can't capture the suspect animal, describe the situation to the local health department/ and the doctor in as much detail as possible. They'll want to know which species was involved, how the animal was behaving, whether or not the attack was provoked, and what type of first aid was administered. Immediate medical care should be sought for the exposed person or pet.
- When you're done, disinfect any surfaces contaminated by the animal's fluids or tissues with a 10% bleach solution (one part chlorine

bleach to nine parts water). Wash your hands and any wounds with soap and water, not bleach—but you may want to mist spray your gloves with the bleach solution. Wear protective gear, especially if using a power washer.

- Clean everything that might have been contaminated before you go to your next job.

Additional guidelines for wildlife

- Quarantine is not an option. There's no conclusive research data on safe quarantine periods. The only way to be sure whether the animal is rabid is to kill it and examine its brain tissue.
- Learn your state's protocols for the treatment of bats found in homes. The greatest concern is when a bat is discovered in a room in which a person has been sleeping, because that person might not know if they've been scratched or bitten while sleeping. Talk to your local health department to determine if the bat needs to be killed and tested for rabies. Don't release or discard any bat found in people's living quarters, unless the possibility of human exposure has been absolutely ruled out.
- Individual bats will sometimes enter a home in the evening, especially during July or August. This doesn't mean there's a roost in that home. The bats may just have wandered in, as wild animals, especially young ones, sometimes do while exploring their territories or feeding. If you're sure that no person or pet has had contact with the bat and it appears healthy, it can be released. (Use a soft-sided container to scoop up the bat after it's landed. Plastic yogurt containers or cardboard boxes are less likely to hurt the bat than a metal coffee can.)

Additional guidelines for domestic animals (cats, dogs, ferrets) and livestock

- There are vaccines to protect cats, dogs, ferrets, and livestock from rabies.
- A vaccinated domestic animal that's been exposed can receive a booster to prevent it from developing rabies.
- An unvaccinated domestic pet that's been exposed can be quarantined for six months (at the owner's expense) and observed to determine whether or not it's infected. The other option is to have the animal killed to avoid the possibility of it developing rabies and exposing people or other animals later on.

Additional guidelines for exotic pets (such as monkeys, gerbils, guinea pigs, hamsters, snakes, iguanas, birds)

- Only mammals get rabies. Not birds, not reptiles (such as snakes, iguanas, and turtles), and not amphibians (frogs, salamanders).

- Even though they're living in a home, these exotic pets are not domesticated species. They're wild species that are native to another country. Again, unfortunately, quarantine is not an option. If an exotic pet has been exposed, the health department may require that it be killed and tested.
- A few species of common exotic pets, such as gerbils and guinea pigs, almost never get rabies.

Prevention—Rabies Vaccinations

WCOs and other people in high-risk jobs should get a rabies pre-exposure vaccine. This consists of three shots. Your doctor should test your blood every two years to determine whether you need a routine booster. For more details about the pre-exposure vaccinations, see "Human Rabies Prevention, US 1999 Recommendations of the Advisory Committee on Immunization Practices," www.cdc.gov/mmwr/preview/mmwrhtml/00056176.htm.

First Aid and Treatment After Exposure

The importance of seeking immediate medical care for people and pets who have been exposed to a potentially rabid animal cannot be overemphasized. This is especially true if you can't capture the suspect animal and have no way to determine whether or not it was infected.

If anyone has been wounded, disinfect the wound by washing it thoroughly with soap and warm water. (You can then apply Betadine, a liquid surgical soap available in many drug stores.) Cover the wound with a sterile bandage, then apply direct pressure to control bleeding.

Have the animal tested for rabies. If the animal was rabid, everyone who was exposed will need treatment (this includes pets and livestock). For someone who received pre-exposure rabies vaccinations, the post-exposure treatment amounts to two shots of rabies vaccine given three days apart.

Unvaccinated people who may have been exposed to the rabies virus will be given six shots in a span of 28 days. Again, treatment must begin as soon as possible. Emphasize to your customers that the shots are now given in the arm, not in the stomach. There's a good chance that your customers have heard and believed scary stories about how horrible the rabies treatment is. That's not true! Calm them down and persuade them to call their doctors right away. Call the local health department and give them the contact information for anyone who has been exposed.

What Your Customers May Want to Know

- How are they most likely to encounter rabies? Through an unvaccinated pet, the most common link between rabid wildlife and people. Even indoor cats should be vaccinated because if an infected bat gets inside the house, the cat will probably chase it. If the bat is sick, the cat has a better chance of catching it and being exposed.

- A nocturnal animal that's active during the day is not necessarily rabid. Healthy female raccoons, for example, sometimes feed during the day, especially during the spring, when they're nursing their young.
- A shabby-looking animal is not necessarily rabid. It could be a nursing female. Her young will sometimes pull at her fur as they feed.
- As with other diseases, exposure to the virus does not automatically mean that you are going to get the disease. But there is no way to test or tell whether or not you were infected, and only post-exposure vaccination will prevent rabies from developing.
- Antibiotics will not treat rabies because the disease is caused by a virus. Antibiotics are used to kill bacteria.

General Guidelines for Dealing With Your State's Rabies Lab (Learn Their Specifics)

- Questions regarding the submission of specimens should be directed to your state department of health. Questions about the handling of animal bites should first be directed to your local health department (see government listings in the phone book).
- **Do not submit live animals.** The lab will only accept approved specimens, and then only the intact head. Exceptions include bats, which should be submitted whole, and livestock (instead of sending the whole head, a sample is prepared: some veterinarians will offer to prepare this specimen).
- Acceptable specimens most likely include only:
  a) rabies suspect mammals for which there has been a reported human exposure (bite, scratch, or contact with saliva or nervous tissue).
  b) all bats encountered in situations where it's probable that people wouldn't realize they had contact with a bat.
  c) rabies suspect mammals with reported bites to (or other "intimate" contact with) a domestic animal.
  d) highly suspect animals that are under surveillance (but there have been no reported contacts), including (i) a member of a rabies vector species showing clear signs of rabies infection, from an area without previous confirmed cases; (ii) a mammal not commonly recognized as a rabies vector, but showing clear signs of rabies infection; and (iii) a domestic animal with a neurological disorder that dies or is euthanized under the care of a veterinarian (if the vet suspects rabies).
  e) specimens associated with approved surveillance programs that are part of wildlife vaccination trials.
- Follow the lab's guidelines for packing and delivering specimens.

- Find out when the lab is open and what their protocol is for emergency and after-hours submissions.
- The lab almost certainly has procedures for sharing the test results. Learn what they are so you don't waste their time with unnecessary phone calls. In some cases, the rabies lab informs the local health department, which then calls everyone else.

For more information about rabies:

- www.cdc.gov/ncidod/dvrd/rabies/

What qualifies as a rabies emergency?

- All cases of direct human contact with a bat, when bite or mucous membrane contact cannot be ruled out.
- Cases in which an animal that is very likely infected with rabies has bitten a person, and the doctor is waiting for the results before beginning treatment. "Very likely infected" is determined by the animal's species, behavior, and the location of its capture.

Review Questions

9. Rabies is usually spread through:
  a) The bite of an infected animal.
  b) Contact with blood.
  c) The air.
  d) Skunk spray.

10. You can tell whether or not an animal is rabid by its behavior. (Circle the correct answer.)
  True False

11. The wild species most likely to carry rabies are:
  a) Dogs and cats.
  b) Raccoons, opossums, and skunks.
  c) Skunks, pigeons, and rats.
  d) Skunks, bats, raccoons, and foxes.

12. If you think you're dealing with a rabid animal, you should:
  a) Wear heavy-duty gloves.
  b) Call your local health department.
  c) Shoot it in the head immediately.
  d) Carefully restrain the animal.
  e) More than one answer is correct.

13. What qualifies as a rabies emergency?
  a) A raccoon walked through the customer's yard during the day.
  b) A person was bitten by a bat.
  c) A skunk is living under the deck.
  d) A bat flew into house through an open window, fluttered around, and flew out a few minutes later.

RACCOON ROUNDWORM

Caused by: A parasite

Most common way people catch it: Dirty hand touches mouth. Contaminated object put directly into mouth.

Worst-case scenario: Death

How common in our region? Although the worm is common in raccoons, few people contract the disease.

Most vulnerable groups: Young children

This disease is caused by a parasite, a roundworm called *Baylisascaris procyonis*. The roundworm larvae cause problems as they travel through the person's muscles and various organs, including the liver, brain, lungs, and eyes. The severity of the infection depends on how many of the parasite's eggs were ingested and where the larvae migrate. Although serious infections are rare, raccoon roundworm can be fatal in people.

Raccoons are the primary host of this roundworm, which is commonly found in their small intestines. The parasite has been found in 90 species of mammals and birds, including mice, squirrels, rabbits, woodchucks, beaver, foxes, coyote, woodrats, quail, herons, mourning doves, owls, and wild turkeys. Livestock and pets have also been affected, including guinea pigs, dogs, chickens, pheasants, and pet parrots.

How Do You Catch It?

Raccoons shed millions of the microscopic roundworm eggs in their feces. It takes about a month for newly deposited eggs to develop to the infective stage. The eggs can only develop into worms when they're in an animal's body, but the eggs are hardy and may survive for years in soil, sand, or water.

People may encounter the eggs through direct contact with raccoon droppings or by touching an object or area that's been contaminated by feces. If they don't wash their hands, they may later transfer the eggs to their mouths. Small children are particularly vulnerable because they tend to put their hands and other objects such as bark, wood chips, toys, soil, or even droppings, into their mouths.

Is it possible to become infested by breathing in raccoon roundworm eggs? Yes, but this is highly unlikely. The parasite's eggs are sticky and relatively big (in comparison to other microscopic disease agents) so they don't easily become suspended in the air. If you shoveled up raccoon droppings with great enthusiasm, tossing dust left and right, then you might breathe in and later swallow some of the parasite's eggs. (Generally, parasites of this type infest people by entering through our digestive tracts, not our respiratory systems.) But you'd need to swallow about 5,000 eggs to receive a dose that could infest you. Inhalation is a minor risk and you may wish to take precautions against it, but the major risk for contracting raccoon roundworm is through hand-to-mouth contact.

Other animals may become infested by eating an infested animal or through contact with the feces of an infested animal.

Symptoms

Symptoms in people may include nausea, skin irritations, tiredness, liver enlargement, loss of coordination and muscle control, blindness, inattentiveness, and coma.

Raccoons rarely show symptoms of the disease. However, the species that don't usually play host to this worm (such as woodchucks, squirrels, birds) tend to show abnormal behaviors when infested. They'll tilt their heads and have difficulty walking or climbing. They may lose their fear of people, circle, roll on the ground, fall over, lie on their sides and paddle their feet, or fall into a coma. In woodchucks, the symptoms of raccoon roundworm can be mistaken for rabies.

Treatment

If someone's been exposed, or even suspects exposure to raccoon roundworm, seek immediate medical care. If the worms can be killed before they migrate through the body, there's a very good chance that the disease will be prevented. But if the condition is not treated early, recovery is less assured. Raccoon roundworm infestations are very difficult to diagnose in people.

Protection on the Job

If you're working in an area that's contaminated with raccoon feces, wear a proper respirator, rubber gloves, rubber boots, and disposable coveralls.

Because the roundworm's outer shell is so sticky, you have to change your clean-up method to make sure the cleaning agent can break through this coating to kill the developing worm. Disinfectants alone will not do the trick. Intense heat is the best option for anything that can withstand open flames. Use a handheld propane torch (weed burner) to flame the droppings, contaminated areas of soil or concrete, and your traps. If that's impractical, give those eggs a "one-two punch" by dousing them with boiling water *and* bleach. The bleach works on that sticky coating, while the boiling water kills the egg. And if that's not feasible, double-bag the materials and bury them deeply.

Contaminated clothing can be double-bagged and discarded or washed in boiling water with bleach. Clean your gear while the feces are still fresh and easier to remove. Scrub rubber boots with bleach and a scrub brush.

Advice for Customers

- Have your pets "wormed" three to four times each year.
- To prevent your children from encountering roundworm eggs, keep them away from areas that are frequented by raccoons.

Cover children's sandboxes. Train them to wash their hands and scrub their fingernails after playing outdoors, especially if they were in your garden or the sandbox.

- Discourage raccoons from living in and around your home (management techniques are discussed in Appendix B).
- Prompt removal and destruction of raccoon feces will reduce the risk of human exposure. Raccoons typically defecate at the base of trees, on fallen logs, large rocks, and wood piles, and in barns or other outbuildings. Raccoon feces may also be found in children's sandboxes, attics, fireplaces, garages, decks, rooftops, haylofts, and compost piles.
- Areas of soil or concrete are best decontaminated by a thorough flaming using a handheld propane torch (weed burner). Wooden decks and patios can be cleaned with boiling water and bleach. Soil can be turned over with a rake or shovel, then flamed. Repeat this process several times. To decontaminate a fireplace or woodstove and chimney, build a roaring fire.

For more information about raccoon roundworm:

- www.cdc.gov/mmwr/preview/mmwrhtml/mm5051a1.htm
  www.michigan.gov/dnr/0,1607,7-153-10370_12150_12220-27261--,00.html

Review Questions

14. You can tell if a raccoon is infested with *Baylisascaris* roundworms without looking inside the raccoon. (Circle the correct answer.)
True False

15. Young children are most vulnerable to this disease because:
a) They're more likely to put contaminated objects and their hands into their mouths.
b) They have small bodies, so the worms reach their brains faster.
c) Raccoons prefer them.
d) Their immune systems haven't been exposed to the disease yet.

16. There were many raccoons living in your customer's backyard. They want to know how to clean it up so their kids can play outdoors again. You tell them:
a) "Forget it, and buy them a computer game."
b) "Clean the deck with boiling water and bleach. Soil should be raked and flamed repeatedly. Replace the sand in the sandbox. Build a roaring fire to clean out the fireplace, stove, and chimney."
c) "Don't worry, the rain will take care of it."
d) "Those full body protection 'bunny suits' are very fashionable—remember the computer commercials?"

HISTOPLASMOSIS

Caused by: A fungus (mold)

Most common way people catch it: The only way it can be caught is through breathing in the mold spores.

Worst-case scenario: Death

How common in our region? Common. The CDC estimates that 80% of people living in areas where the mold spores are common have been exposed to histoplasmosis. Most cases are mild. Histoplasmosis causes about 800 deaths in the U.S. each year.

Most vulnerable groups: Very young or very old people, and people with compromised immune systems, because they're more likely to develop the most dangerous form of the disease. Construction workers, WCOs, and cavers, because they're more likely to encounter the spores.

Histoplasmosis is a lung infection caused by inhaling the spores of the fungus, *Histoplasma capsulatum*. There are three forms of this disease.

Most people experience its mildest form, usually showing no symptoms or suffering a minor, flu-like illness that gets better on its own. In fact, many people are unaware they've been infected. The disease can develop into a chronic form that resembles tuberculosis, with the patient's condition worsening over months or years. The most serious version, which affects the fewest people, is called "disseminated histoplasmosis." This means the fungal infection has spread to other organs, and unless treated, it's usually fatal.

Why do some people become so much sicker than others? It probably depends on the number of spores inhaled and the person's age and health. Young children, long-term smokers, and the elderly, especially those already suffering from a lung disease, are more likely to show symptoms of histoplasmosis. People with compromised immune systems (such as those with diabetes, cancer, HIV or AIDS, and transplant recipients) are at the greatest risk for developing the most dangerous form of this infection. Histoplasmosis is one of the most frequent opportunistic infections afflicting people who are HIV-positive.

The *Histoplasma* fungus is common in the central and eastern U.S., especially along the Ohio, Mississippi, and St. Lawrence River Valleys and the Rio Grande. The CDC estimates that as many as 80% of the people living in these areas have been exposed to histoplasmosis. And 10–25% of those residents who are HIV-infected develop the most severe form of the infection. Overall, each year in the U.S., there are about 500,000 infections, 5,000 people hospitalized, and 800 deaths due to histoplasmosis.

How Do You Catch It?

You must inhale the mold spores to catch histoplasmosis. (If symptoms occur, they usually begin about 10 days after exposure.) Spores are often encountered in old or abandoned bird or bat roosts, especially those that are outside or exposed to rain; in chicken coops; and in caves. The spores are airborne. They may be inhaled when contaminated soil or droppings are disturbed.

The disease is not contagious; it's not spread from person to person, or by other animals. Birds and bats do not carry this disease but they are associated with it because their droppings enrich the soil and promote the growth of the fungus.

The fungus is naturally occurring and can grow in various soils, with or without droppings. Although it's almost always associated with soil, the fungus has, in rare cases, been found in droppings alone. It grows readily in the soil beneath bird roosts, but it cannot form spores under the acidic conditions of fresh droppings. This means that an active roost may only give off a few spores. The droppings must dry out and then get wet before spores can be released. Generally, droppings need to accumulate for three or more years underneath a roost before the spores increase to significant levels.

If the soil is stirred up under dusty conditions, massive numbers of spores may be released. Severe epidemics have occurred in association with bird roosts that were bulldozed during construction projects. Once airborne, spores can easily be carried long distances by the wind.

Treatment

Antifungal medications are used to treat severe cases of histoplasmosis. Mild cases usually resolve themselves without treatment.

Protection on the Job

Precautions should be taken when working around old or abandoned roost sites, especially if they've been exposed to rain. Avoid stirring up dust and inhaling spores. If at all possible, schedule jobs when the weather is cool and damp. Wear a proper respirator, disposable coveralls, goggles, gloves, and disposable shoe coverings. If materials are likely to fall from overhead, wear a hood.

Ventilate the area, if possible. Dampen contaminated materials to reduce the amount of dust, and keep them damp as you're working. The droppings can be wiped up with a damp sponge. Double-bag them for disposal. Any dead birds or bats should also be sprayed with a household disinfectant, then double-bagged. An industrial vacuum with a high-efficiency filter can also be used to clean up the site.

Advice for Customers

- If you live in one of the areas in which the fungus is common, it's smart to protect yourself before going near bird or bat droppings. If you're concerned about exposure, avoid dusty places that might be contaminated with the mold, such as construction and demolition sites, caves, attics, and poultry barns with dirt floors.
- Sanitize the soil beneath bat and bird boxes annually. Remove the droppings or guano using proper precautions.
- If you develop flu-like symptoms anywhere from 3–18 days after a potential exposure to the spores, you may want to see a doctor for testing.
- The best way to prevent exposure to histoplasmosis is to discourage birds and bats from roosting within buildings.

For more information about histoplasmosis:

- www.cdc.gov/ncidod/dbmd/diseaseinfo/histoplasmosis_g.htm
- www.cdc.gov/niosh/97146eng.html ("Histoplasmosis: Protecting Workers at Risk")
- www.nlm.nih.gov/medlineplus/ency/article/001082.htm
- www.aegis.org (AIDS education global information system)

Review Questions

17. People catch histoplasmosis from:
   a) Sexual contact with an infected person.
   b) Breathing in the mold spores.
   c) Rubbing their eyes after they've touched an area contaminated with bat droppings.
   d) Cleaning the cat's litter box.
   e) Birds.

18. Histoplasmosis is common in the eastern U.S. (Circle the correct answer.)
True False

19. Typical hot spots for this disease include:
   a) Caves.
   b) Attics that are inhabited by rodents.
   c) Old, abandoned bird or bat roosts.
   d) Poultry barns.
   e) Answers "a," "b," and "c" are correct.
   f) Answers "a," "c," and "d" are correct.

TOXOPLASMOSIS

Caused by: A parasite

Most common way people catch it: Touch contaminated object, then unwashed hands touch mouth. Put contaminated object directly into mouth.

Worst-case scenario: Can cause miscarriages or serious birth defects

How common in our region? Common

Most vulnerable groups: People with compromised immune systems; pregnant women; fetuses. Most infections are mild. Pregnant women can transfer the disease to their fetuses, but it usually doesn't cause problems. Miscarriages and severe birth defects are possible.

Toxoplasmosis is an infection in mammals caused by a microscopic parasite, a protozoan called *Toxoplasma gondii.* Usually, the disease is mild and is often mistaken for a simple cold or viral infection.

There are several ways in which toxoplasmosis resembles histoplasmosis: most people who are infected never realize it; it often causes flu-like symptoms; people with compromised immune systems are at greater risk for developing a much more severe infection; and severe infections may result in brain damage or death.

But there's a significant difference between the two diseases. A pregnant woman can pass the toxoplasmosis infection to her unborn baby. Miscarriages, stillbirths, and severe birth defects, including blindness, cerebral palsy, and mental retardation, are possible. (The disease is more serious if passed on to the fetus in the first three months; however, it's more commonly transmitted later in the pregnancy.)

On average in the U.S., one out of a thousand babies is born with toxoplasmosis each year. But that doesn't tell the whole story. The March of Dimes reports that up to 90% of infected babies appear normal at birth, and 55–85% of them develop symptoms months to years later, suffering from eye infections, hearing loss, and learning disabilities.

Toxoplasmosis is common, but very few people have symptoms because their immune systems usually keep the parasite from causing illness. Symptoms in adults include swollen lymph glands, mild fevers, muscle aches, headaches, tiredness, confusion, and pains that last for a few days to several weeks. Severe infections can result in brain damage and damage to the eyes, and may become chronic.

How Do You Catch It?

People catch toxoplasmosis by eating or handling raw or undercooked meat that's infected with the parasite's eggs (especially pork, lamb, or venison) or through direct contact with infected feces (usually from cats) or contaminated soil. Most people are likely to become infected after cleaning a cat's litter box or gardening. They may touch contaminated soil or other fouled objects, forget to wash their hands, and then transfer the eggs to their mouths. Rarely, people contract toxoplasmosis through organ transplantation or transfusion.

Cats acquire the *Toxoplasma* parasite by eating infected wild animals or raw meat. Most mammals can be infected with this parasite. The eggs take about two days to become infective.

According to the American Veterinary Medical Association, infected cats only shed the eggs for one to two weeks of their lives, right after their first exposure to the parasite. Like humans, cats rarely have symptoms when first infected, so most people don't know if their cat has been exposed to *Toxoplasma*. There are no good tests available to determine if your cat is passing *Toxoplasma* in its feces.

Treatment

In an otherwise healthy person who is not pregnant, treatment is often not needed. Symptoms will usually go away within a few weeks. For pregnant women or people who have weakened immune systems, drugs are available to treat toxoplasmosis. There are tests to determine if a fetus is infected; if so, medication may prevent or reduce the severity of the effects of the infection.

Protection on the Job

Even WCOs who never handle a nuisance complaint involving cats might encounter this disease. Cats and wild animals frequent some of the same areas. Use your standard precautions to avoid contact with cat feces, and, as always, wash your hands well with soap and warm water, especially before you eat, smoke, or prepare any food.

Advice for Customers

- Relax, and consider this in perspective. First, toxoplasmosis is more of a concern if you're pregnant or have a compromised immune system. Even if you fit either of those categories, you can still wash your hands! That's a pretty simple way to avoid contracting this disease.
- There's no need to get rid of pet cats or avoid adopting a cat. To prevent infections, keep cats indoors and feed them dry or canned cat food.
- Feral cats, stray cats, and pets who are allowed to roam outdoors may eat infected small mammals and defecate on your property. Remove food sources to discourage stray cats.
- Avoid handling stray cats, especially kittens. If you want to adopt a stray, or don't know whether a cat you'd like to adopt roamed outdoors or was fed raw meat, talk to your veterinarian before you bring the cat into your home.

- Cover sandboxes, and carefully wash your hands after working in your garden (or wear gloves).
- Clean the litter box daily. If this is done, then any eggs that are present would still be in the non-infective stage.
- Ideally, someone who's healthy and not pregnant should clean the litter box and handle raw meat. If this isn't possible, wear gloves while doing either activity, then wash your hands well with soap and warm water.
- Remember to wash any cutting boards, sinks, knives, and other utensils that might have touched raw meat or unwashed vegetables.
- Cook all meat thoroughly, at least to 150 degrees (no longer pink in the center, or until the juices run clear). Don't sample meat before it is fully cooked. Red meat that's been smoked, cured, or frozen for at least 24 hours is also safe from this parasite. Chicken, other fowl, and eggs almost never contain this parasite, according to the CDC.

For more information about toxoplasmosis:

- www.cdc.gov/ncidod/dpd/parasites/toxoplasmosis/factsht_toxoplasmosis.htm
- www.cdc.gov/hiv/pubs/brochure/oi_toxo.htm
- http://www.marchofdimes.com/professionals/14332_1228.asp
- www.avma.org/careforanimals/animatedjourneys/pethealth/pethealth.asp#6

Review Questions

20. Toxoplasmosis is like histoplasmosis because:

a) They're both caused by a parasite.
b) Both cause flu-like symptoms that are often mistaken for something else.
c) With both diseases, people with compromised immune systems are more likely to develop a severe infection than others.
d) A pregnant woman can pass either disease to her unborn child.
e) Answers "b" and "c" are correct.
f) Answers "a" and "c" are correct.

21. How do people catch toxoplasmosis?

a) Through direct contact with cat feces.
b) By eating or handling raw or undercooked meat.
c) By breathing in the spores.
d) Through gardening.
e) Answers "b" and "c" are correct.
f) Answers "a," "b," and "d" are correct.

22. WCOs just need to know how to answer customers' questions about this disease, because they're not likely to be exposed to it. (Circle the correct answer.)

True False

HANTAVIRUS

(disease name: "Hantavirus pulmonary syndrome")

Caused by: A virus

Most common way people catch it: Cleaning, working, or living in an area that was infested by rodents

Worst-case scenario: Death

How common in our region? Fairly uncommon

Most vulnerable groups: People in close contact with rodent nesting areas (such as those whose homes are infested, agricultural workers, WCOs, and people who clean out buildings that are only used seasonally)

There are several hantaviruses that cause a respiratory disease in people. The disease, called "Hantavirus pulmonary syndrome," was first officially recognized in the U.S. in 1993, but it's not new. Medical records confirmed a 1959 case and, much earlier, Navajos identified a similar disease that might have been hantavirus. So far, it's fairly uncommon and the chances of becoming infected are low. But it's potentially deadly.

The first signs of sickness, especially fever and muscle aches, appear one to five weeks after exposure, followed by shortness of breath and coughing. Once this phase begins, the disease progresses rapidly and hospitalization is often needed within 24 hours.

How Do You Catch It?

In the Northeast, hantavirus is spread by the deer mouse, *Peromyscus maniculatus,* and the white-footed mouse, *Peromyscus leucopus.* Both may invade homes. In the Southeast, the deer mouse, the cotton rat, *Sigmodon hispidus,* and the rice rat, *Oryzomys palustris,* carry the virus. The deer mouse is the primary vector in the U.S.

Rodents shed the virus in their urine, droppings, and saliva. The virus can then become airborne, and people catch it by breathing in these microscopic particles. It's also possible, although less common, for someone to catch the disease from the bite of an infected rodent. If you touch an object that's contaminated with rodent urine, droppings, or saliva and then touch an open wound or your eyes, you could also become infected. It may be possible to catch the disease by eating food or drinking water that's been contaminated by mice. Hantavirus is not transmitted from person to person in the U.S.

WCOs are in a higher risk category than many other people because they frequently enter areas that are infested by mice, such as crawl spaces and attics. They're more likely to disturb the materials that contain the virus, too.

Other hot spots for the disease include buildings that are only used seasonally. When people open up cabins, sheds, barns, garages, or storage facilities for farm and construction equipment, they are likely to encounter rodent infestations. Rodents also infest homes, and sometimes people are unaware of their presence. Sharing a home with rodents puts the residents at higher risk for this disease.

Rodents catch hantaviruses by touching other rodents. Transmission is probably associated with fighting. Rodents are a vector; there's little evidence that they suffer from the disease. No other animals are known to have a direct role in the transmission of hantaviruses.

Treatment

Patients receive supportive treatment, often in the intensive care unit, and antibiotics to prevent secondary bacterial infections.

Protection on the Job

Ventilate the area before entering, if possible. This is particularly important if the building has been closed for a while, such as a seasonal cabin.

Wear rubber gloves and a proper respirator while handling traps containing rodents or cleaning up their droppings, urine, or nest materials. Avoid stirring up dust because it may contain hantavirus. Wet contaminated materials with a 10% bleach and water solution or household disinfectant. Let this soak in, then wipe up with a damp towel or sponge. Mop or sponge the area with disinfectant. Double-bag contaminated materials for disposal. (Any dead rodents should also be sprayed with disinfectant, then double-bagged for disposal.) Disinfect your gear and traps following standard guidelines.

Advice for Customers

- Toys, silverware, and such items can be disinfected, but it's best to discard contaminated foods, drinks, napkins, paper plates, or cups.
- There are several good reasons to exclude mice from your home. They spread diseases, such as hantavirus. They chew on wires and sometimes make their nests in fans or vents, which could lead to fires. They may destroy insulation and cause your heating and cooling bills to increase. Once they're removed, repairs may be needed to ensure that they can't get back in.

For more information about hantaviruses:

- www.cdc.gov/ncidod/diseases/hanta/hps/index.htm

Review Questions

23. Hantavirus is primarily spread by deer mice. (Circle the correct answer.)
True False

24. Hot spots for hantavirus include:

a) Crawl spaces and attics.
b) Buildings that are only used seasonally, such as farm equipment storage sheds, summer camps, and barns.
c) Abandoned bird and bat roosts.
d) Answers "a" and "b" are correct.

25. To protect themselves from exposure to hantavirus, many WCOs wear respirators. Which other diseases are **commonly** caught by airborne transmission?

a) Histoplasmosis.
b) West Nile virus.
c) Rabies.
d) Toxoplasmosis.
e) Raccoon roundworm.

MANGE and DISTEMPER

These two diseases aren't a serious medical concern for people, but they do commonly affect wildlife and pets. People can be infected with mange and canine distemper, but canine distemper doesn't cause symptoms in humans. So why are we including these diseases in this book?

You could pick up the mites that cause mange from contact with a mangy animal or contaminated equipment. If you were careless, you could then give those mange mites a free ride into a new area. Mange can be fatal to wildlife, so try to avoid spreading the disease. If you deal with a mangy animal, clean your equipment carefully before moving on to your next job (handle carcasses with care, too).

Distemper can be mistaken for rabies so it's helpful to know something about the disease.

MANGE

The term "mange" applies to several skin diseases in mammals that are caused by microscopic burrowing mites. It's a contagious infestation that affects many species, including wildlife and domestic animals. Two forms, "sarcoptic mange" (also known as "scabies") and "chorioptic mange" can be transmitted to people.

New hosts pick up the parasites from direct contact with an infested animal or its nesting area, or when the mites leave an abandoned nest in search of a new host. Woodchucks commonly pick up mites when they move into an abandoned fox den. Your customers might become infested if raccoons were removed from their attic but no one cleaned up afterwards, causing the now-hungry mites to wander into the home looking for another source of food and shelter (a new host).

Mange is almost always fatal to red foxes and takes a heavy toll on coyotes, too. It may be a significant cause of death among squirrels during the winter. Although rare in well-fed, well-kept cats, mange is a problem in domestic dogs.

Sarcoptic mange in wildlife and domestic animals usually causes itchiness, hair loss, crusty scabs (often seen first on the head), and thick, wrinkly skin. The skin changes can cause blindness and interfere with the animal's ability to hear or eat. In advanced cases, the animal may be weak and emaciated and smell foul, a result of secondary infections caused by scratching. If untreated, the animal may die of exhaustion, dehydration, or a secondary infection.

In people, this infestation causes a rash that usually looks like pimples, but may look like blisters or an inflammation; it usually appears on the forearms, thighs, and abdomen. Reactions vary. Generally, the mites die off, because they don't do as well on people. Although doctors don't often attempt to kill the mites with drugs, they may offer patients medication to control the itchiness so the patient doesn't scratch constantly, which could invite other infections.

Protection on the Job

- Wear disposable gloves.
- Minimize contact with mangy animals, by using restraining devices. Wash promptly afterwards.
- Minimize contact with potentially infested gear or clothing. Clean your gear if you think it may have picked up mites.
- Freezing kills the mites and may be useful for decontaminating clothing or other gear.

Advice for Customer

- If your pet is scratching a lot or developing bald spots, take it to a veterinarian. Mange is treatable. Treatment for pets usually includes clipping, medicated baths or sprays, and medication.
- The dog's bedding, and household furniture, may need to be treated, removed, or burned.

DISTEMPER

"Distemper" is often used to refer to two diseases that show some of the same symptoms but are caused by different viruses. These diseases affect dogs and cats and many carnivorous species of wildlife, such as raccoons, foxes, and coyotes. *The symptoms can be confused with rabies.* In fact, many raccoons that are tested for rabies actually prove to be suffering from canine distemper. Neither canine nor feline distemper has been shown to cause illness in people.

Dogs suffer from "canine distemper," characterized mostly by respiratory and neurological symptoms. The cat version is called "feline distemper" or "panleukopenia" (previous names included "cat fever" and "cat typhoid"). Members of the raccoon and weasel families may also catch feline distemper.

Panleukopenia symptoms are usually associated with diarrhea and anemia. The term "panleukopenia" means "total loss of white blood cells." That should give you a sense of the seriousness of this disease. Panleukopenia is a complex disease. Its symptoms can range from very mild to extreme and are sometimes atypical. It has been mistaken for poisoning or swallowing a foreign object.

Canine distemper is the greatest single disease threat to the world's dog population. It's highly infectious and it's nearly impossible for a dog to avoid exposure to this virus. More than 50% of the adult dogs and 80% of puppies that contract it die. Other effects include permanent problems with vision, hearing, sense of smell, and partial or total paralysis. Younger dogs and puppies are the most vulnerable (in wildlife, most cases are seen in the spring and summer, when wild canids, such as foxes and coyotes, bear their young).

Panleukopenia is also a highly contagious and common viral disease. Cats may catch distemper from contact with the blood, urine, droppings, nasal secretions, or fleas of infected cats or from contact with contaminated bedding, cages, food dishes, and the hands or clothing of handlers. Young kittens, sick cats, and indoor cats that have not been given boosters recently are most susceptible; older cats are more likely to have acquired some immunity. Pregnant female cats that contract the disease, even in its mildest form, may give birth to kittens with severe brain damage. Kittens younger than 16 weeks of age may die at a rate of about 75%, and others at a rate of 50%.

Advice for Customers

- Vaccinate your pets. Vaccinate your pets. Vaccinate your pets. (Think they'll get it?)
- Both diseases are so common and the symptoms are so varied that any sick young dog or sick cat should be taken to a veterinarian.
- The virus that causes feline distemper is resistant to many chemicals and may remain infectious at room temperature for as long as one year.
- Indoor cats can catch this disease because you may bring the virus into the house if you've petted an infected cat or if it's brushed against your clothing. Hand-me-down cat carriers, beds, or food dishes could be contaminated.

For more information:

Mange

- http://edis.ifas.ufl.edu/scripts/htmlgen.exe?DOCUMENT_MG118
- www.michigan.gov/dnr/0,1607,7-153-10370_12150_12220-26949--,00.html

Canine Distemper

- www.avma.org/careforanimals/animatedjourneys/pethealth/canine.asp#3
- www.michigan.gov/dnr/0,1607,7-153-10370_12150_12220-26505--,00.html

Feline panleukopenia

- www.avma.org/careforanimals/animatedjourneys/pethealth/feline.asp#4
- www.michigan.gov/dnr/0,1607,7-153-10370_12150_12220-26505--,00.html#Feline

Review Questions

26. People can catch mange, but they can't catch distemper. (Circle the correct answer.)
True False

27. Distemper can be confused with rabies because the symptoms are similar. (Circle the correct answer.)
True False

WEST NILE VIRUS

(disease name: West Nile virus-associated illnesses, including West Nile encephalitis, West Nile meningitis, and West Nile fever)

Caused by: A virus

Most common way people catch it: Mosquito bite

Worst-case scenario: Death

How common in our region? The range of the virus is expanding. Because it's mosquito-borne, many people can be exposed to the virus. In 2005, 3,000 cases were reported in the U.S., with 119 fatalities. Of those cases, 43% resulted in the more serious forms of disease: encephalitis or meningitis. In the northern U.S., the epidemic peaks in the late summer through the fall, while in warmer regions, cases are seen throughout the year.

Most vulnerable groups: The elderly and those with compromised immune systems. People over the age of 50 are at higher risk for severe disease.

Please note: Researchers are currently seeking answers to some important questions about this virus.

West Nile virus affects more than 70 species of domestic and wild birds (especially American crows, jays, hawks, and owls) and mammals (especially people and horses). It's also been found in bats, chipmunks, raccoons, skunks, squirrels, domestic rabbits, dogs, cats, mountain goats, and reindeer—and was recently identified as the cause of death of a few American alligators, although we don't yet know how susceptible reptiles are to this virus. Most mammal species do not become sick if infected with West Nile virus.

The virus causes a variety of symptoms, which usually appear in 3–14 days (although the precise incubation period is not yet known). Most common are a group of mild, flu-like symptoms including fever, rash, tiredness, and loss of appetite (West Nile fever). Some people develop a serious infection of the nervous system: West Nile encephalitis (inflammation of the brain) or West Nile meningitis (inflammation of the spinal cord and the membrane around the brain). A smaller group suffers paralysis or death.

The disease was first reported in the eastern United States in the summer of 1999, although it may have been present before then. It's now believed to be permanently established in the U.S. There's good evidence that birds spread the disease as they migrate.

How Do You Catch It?

People primarily catch West Nile virus from the bite of certain species of mosquitoes (if that mosquito was infected). The disease lives in the insects' salivary glands. Mosquitoes are vectors of West Nile virus, which means that they can transmit the disease, but do not suffer from it themselves. The northern house mosquito, *Culex pipiens*, is an important vector of West Nile virus but we're not sure what role it may play in the transmission of the virus to mammals. In the eastern U.S., this is the most common mosquito around the home. There are other ways a person might be exposed to West Nile virus, such as from blood transfusions, organ transplants, and breast milk, or by handling infected birds (the virus may be present in feathers, feces, or dander).

Normally, the virus cycles between mosquitoes and birds. When an infected mosquito bites a bird, it transmits the virus to the bird. The virus circulates in the bird's blood for a few days. Uninfected mosquitoes that bite an infected bird may then pick up the virus, which reproduces in the mosquito's body. Once there's a certain amount of virus in the mosquito's body, it can transmit the virus to another animal.

Mosquitoes don't bite only birds, of course. When an infected mosquito bites a person, horse, or other mammal, it transmits the virus to that animal. The virus then multiplies and may cause illness. But the virus doesn't reproduce effectively in mammals, so an uninfected mosquito that bites an infected mammal probably can't pick up the virus. That's why mammals are currently considered "dead-end" hosts for the West Nile virus.

Some species of mosquito prefer to feed on birds, some tend to bite large mammals such as horses, and others bite people. Which mosquitoes find which animals tastiest? Do their preferences change during the season? Researchers are trying to answer these questions, which will help us better understand how, when, and where people are most likely to catch West Nile virus.

Right now, West Nile virus is infecting many birds and horses. Some wildlife biologists are concerned that West Nile virus could seriously affect the populations of some rare or endangered birds. When dogs and cats are infected, the virus doesn't appear to cause serious illness in them.

Treatment

People infected with West Nile virus receive general supportive treatment. There is no vaccine for people yet, but specific medications and a vaccine are under development. There is a West Nile virus vaccine for horses produced by Fort Dodge under a conditional license, pending results from studies of its effectiveness and potential side effects. Other equine vaccines are under development. Horses vaccinated against other related viruses, such as eastern equine encephalitis, western equine encephalitis, and Venezuelan equine encephalitis, are not protected against West Nile virus.

Protection on the Job

Until we have a better understanding of which mosquitoes are most likely to transmit the virus to people, we can't offer a strategy that takes advantage of the particular habits of the vector. For example, some mosquitoes feed during the day while others are active from dusk to dawn. They're found in different habitats. And mosquitoes may not be the only vector of West Nile virus. Scientists are currently studying whether or not the bird louse can transmit the virus to people.

With all these unknowns, what should you do? Avoid mosquito bites and don't handle dead birds with your bare hands (wrap them in layers of plastic instead). Wear loose-fitting clothing or mosquito repellent to reduce your chance of being bitten by an infected mosquito. Light colors are less likely to attract mosquitoes. Long sleeves aren't a bad idea, either. And stay tuned!

Advice for Customers

- If you notice conditions at your customer's home that would encourage mosquito breeding, you may wish to point these out. The *Culex* mosquito has a small home range. It usually doesn't fly more than 300 feet from its breeding site, so it's possible to dramatically reduce the number of *Culex* mosquitoes near the home. These mosquitoes breed in small pools of water that contain leaves or other debris.
- Dump out standing water from rain gutters, recycling bins, tires, tarps, children's pools, garbage can lids... wherever it collects.
- Turn over items that might collect water, like wheelbarrows and pools.
- Change the water in birdbaths at least once a week.
- Clean the debris from rain gutters early in the spring. Clean up any garbage piles.
- Keep roadside ditches clear, so water can flow freely through them.
- Remove decaying plants and any other floating debris from garden ponds.
- Stock ornamental ponds with mosquito predators such as dragonflies and goldfish to reduce mosquito populations. Or aerate the pond by adding a fountain. There are a variety of pesticides that will kill mosquito larvae in water; read the labels.

For more information about West Nile virus:

- www.cdc.gov/ncidod/dvbid/westnile/index.htm
- www.cfe.cornell.edu/ERAP/WNV

Review Questions

28. People generally catch West Nile virus from:

a) Birds.

b) Bats.

c) Mosquitoes.

d) Breathing in spores.

29. WCOs are not at greater risk for this disease than the average dog owner. (Circle the correct answer.)
True False

30. This disease is more deadly than people realize. (Circle the correct answer.)
True False

LYME DISEASE

Caused by: A bacterium

Most common way people catch it: Bite from the black-legged tick (a.k.a. "deer tick") is the primary way it is caught

Worst-case scenario: Can lead to disabling conditions including arthritis, problems with the heart and nervous systems, and personality changes

How common in our region? Common. (In 1999, for example, 92% of the cases reported to the CDC were from Connecticut, Rhode Island, New York, Pennsylvania, Delaware, New Jersey, Maryland, Massachusetts, and Wisconsin.)

Most vulnerable groups: People who spend more time outdoors and those who work in landscaping, forestry, and related jobs

Lyme disease is caused by a bacterium, *Borrelia burgdorferi.* The bacteria are transmitted to people by bites from the black-legged (or "deer") tick, *Ixodes scapularis.* There are more than 16,000 infections in the United States each year. If untreated, Lyme disease can cause arthritis, especially in the knees; heart problems; cognitive disorders; sleep disturbance; fatigue; and personality changes. Infrequently, Lyme disease may be severe, chronic, and disabling, but it's rarely if ever fatal.

Some infected individuals show no symptoms, or only mild, flu-like symptoms such as fever, headache, fatigue, and muscle pain. Symptoms often begin 7 days after infection, although they may be noticeable anywhere from 3–30 days afterward.

Early symptoms include a flu-like illness with a headache, slight fever, muscle or joint pain, neck stiffness, swollen glands, jaw discomfort, and inflammation of the eyelids. A diagnostic rash appears in 65–75% of cases. This red rash starts at the site of the tick bite. It often has a bull's-eye appearance. The rash grows quickly, often forming a nearly circular lesion of about 1–8 inches, and may burn or itch. Additional smaller skin lesions may appear at other sites of the body and may last for days or weeks.

Additional symptoms may develop in untreated patients. These often include arthritis, especially of the knees. The joint pain and swelling usually occur one or more months after infection, may involve one or more joints, and may recur in different joints. Problems might develop with the heart, muscles, and nervous system, causing cognitive disorders, sleep disturbance, fatigue, and personality changes.

The disease is more common in coastal areas with sandy soils. People who spend more time outdoors and those who work in landscaping, forestry, and related jobs are at higher risk for the disease. Gardening, hiking, camping, fishing, and hunting may expose you to ticks with Lyme disease. WCOs are not necessarily in a higher-risk category, but should take precautions, especially if working in an area favored by ticks.

The disease's name comes from its discovery in Lyme, Connecticut, in 1977.

How Do You Catch It?

People contract the disease when an infected tick feeds on them—if the tick has remained attached to their bodies long enough to transmit the disease. That usually takes about two days. If the tick is located and removed within 24 hours, infection is unlikely.

For Lyme disease to exist in an area, at least three things must be present outdoors: the Lyme disease bacteria; black-legged ticks that can transmit the bacteria; and host mammals such as mice and deer that provide a blood meal for the ticks through their various life stages.

Lyme disease has a more complicated transmission cycle than the wildlife diseases discussed earlier because the life cycle of the vector, the black-legged tick, takes two years to complete. Although the black-legged tick may be infected with the Lyme disease bacteria in three of its four life stages, it's almost always a nymphal tick that transmits the disease to people. That's because larval ticks are rarely infected when they're feeding. And adult ticks are much larger, so they're more likely to be removed before they've had a chance to transmit the bacteria to the person.

Lyme disease affects humans and dogs but doesn't cause illness in deer.

Treatment

Patients treated with antibiotics during the early stages of the disease usually recover quickly and completely. Even patients treated during later stages generally respond well and recover. A vaccine is available for dogs, but there's no reliable vaccine for people yet.

Protection on the Job

When possible, avoid tick-infested areas, especially from May through July. These ticks prefer deciduous forests and habitats that contain leaf litter. They need moist cover from the wind and other elements. Suburbs often contain prime habitats (forested areas interspersed with residential development); ticks also prefer overgrown brushy areas, particularly along forest edges.

To find ticks more easily on your clothing, wear light colors. Tuck your pant legs into your socks. You could also tape the tops of your socks over your pant legs with duct tape for added protection. If you decide to use tick repellent, follow the label's instructions carefully.

Check yourself for ticks while you're outdoors, and remove any you find. (Try doing this during your lunch break.) Once indoors and naked, inspect your body carefully and remove ticks. Make sure to do this by bedtime. Showering, with vigorous scrubbing, may also help to remove ticks. After outdoor activities, remove and wash your clothing promptly and dry it at a high temperature.

If a tick has been attached to your body for more than 24 hours, place the tick in alcohol or a vial for identification. Note the date of removal. If it proves to be a black-legged tick, see a doctor.

Advice for Customers

- Research has shown that tick populations can be reduced by 72–100% when leaf litter is removed. Clean up your yard and keep it well trimmed. Remove brush piles and wood piles from areas right around the house.
- Carefully inspect yourself and your pets for ticks soon after going indoors, and remove any ticks right away.
- To remove a feeding tick, dab it with alcohol. If feeding has just started the tick may release easily. If it doesn't let go in a few minutes, take a pair of tweezers with a pointed tip and grasp the tick close to your skin. Pull steadily until the tick is removed.
- Don't be alarmed if the tick's mouthparts remain in the skin. The bacteria that cause Lyme disease are contained in the tick's midgut, not its mouth. Just cleanse the area with an antiseptic to prevent infection.
- Don't use petroleum jelly, a hot match, nail polish, or other products to remove a tick. Grasping the tick at its back end, or heating it, can force disease organisms into the wound.
- Hunting deer won't significantly reduce the risk of exposure to Lyme disease because a typical annual harvest doesn't reduce deer density enough to significantly reduce the tick population. The primary reservoir for Lyme disease includes small rodents, not deer.
- Ticks do not like sunny, dry environments. Use bark chips or gravel to create a 3-foot-wide barrier between woodland areas and lawns to keep ticks away from the areas you use. Keep decks and children's play equipment more than 10 yards from brushy or woodland edges.

For more information about Lyme disease:

- www.cdc.gov/ncidod/dvbid/lyme/index.htm

Review Questions

31. What's the best way to remove a tick from your body?

a) Hold a lit cigarette or hot match next to the tick.
b) Smother it with petroleum jelly ("Vaseline").
c) Dab it with alcohol, then pull it out with a pair of tweezers.
d) Shower and scrub furiously.

32. If you remove the tick soon enough, you're not likely to get the disease. (Circle the correct answer.)

True False

Basic tips to reduce risks from wildlife diseases

What you need to know:

- Which wildlife diseases are present in your area.
- How they're spread.
- Symptoms in wildlife and in people.
- How to recognize disease hot spots.
- Recommended safety gear for different situations.

Before you start a job:

- Get pre-exposure rabies vaccinations and tetanus shots. Stay current.
- Have emergency phone numbers handy—local police, animal control, department of health, state wildlife department, your doctor.
- Vaccinate your pets.
- Wear protective gear when needed, and know how to use it properly.

While you're working:

- Wash your hands thoroughly and often, especially before you eat, drink, or smoke.
- Keep your gear clean, too.
- Record all animal contact in your daily log.
- Be even more careful when handling a sick animal or one that's behaving oddly.
- If you've been bitten or scratched, or are sick, go to the doctor promptly. Tell your doctor about your work.
- Dispose of animals and any contaminated materials safely.

After you're done for the day:

- Clean your gear, truck, and clothing.
- Shower and scrub well.

What Do You Do With Contaminated Materials and Dead Animals?

Most WCOs do a certain amount of clean-up of the site as part of their service. If the site presents a formidable mess, some WCOs contract for clean-up separately while others recommend a cleaning service. Whether you do a little or a lot of cleaning on site, you'll still need to clean your gear and your truck.

Clean and disinfect your equipment with a commercial disinfectant or a 10% chlorine bleach solution—that's one part bleach to nine parts water. This doesn't last long, so mix up a new batch every day. Many WCOs keep a 10% bleach solution in a quart-sized spray bottle in the truck for quick, small cleaning jobs. (Always read the label on any cleaning product or disinfectant before using it.)

Never mix bleach and ammonia! A toxic gas will result. A similar gas may be produced when bleach is applied to bird or reptile droppings. Animal urines, and the droppings of other wildlife, are generally too low in ammonia to cause this reaction.

Remember, if you're working around bird, bat, or rodent droppings or nest materials, don't stir up dust. Wear the proper safety gear. Use disinfectant to wet down anything that might have been contaminated, including any dead animals. Wipe up with a damp towel or sponge, or use a commercial, heavy-duty vacuum.

Cleaning up after raccoons is a bit trickier. The eggs of the parasite that causes raccoon roundworm are resistant to disinfectants. Contaminated materials, feces, and soil should be burned, if possible. Flame your metal traps using a propane torch, or use boiling water and bleach. For those materials that can't be burned, either wash with boiling water and bleach (a good option for decks, porches, and contaminated clothing) or double-bag the material and bury it deeply.

To refresh your memory for safety protocols, here are the diseases you're more likely to encounter when working with:

Birds:	Histoplasmosis
Bats:	Histoplasmosis, rabies
Rodents:	Hantavirus (squirrels often get mange)
Raccoons:	Rabies, raccoon roundworm, feline distemper, canine distemper
Foxes, coyotes:	Rabies, mange, canine distemper

All of these species attract a variety of parasites, too. And although they don't often show symptoms, woodchucks, squirrels, and birds may be infected with raccoon roundworm. Any mammal can be infected with rabies. West Nile virus may be transmitted to people who handle the carcasses of infected birds.

Carcasses and other potentially contaminated materials (gloves, protective clothing, nesting materials) must be disposed of properly, because they can also spread diseases. Your health department may offer specific disposal instructions. Otherwise, these items may be buried, burned, or sent to a landfill.

Here are some points to consider for each disposal method, based on recommendations proposed by the National Wildlife Control Operators Association:

Burial
- Secure the landowner's permission.
- Pick a location that will protect both surface water and groundwater from contamination. The grave should be at least 200 feet away from any wells used to supply drinking water.
- Ideally, the carcass should be covered with two feet of soil within a day of burial. (This can be extremely difficult during the winter, so you may need to switch to another method.)
- If using a common grave (for multiple animals or species), it should not be located within a 100-year floodplain zone or a wetland area.

Incineration
The incinerator must have approval from both state and local authorities to burn pathological wastes. A local veterinary hospital or pet cemetery may be willing to dispose of your carcasses, but may charge a fee.

Removal to a standard landfill. (Type II-licensed solid waste disposal facility, a.k.a. "the dump")
- Double-bag the carcass and any contaminated materials in heavy-duty plastic garbage bags, or use other suitable airtight containers.
- Follow state and local regulations for solid waste disposal.

To learn more...

- Develop a short talk about rabies prevention, and offer it during a community safety fair (police departments often sponsor these events). Speak to your local county health department's rabies coordinator for help in preparing your talk. They may be able to offer brochures or videos.

- Write a short piece for a local newspaper (often known as an editorial, or "op-ed') explaining why it's important to vaccinate your pets. Check with a local veterinarian or the state's Veterinary Medical Society for more information on pet vaccination.
- Create tip cards for your customers that explain which diseases are carried by a particular species, how those diseases are spread, and how to lower your chances of catching the disease. For example, the raccoon card might list rabies, canine distemper, feline distemper, and raccoon roundworm, give the CDC and rabies lab websites for more information, explain how people and pets catch these diseases, and then how to clean up afterwards.

Review Questions

33a. You've been hired to remove a raccoon from the rafters of a warehouse that's located in a city. Which safety issues should you consider when setting up the job? (Check all those that apply.)

___ Mange
___ Tetanus
___ Rabies
___ Bites
___ Toxoplasmosis
___ Raccoon roundworm
___ Falls
___ Distemper
___ Heat stress
___ Histoplasmosis
___ Hantavirus
___ Lyme disease
___ West Nile virus

33b. And what safety equipment would you use to protect yourself in this situation? (Check all that apply.)

___ Respirator
___ Helmet
___ Gloves
___ Restraining device
___ Water bottle
___ Trap
___ Ladder brace
___ Ropes or bungee cords
___ Disposable clothing

34. There are no animals currently present, but which of these wildlife diseases might you still encounter while working in an attic? (Check all that apply.)

___ Mange
___ Tetanus
___ Rabies
___ Toxoplasmosis
___ Raccoon roundworm
___ Distemper
___ Heat stress
___ Histoplasmosis
___ Hantavirus
___ Lyme disease
___ West Nile virus

35. While working on a roof trapping squirrels, you notice children taunting a dog on the street below. Even from this distance, you can see that the dog is skinny, its fur is patchy, and it's uncoordinated. The dog runs into a fire hydrant. You have a bad feeling about this situation. Who would you call?

a) A university professor.
b) A veterinarian.
c) An animal control officer.
d) The health department.
e) The police.
f) A doctor.
g) Another WCO.
h) An animal shelter.
i) A wildlife rehabilitator.
j) A TV station—this could be your fifteen minutes of fame!

36. A business owner hires you to remove pigeons that have been fouling cars in her parking lot. She's worried that her customers might be exposed to rabies. You tell her:

a) "The customers probably didn't get too close to the pigeons, so the chance of exposure is low. I wouldn't worry about it."
b) "Birds don't carry rabies."
c) "Pigeons are nocturnal, so the customers won't be nearby when the birds are active."
d) "I drive past this lot every day and hardly ever see a car. And by the way, can I have a deposit?"
e) "To play it safe, call the health department. They'll probably want to interview all of your customers."

37. Airport staff are worried about risks posed by wildlife. They've written the following list. Briefly explain how one of these problems might happen.

- Financial: potential damage to airplanes.
- Safety: plane crashes; failure of traffic control system.
- Health: sparrows flying inside buildings—is this a disease threat? Any worker safety issues from mouse droppings in kitchen?

Zoonotic Diseases in the Northeast

Transmission routes (how people contract the disease):

Fecal-oral: Person touches contaminant and then, with unwashed hands, touches mouth.
Ingestion: Person eats or drinks contaminated product.
Inhalation: Person breathes in disease agent.
Wound/eye/nose/mouth: The contaminant invades a person's wound, eyes, nose, or mouth.

Sample listing:

Disease name
Agent (what causes the disease)
Vectors (animal carriers in the Northeast)
Route (how people contract the disease)—see above
Symptoms in people

Ancylostomiasis, Cutaneous larval migrans, Creeping eruptions

Agent: Parasite (hookworm)

Vectors: People, dogs, cats

Route: Worm burrows into skin; ingestion

Symptoms: Meandering, itchy red lines on skin

Anthrax

Agent: Bacteria (spores)

Vectors: Most mammals, mostly food animals, rabbits, mice

Route: Inhalation; ingestion; wound/eye/nose/mouth via contact with infected carcass

Symptoms: (Two forms) flu-like, pustules on skin, cough, respiratory failure, vomiting, diarrhea. Can be fatal.

Aspergillosis

Agent: Fungal spores

Vectors: None, but grows in soil enriched by bird and bat droppings (especially pigeons, poultry)

Route: Inhalation

Symptoms: Mild, flu-like pneumonia

Babesiosis

Agent: Parasite (protozoa)

Vectors: Wild and domestic animals, white-footed mouse, meadow vole, white-tailed deer (via tick)

Route: Tick bite

Symptoms: Irregular fever, chills, headache, muscle pain, fatigue

Botulism (food poisoning)

Agent: Bacterial toxin

Vectors: Birds, mammals

Route: Ingestion

Symptoms: Nausea, vomiting, disturbed vision, weakness

Brucellosis

Agent: Bacteria

Vectors: Hoofed animals

Route: Ingestion (milk); fecal-oral; wound/eye/nose/mouth

Symptoms: Intermittent fever, chills, headache, body aches, weakness, weight loss

Campylobacteriosis

Agent: Bacteria

Vectors: Hoofed animals, feral cats and dogs, birds, young animals

Route: Fecal-oral

Symptoms: Sudden gastrointestinal disease, diarrhea, abdominal pain, fever, pseudoappendicitis

Cat scratch disease

Agent: Bacteria

Vectors: Feral cats, dogs, squirrels

Route: Cat bites, scratches, or licks person

Symptoms: Fever, malaise, swollen lymph nodes

Cryptococcosis

Agent: Fungal spores

Vectors: Pigeons, starlings

Route: Inhalation; ingestion

Symptoms: Meningitis. Can be fatal.

Cryptosporidiosis

Agent: Parasite (protozoa)

Vectors: Rodents, birds (poultry), ruminants, fish, reptiles, domestic animals

Route: Fecal-oral; waterborne; wound/eye/nose/mouth

Symptoms: Enteritis, diarrhea, low fever, malaise, nausea, vomiting, abdominal cramps

Dermatomycosis

Agent: Fungus (ringworm)

Vectors: Dogs, cats, cattle, horses, coyote, foxes, other mammals, birds

Route: Wound/eye/nose/mouth

Symptoms: Ring-like lesion, scaling, redness

Encephalitis (several related diseases, including West Nile, Eastern, Western, and St. Louis encephalitis)

Agent: Viruses

Vectors: Birds, bats, some rodents, jackrabbits (via mosquitoes)

Route: Mosquito bites

Symptoms: Mild flu-like to severe. Can be fatal. (There's a high fatality rate in horses afflicted with Eastern equine encephalitis.)

Ehrlichiosis (two forms)

Agent: Bacteria (rickettsia)

Vectors: Dogs, cattle, other mammals (probably via tick)

Route: Tick bite

Symptoms: Flu-like, sometimes a rash appears

Giardiasis

Agent: Parasite (protozoa)

Vectors: People, dogs, beavers

Route: Fecal-oral; ingestion; environmental contamination (water)

Symptoms: Diarrhea; greasy, frothy, and smelly feces; cramps; flatulence

Hantavirus pulmonary syndrome

Agent: Virus

Vectors: Deer mice, white-footed mice

Route: Inhalation, rodent bite

Symptoms: Flu-like. Can be fatal.

Histoplasmosis

Agent: Fungal spores

Vectors: None, but grows in soil enriched by bird and bat droppings (especially pigeons)

Route: Inhalation

Symptoms: Mild, flu-like pneumonia. Can be fatal.

Leptospirosis (Weil's disease)

Agent: Bacteria

Vectors: Rodents, rabbits, foxes, skunks, raccoons, opposum, deer

Route: Fecal-oral; ingestion; wound/eye/nose/mouth

Symptoms: Fever; jaundice; pain in stomach, joints, or muscles; nausea. Can be fatal.

Lyme disease

Agent: Bacteria

Vectors: Wild rodents (some mice, chipmunks), raccoons, deer, rabbits, birds (via deer/black-legged tick)

Route: Tick bite

Symptoms: Flu-like, rash, arthritis

Lymphocytic-choriomeningitis

Agent: Virus

Vectors: House mouse and rats

Route: Inhalation; fecal-oral; rodent bites

Symptoms: Flu-like, inflammation of testes, joint pain, paralysis, coma

Mange

Agent: Parasites (mites)

Vectors: Dogs, cats, foxes, coyotes, rodents, other mammals

Route: Mite bites; wound/eye/nose/mouth

Symptoms: Crusty, red rash between fingers and on back of hands and arms

Psittacosis

Agent: Bacteria

Vectors: Birds (especially pigeons), parakeets, poultry, waterfowl

Route: Inhalation, especially from dry feces

Symptoms: Flu-like, anorexia, pneumonia. Can be fatal.

Q (query) fever

Agent: Bacteria

Vectors: Mammals (especially sheep, goats), birds

Route: Wound/eye/nose/mouth

Symptoms: Flu-like, high fevers, pneumonia, hepatitis. Can be fatal (rarely).

Rabies

Agent: Virus

Vectors: Mammals, especially raccoons, striped skunks, foxes, bats

Route: Animal bite or scratch; wound/eye/nose/mouth; rarely, inhalation

Symptoms: Paralysis, convulsions, coma. Nearly always fatal.

Raccoon roundworm

Agent: Parasite (roundworm)

Vectors: Raccoons

Route: Fecal-oral

Symptoms: Nausea, tiredness, enlarged liver, loss of coordination, coma. Can be fatal.

Rickettsial pox

Agent: Bacteria (rickettsia)

Vectors: House mice

Route: Mite bite

Symptoms: Resembles chicken pox

Rocky Mountain spotted fever

Agent: Bacteria (rickettsia)

Vectors: Wild rodents, rabbits, hares, carnivores, birds (via tick)

Route: Tick bite

Symptoms: Rapid onset fever, headache, muscle aches, nausea, vomiting, abdominal pain, rash, loss of muscle control. Can be fatal.

Salmonellosis (food poisoning)

Agent: Bacteria

Vectors: Rodents, swine, cattle, wild birds, poultry, pet reptiles (turtles, iguanas, etc.)

Route: Ingestion; wound/eye/nose/mouth

Symptoms: Sudden headache, fever, abdominal pain, nausea, diarrhea, vomiting

Schistosomiasis (swimmer's itch)

Agent: Parasite

Vectors: Rodents, waterfowl

Route: Waterborne (can penetrate intact skin)

Symptoms: Rash

Sporothrichosis

Agent: Fungal spores

Vectors: Cats, dogs, horses, soil, plants

Route: Wound/eye/nose/mouth via soil or plants

Symptoms: Skin lesions, spreads along lymph nodes

Toxocariasis, Visceral larval migrans

Agent: Parasite (roundworm)

Vectors: Dogs, cats, raccoons, other mammals

Route: Ingestion of eggs (soil, contaminated items)

Symptoms: Fever, cough, wheezing, eye problems, brain damage

Toxoplasmosis

Agent: Parasite (protozoa)

Vectors: Cats

Route: Fecal-oral; ingestion (of contaminated eggs in meat)

Symptoms: Fever, malaise, muscle pain, headache. Can be fatal.

Tularemia

Agent: Bacteria

Vectors: Wild rodents, rabbits, hares, muskrats, carnivores, birds, hoofed animals (deer), bull snakes

Route: Bites from ticks or biting insects; inhalation; ingestion. Can penetrate intact skin.

Symptoms: Mild to severe; pneumonia, ulcer at bite site, swollen lymph nodes, death

Typhus

Agent: Bacteria (rickettsia)

Vector: Rodents (including rats, mice, squirrels), opossum (via rat flea or body louse)

Route: Flea or lice bites; inhalation; wound/eye/nose/mouth

Symptoms: Fever, severe headache, chills, general pains, possible rash

In our region, these are the diseases you'd most likely be exposed to by...

MAMMALS

- Rabies (any mammal, but most commonly from skunks, raccoons, foxes, bats)
- Raccoon roundworm (mostly raccoons; also mice, squirrels, woodchucks, rabbits)
- Histoplasmosis (from bat roosts and caves, not directly from bats)
- Hantavirus (deer mouse, white-footed mouse)
- Mange (foxes, coyotes, squirrels)
- Toxoplasmosis (contact with contaminated meat, cat feces, or soil)

BIRDS

- Histoplasmosis (from bird roosts, poultry barns, not directly from birds)
- Possibly West Nile virus from contact with the carcasses of infected birds

REPTILES

- Salmonellosis

PARASITES AND INSECTS

- West Nile virus (mostly from mosquito bites)
- Lyme disease (from ticks; people are most likely to be infected from May through July)

HEALTH ISSUES FROM OTHER WCO JOB HAZARDS

- Falls
- Car accidents
- Heat-related illnesses
- Tetanus infection (puncture wound, contact with infected nail) or other infections from wounds
- Exposure to toxic substances such as pesticides and fiberglass, which are often found in attics
- Electrocution and other hazards caused by dangerous sites
- There are some venomous snakes in the eastern U.S.

Chapter 5
Best Practices, a Step-by-Step Guide

Overall learning objectives for this chapter

- Understand the best practices approach to solving wildlife conflicts.
- Know how to interview people and inspect sites to figure out what caused the problem.
- Describe the different management options available to your customers.
- Be familiar with the techniques used to remove wildlife from buildings and outdoor areas.
- Identify the different euthanasia methods, and know when and why you might use each one.
- Understand how to prevent wildlife damage.
- Know why it's important to evaluate your practices and what you can gain from this effort.

SECTION ONE: Introduction

LEARNING OBJECTIVES

5.1 List the five steps involved in the best practices approach to solving wildlife conflicts.

5.2 Give one example of why it's crucial to properly identify which species is causing the problem.

As a wildlife professional, we hope you'll help your customers understand that they can protect their homes and businesses from wildlife damage without banishing all wildlife from the community. Whenever possible, use best practices and encourage your customers to solve their problems for good—not just temporarily.

In Chapters One and Two, we described a best practice as an effective method for solving a nuisance wildlife problem that also minimizes risks to the environment and our health and well-being. This decision-making strategy balances concerns about safety; the humane treatment of wildlife; practicality; landowner rights; the protection of wildlife populations and habitats; and ethical, legal, financial, and aesthetic issues. Often, the most effective long-term solution involves the use of several best practices, such as a combination of removal and exclusion.

This book focuses on best practices but in some circumstances, legal techniques that aren't described in this book may be appropriate. For example, in an emergency, the need to ensure safety may be so pressing that a technique which doesn't rate as a "best practice" is considered the best choice for that situation. That's one of the strengths of the best practices approach: it's flexible and offers many options (Figure 5-1).

FIGURE 5–1 One size does not fit all! Your chosen strategy for dealing with these beavers might change dramatically depending on the site and the needs of the customer. For example, beavers may not be tolerated at all in a reservoir used to supply the town's drinking water, while the same level and type of activities might not bother a private landowner. Each nuisance wildlife situation presents different challenges and requires a customized solution. *(Paul D. Curtis)*

Some methods that seem questionable today could be perfected and achieve the status not only of best practice, but also become standard operating procedures. For that reason, we have included discussion of some practices that have not yet been well studied, but seem promising. If you're wondering about the merits of a tool or technique, seek information from a trustworthy and current source.

Now, with your understanding of the best practices decision-making strategy, and the legal and safety issues you may confront on the job, we're ready for the details, the tools and techniques that form best practices. This is when you get to play detective. Investigate the situation, and then use your expertise to solve the puzzle.

The best practices approach to solving a wildlife damage problem includes five steps:

1. Assess the situation
2. Choose management options
3. Do it (tools and techniques)
4. Prevent future problems
5. Evaluate success

Sounds like a lot of trouble! Is it worth it? Yes. Here's a real-life example that shows why. In one case, night herons were raiding a fish hatchery. Researchers wanted to know if they could deceive the birds, so they played a tape of a propane cannon explosion to drive them off (that's much easier than using the real thing). But six nights later, the birds had gotten used to the noise and had settled back down to their dinner. Then the scientists tried a recording of night heron distress calls. Bingo! More than 80% of the herons left the pond, and six months later, this technique still worked. Here's the crucial bit: most birds only react to distress calls from their own species. So if the researchers hadn't bothered to properly identify which species was causing the damage, they might have used the wrong sounds.

Maybe you don't handle agricultural problems, but don't worry, you'll still encounter many cases of mistaken identity. Some customers may confuse moles for voles, raccoons for badgers, or woodchucks for muskrats. Is it a young Norway rat or an adult house mouse? The techniques used to deal with those animals differ. Proper identification is the first step to identifying the source of the problem.

Resist the temptation to jump to conclusions, too. Just because an animal is seen at the "scene of the crime" doesn't mean it's the culprit. For example, turkeys are sometimes blamed for crop damage that was actually caused by raccoons or squirrels. Why? Turkeys are active during the day, so people are more likely to notice them in the fields and vineyards. Raccoons are primarily nocturnal so fewer people are aware of their activities. Squirrels are much smaller and less visible than a flock of turkeys.

To learn more...

- Start your own "wildlife: who is it?" file, for those species that customers often misidentify, such as moles and voles. Jot down a few quick questions you could ask over the phone to help identify the species correctly.

Review Questions

1. Your customer believes there are rats in their apartment. You find smudge marks on the wall and dry, hard droppings that crumble when you touch them. The table lamp's wire is frayed. You want to put down non-toxic tracking powder to determine if there's still an infestation, and if so, whether it's rats or mice. This upsets your customer, who wants you to put down traps immediately. How do you explain your strategy?
   a) "My traps are expensive, so I don't take them out unless I'm sure they're needed."

b) "To do the best job for you, I have to be sure whether it's rats or mice because the techniques for catching them are different. For example, I'd use a smaller trap for mice."
c) "I charge more for rats, so I'd need to know before I write up the contract."
d) "I'm writing a book and need a good photograph of rat tracks. Do you mind?"

2. Pick the five steps involved in the best practices approach to solving wildlife conflicts:
___ Assess the situation.
___ Order pizza for your crew.
___ Take photos of the whole site for your records.
___ Choose management options.
___ Do it (use those tools and techniques).
___ Clean-up all the signs of the animal's presence.
___ Kill the animals.
___ Prevent future problems.
___ Evaluate your success.

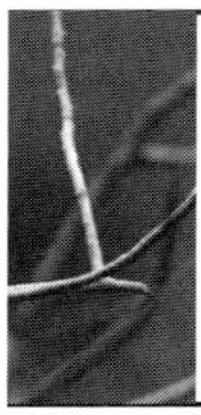

STEP ONE: Assess the Situation

LEARNING OBJECTIVES

5.3 Describe two benefits of doing a thorough inspection.

5.4 Give an example of a question you'd ask to figure out whether a problem was caused by a raccoon or a squirrel.

5.5 Describe four kinds of animal sign that you'd look for during an inspection.

5.6 Your knowledge of wildlife habits helps you estimate how many nuisance individuals may be present—before you even get to the site. Describe two biological facts that help you estimate animal abundance.

5.7 You've just noticed something that's going to change the way you deal with this situation. What is it?

The very first thing you need to do is find the source of the customer's problem. That means you'll need to understand the type of damage and how bad it is. You'll have to identify the culprit. You'll look for clues that will help you figure out what attracted the nuisance animals to the site. (Remember the two key enticements, food and shelter.)

Wild animals usually provide numerous signs of their presence. Once you've gained experience in "reading" these signs, the clues you gain from your site inspection and customer interview should help you identify the species, estimate the number of animals present, and find the areas where they're most active.

Signs of Wildlife Presence

- **Visual sighting.** This is one of the easiest ways to identify the species (if you can trust the observer). You may also find carcasses. If nocturnal animals are often seen during the day, this may mean that the animal has young and is feeding more often, or that the local population is high, especially with rats and mice. If dealing with a bat colony, you may quickly identify the species but have a harder time locating the entry holes. Stand outdoors at dusk or dawn, and watch where the bats enter or leave the building. There's the hole!
- **Sounds.** Various squeaks, growls, cries, hisses, chitters, screeches; gnawing; or clawing, scampering, and climbing inside the walls, above the ceiling, between the floors, or underneath cabinets. Learn to tell the sounds of adults from those of young.
- **Odors.** You may smell the droppings, fermenting urine, or body oils of wildlife that are living indoors. With a little experience, you can tell the odor of a house mouse from that of a rat. Skunks have a well-known scent, but woodchucks can also be identified by their odor. Dens of other animals, including raccoons, have their own perfume.
- **Droppings** may be found along runways, near shelters, in piles near an entry hole (Figure 5-2), or in other places used often. Fresh droppings are shiny and soft in texture, while old ones are dry, lighter in color, and hard. Bat droppings always crumble, while rodent droppings don't.
- **Urine.** You can see rodent urine using an ultraviolet light—it glows blue-white. Unfortunately, other materials also do this, which can be confusing until you become familiar with the typical background fluorescence of a home or office. You may also notice discoloration on building materials, often in attics or crawl spaces. That's caused by a large amount of urine, which could indicate the presence of raccoons, flying squirrels, or a large bat colony (Figure 5-3).
- **Nests and food caches** can sometimes be found when cleaning garages, attics, basements, closets, and other storage places. Rats, squirrels, and other rodents often store food in attics.
- **Entry sites** (holes, cracks, loose siding). The location, size, and condition of the entry sites are important clues to the species involved (Figure 5-4).

FIGURE 5–2 You'll often see a pile of bat droppings under the main entry hole. *(Connecticut Nuisance Wildlife Control Operators Association, Inc.)*

FIGURE 5–3 Notice the urine stains on the chimney from bats. *(Connecticut Nuisance Wildlife Control Operators Association, Inc.)*

FIGURE 5–4 Animals often gain entry under the eaves, as did a squirrel in this home. *(Connecticut Nuisance Wildlife Control Operators Association, Inc.)*

- **Burrows.** Woodchucks, chipmunks, moles, voles, and Norway rats burrow, and you can learn to tell their burrows apart. (Other animals, including raccoons and skunks, will use burrows but they don't make them.) The location of the burrow, its size, the type and number of entrances, and objects located near the burrow will help you identify the species.
- **"Leftovers."** You'll sometimes find the remains of a meal near an animal's den. You may be able to identify what the animal was eating, and that can help you identify the animal using the den. For example, you'll sometimes find a fair amount of prey remains, such as bones, feathers, or fur, near the den of a fox or coyote. If there's no sign of prey, then you're probably dealing with a herbivore, such as a woodchuck.
- **Runs.** Smooth or worn trails may be found next to walls, along fences, or under bushes and buildings. Runs within buildings may be well-polished trails that are free of dust. Trails through insulation are common.
- **Smudge marks** are often seen in the animal's run where it rubs against a surface during its travels, leaving behind dirt and oil from its fur (Figure 5-5). Look on pipes, beams, against walls, and on the outside edges of holes (Figure 5-6).

FIGURE 5–5 Buildings are vulnerable at their joints. Check carefully for smudges that show an animal's been going in and out. *(Connecticut Nuisance Wildlife Control Operators Association, Inc.)*

FIGURE 5–6 A raccoon climbed this pipe to gain entry to the attic. See the marks? *(Connecticut Nuisance Wildlife Control Operators Association, Inc.)*

- **Tracks and claw marks.** Footprints, tail marks, and wing prints may be found in dusty surfaces, sand, soft soil, and in snow. If the surface doesn't show tracks well, you can sprinkle nontoxic tracking dust (such as chalk powder, flour, or unscented talc) in a likely area, then return later to see if there are any tracks. When used outdoors, the dust must be protected from wind and rain. You may find claw marks on woodwork, trees, or in dust. Consider photographing and labeling the images (after you're sure you've properly identified the species).

- **Hair, feathers, or shed skins.** You may find tufts of hair on a fence or baseboard, feathers in an attic or above a dropped ceiling, or, less often, the shed skin of a snake. With practice, you may be able to identify the species from this sign. To improve your identification skills, consider making some hair sample charts. Clip a tuft of hair from a nuisance animal you've dispatched, and attach it to the chart with the species identified. (Please note that you may not possess any part of a threatened or endangered species without a special federal permit. Even if you found a shed eagle feather while you were out hiking, you cannot legally keep it without that permit.)
- **Gnawing** (wood chips, tooth marks, holes, shredded fabrics, frayed wires). Look for evidence of chewing wherever wildlife might try to enlarge a crack or enter something. Wood chips may be seen near baseboards, doors, basement windows, kitchen cabinets, furniture, and stored materials (Figure 5-7). You could find shredded clothing, or see toothmarks on pipes. Rodents and raccoons often chew on the insulation around wires. The size of the toothmarks (or of a hole chewed in a baseboard) will frequently help you tell whether you're dealing with rats, squirrels, or mice (Figure 5-8).
- **Pets become excited.** When cats or dogs hear or smell rodents in some inaccessible spaces such as a wall, they may become extremely interested, whining, sniffing, and scratching at the spot.
- **Access routes.** Walk around outside and try to imagine the route the animal might have used to gain entrance to the building. Are there trees or utility lines near the roof? Could it have crawled under a porch, up a chimney, or along a downspout? Is there an attached garage that might have been left open? These clues point to likely culprits. Skunks, for example, aren't going to jump from a tree branch onto the roof, and tree squirrels aren't as likely to wriggle in underneath a porch.

FIGURE 5–7 What happened here? A squirrel trapped in a basement tried to chew its way out. The squirrel tried to squeeze underneath the door that led to the basement, gnawing and scratching at the bottom of the door and the rug. *(Connecticut Nuisance Wildlife Control Operators Association, Inc.)*

FIGURE 5–8 See the bits of wood and shredded fabric? The squirrel also chewed on the window in the basement, hoping to escape that way. It chewed most of the mullions down to bare wood. See the bits of wood on the ledge? *(Connecticut Nuisance Wildlife Control Operators Association, Inc.)*

Your knowledge of the habitat preferences and behavior of wildlife will help you estimate how many animals might be present. For example, if your customer complains of noises in the attic in March, you're probably expecting to find squirrels, and not just one—probably three to eight, because that's when female squirrels are raising their young. (If you see a female mammal, you may be able to tell if she's nursing because her nipples would be larger. This is hard to see on some of the small, fast-moving animals like flying squirrels. So if it's the breeding season for that species, assume that young may be present. And be careful, because individual animals may give birth outside of the breeding season as defined by textbooks.)

Other aspects of the animal's lifestyle that help you answer the question, "how many?" include its social habits (generally found alone, or in a group?), daily movements, and whether or not it hibernates or migrates. How fresh are the signs? Is this problem new or is it well established? How large is the property, and how many individuals of that species would you expect it to support?

But you're more than a wildlife detective. You also have to nose around in your customer's home, and ask questions to find out if the inhabitants (or their neighbors) are causing the problem. People sometimes feed or house wildlife without realizing it. Squirrels, for example, may be attracted to spilled seed at a bird feeder. This may not be a big problem. But once nearby, the squirrels may run across the roof and find a hole that gives them access to the attic. Perhaps they decide that's an ideal place to raise their young. To solve this conflict, you'll probably need to remove the food source, ensure that the female and all of her young are no longer inside, and repair the building.

There are other things you need to know about your customers before you choose a strategy, such as whether there are any pets roaming freely in the area. So, Sherlock, how do you proceed? Most WCOs interview their customers and inspect the premises to look for clues.

Here are the questions you'll hope to answer with clues gained from the interview and the inspection:

- Which wildlife species is causing the problem? Is there more than one species involved?
- How many individual animals might be present?
- Could there be young present?
- Is this the first time they've had a wildlife problem? If not, what happened before?
- How long has this problem existed?
- Are the culprits readily accessible or hard to reach?
- How risky is this situation for you, your customers, and their neighbors?
- Can you target only the offending animals?
- Will it be easy to be discreet or would your control activities be in public view?
- If you repel or exclude the animals, where might they go?
- Does this look like a short-term problem or one that's likely to happen again and again?
- Given the location of the site, are there any local laws that would affect your management options?
- In addition to this problem, do you see signs that could predict future problems?

Interview

First, it's best to talk to the right person—whoever knows the most about the problem and the property. Ask both general questions and some very specific ones. A general question might be, "Please describe what you know about the problem," or "Have you noticed any problems since my last visit?" This type of question gives customers a chance to share their information. That could trigger other questions that might guide your inspection.

You'll develop your own interviewing style, but here are some questions to probe for specific information. Ask the customer, when did you first notice this problem? How often does it happen? Did you see any animals or any signs of animals? You might want to prompt them by asking whether they've heard noises, for example. Then be sure to find out if the noises are heard during the day or at night. Can they locate the noise? You may want to ask several questions about the noises they're hearing, because sometimes people mistake mechanical sounds such as the beeping of a smoke alarm that has a low battery, or a swaying utility line, for animal noises.

Remember to ask if they've ever had any trouble with wild animals before. If they say yes, have them describe that situation in more detail. When did it happen, did they identify the culprits, and what did they do?

There are also a few important questions to ask about the household. Are there any children or pets present? That might limit which wildlife control techniques you choose. Did anyone have the type of contact with the wild animal or its droppings and nesting materials that might have exposed them to a disease? What about pets? This is especially important if dealing with bats, raccoons, skunks, and foxes, the species in our area that are most likely to carry rabies. If there's been contact, you'll have to follow health department guidelines. That's the law.

Inspection

Before you can confirm the problem and contemplate solutions, you'll usually need to inspect the property. Take a good look at the whole landscape and the neighborhood. Are there signs that the person's behavior is contributing to the problem, such as overflowing garbage cans, a pet dish left outdoors, or a large bag of open bird feed on the porch? This is the big picture that could explain why the problem developed. With the added information from an inspection, an experienced observer can provide management options for problems associated with uninvited animal guests (Figures 5-9a and 5-9b).

Follow appropriate safety precautions throughout the inspection. Before you start your inspection, think about the space you're entering and the typical hazards associated with it. In an attic, you might worry about heat stress; ladder accidents; airborne diseases stirred up from rodent, bat, or bird droppings; and scratches from nails that might expose you to tetanus. Heat stress and airborne diseases are also concerns when entering a crawlspace, but you may not encounter nails or need to use a ladder. (See Chapter Four.)

Find out where the property lines are, and whether the caller owns or rents. If it's a rental unit, talk to the property owner or manager to secure

FIGURE 5–9a You can often turn a vulnerable site into a darn good fortress by installing a barrier around it. A flat rooftop and satellite dish are protected under netting. *(Bird Barrier, Inc.)*

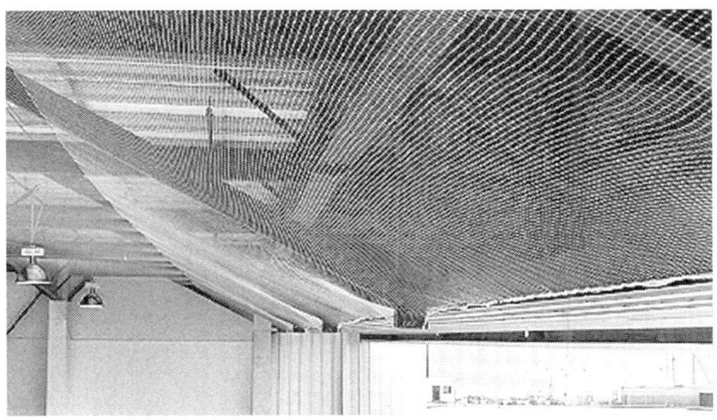

FIGURE 5–9b Netting works equally well inside the warehouse to keep birds from roosting on the rafters. If your customer worries about how things will look after you're done, netting may be a good option because it's hard to see, especially from a distance. *(Bird Barrier, Inc.)*

permission. (If the renter refused to pay, the managers might not be willing to if they didn't authorize the work.)

During your inspection, you might focus on the specific problem area, but there are some benefits to inspecting the entire building or property. A "problem-specific" inspection takes less time and addresses your customer's immediate concern, but a more thorough inspection may identify other problems that your customer doesn't know about. For example, you might be called for raccoons in a chimney and find that the animals have also entered the attic. Or maybe they haven't made it into the attic, but you suspect it's only a matter of time because the type of roof vent on this house is very vulnerable to raccoon damage. Rotted soffits, trim, and roofs are also vulnerable to animal entry.

A thorough inspection will also reduce the likelihood of overlooking some aspect of the current wildlife problem. What if there's more than one species involved? Even if the customer didn't ask you to look for that, they might not be happy if you missed it.

"I had a job where a raccoon was living in one of two spaces in the attic of a large house. If I'd limited my inspection to the area where the customers regularly heard the raccoon making noise, I would have missed the actual entry site. This raccoon got into the building through a hole that wasn't visible from the outside. Then it traveled through a small wall void into the second crawlspace area, where it was most active."

—Lynn Braband, former WCO in New York

Take pictures during the inspection, using a digital or instant camera so the images are available right away. It's much easier to explain the situation to your customers if you can show them pictures of the damage, the animal sign, and the entry holes. In fact, you may want to create a photo album that shows structural problems, typical animal damage, management options, and prevention techniques. That's a handy resource to keep in the truck. Consider offering fact sheets to customers, too. This could save you time while reducing the odds of misunderstandings. (Make sure the information

Detective tools (inspection equipment)

- Good flashlight with extra batteries and bulb
- Small, flat pry-bar
- Stepladder, extension ladders
- Binoculars (useful for tall buildings)
- Mirror, TV camera, or fiber optic scope attached to a long, flexible pole (for viewing hard-to-see areas such as wall voids)
- Ultraviolet light for detecting rodent urine stains
- Digital or instant camera (so you can show your customer exactly what you found)
- Inspection form for recording information
- Identification guides, including those that show animal sign (tracks, scat, nests)
- Plastic bags for collecting animal sign (to show evidence to customers)

is reliable! Dependable sources include the state wildlife agency, health department, and Cooperative Extension.)

It's just as important to take notes during the inspection. Write down your management recommendations, either on the inspection form or the contract. Give the customer a copy of the inspection report and remember to keep one for yourself. Once you've reached agreement on the approach, you and your customer may want to sign a contract that clearly describes the work, your fees, and terms of payment. Many WCOs also collect a deposit at the beginning of a job, although this is usually most practical with private individuals. Different arrangements are often made with corporate accounts.

Another tip that may save you time later is to be systematic as you conduct your inspection. It's easier to remember whether you've checked a particular area if you always work in the same pattern. After you've done a few inspections, you'll develop your own style.

Here's an example of a thorough inspection of a building for existing or potential wildlife problems. Start inside the building at the top and work your way to the bottom. Within each room, move either clockwise or counterclockwise. Pay special attention to the corners, and spaces underneath and behind furniture. If there are suspended ("dropped") ceilings, push up the panels in a few places to check above the ceiling. Inspect attics; basements; the areas underneath sinks; the places where pipes, cables, and wires enter the building; and crawl spaces. Look inside closets and built-in drawers.

You may be able to use light to your advantage, especially in older buildings. During the day, turn off the lights and look for places where the daylight shines through walls and floors. Some of those cracks might provide access points for small animals. Once you're done with the inside of the building, move outdoors. You can use light to your advantage here, too. At night, turn the lights on, then go outside to see if there are places where the lights shine through the walls or roof. Use binoculars for a good view of the roof and search methodically, from the top all the way down. Most modern buildings have ridge and soffit vents that allow air and light into the attic, so this trick won't work in them. Light may slip behind chimney flashing that isn't permeable to animals, too, but that's always a good spot to check because it is a vulnerable area.

What's the condition of the eaves, dormers, windows, vents, ledges, chimneys, and roof corners? Animals often gain entrance at joints and places where different building materials meet. Pay careful attention to the foundation because that's a vulnerable area. Check beneath decks, porches, and crawl spaces. Don't forget the garage; barns; sheds; the places where they store garbage cans; dumpster areas; compost heaps; and piles of firewood, lumber, or junk.

To learn more...

- Ask a friend to pretend to be one of your customers. Interview your friend about the problem, then inspect the premises and share your advice.
- Jot down a few questions you might ask customers that would help you identify which animal is causing the damage.
- Go to a bookstore or library to browse through the latest field guides to animal signs.
- Visit a zoo or nature center that exhibits some of the animals you might encounter in your job. Spend some time watching them. Close your eyes, and pay attention to the odor. Talk to staff about the species' habits. How would they know if that critter was present?
- Start your own wildlife sample charts to help you identify species by their fur, feathers, or shed skins. You can collect samples from the animals you trap.

Review Questions

3. You're called to a home to deal with a rodent problem. Your customer has three dogs and a toddler. That makes you think:

a) "No wonder the place smells."
b) "I should attach the mouse traps to beams, where the dogs and the child can't reach them."
c) "Why aren't these lazy dogs taking care of the mice?"
d) "I could set out live traps, which would be safer for the kid."
e) Answers "b" and "d" are correct.
f) Answers "b," "c," and "d" are correct.

4. When you first talk to your customer to gain information about the nature of the wildlife conflict, you might ask:

a) "What can you tell me about this problem?" and "Why didn't you call earlier?"
b) "What's going on?" and "Can you pay for this?"
c) "What have you seen and heard?" and "When do you notice the problem?"
d) Very little, because people usually misidentify wildlife.

5. There's a noise in the dropped ceiling. Your customer isn't sure what it is, but you know it could be a rodent or a bird. Your truck is packed tight, so you want to make sure you bring the right equipment for the job. What question do you ask her?

a) "When do you hear the noises, during the day or at night?"
b) "How loud is it?"
c) "Where are the noises coming from?"
d) "You mind if I leave this skunk in your driveway while I finish up my route? That would make it easier for me to fit everything into my truck without going back to the office."

6. Your customer just bought an old house. They've been told there's a bat or two in the attic, but they were too scared to check. They want to know how many are up there, and have them removed before they move into the house, which they're planning to do the first week in July. Because you're the wildlife Sherlock Holmes, you tell your customer you'd be glad to help, but:

a) This time of year, there's a good chance there's more than one or two bats in the attic. There could be a whole colony of females raising their young.
b) You can start now, and make sure that the bats can't get from the attic into the living spaces. But you'd like to let this batch of young leave on their own, which they'll do by mid-August. Once they're gone, you can finish bat-proofing the house.
c) The bat's probably long gone. Don't worry about it. You'll never go into the attic, anyway.
d) Answers "a" and "b" are correct.

7. You think there's a mammal in the house. While inspecting the kitchen, you'd look for these signs (check all that apply):
    ___ Gnaw marks on the cabinets
    ___ Droppings in drawers and by the kickboards
    ___ Smudge marks on the baseboards
    ___ Urine stains on walls and woodwork
    ___ Holes chewed in cereal boxes and other stored foods
    ___ Burrows
    ___ Piles of nuts and seeds in a cabinet that's not used often
    ___ Odors
    ___ Fur
    ___ Tracks
    ___ Noise
    ___ The creature
    ___ Pets scratching at a wall
    ___ Livestock remains
    ___ Broken egg shells
    ___ Nest
    ___ Crop damage

STEP TWO: Choose Management Options

LEARNING OBJECTIVES

5.8 Describe the options for managing a wildlife conflict.

5.9 Explain one way you'd make an area less attractive to nuisance wildlife by actually changing the environment, and one way you'd do this by changing people's habits.

5.10 List two of the factors that influence the effectiveness of repellents.

5.11 Which management option would almost always be a part of your solution? Why?

At this point in our story, Sherlock would retreat to his apartment, pick up his pipe, and think. So many clues, so many possible solutions: how do you choose?

With a proper species identification, the information you learned during your interview and inspection, and your knowledge of that animal's life history, you can estimate how many individuals might be present. Again, this will depend on the time of year (is the species breeding, rearing young, hibernating, or migrating?) and the animal's social habits. Woodchucks are generally solitary, while commensal bats are social and often found in colonies. Is this animal territorial? Will it drive away others at certain times of the year? Does it have a large home range? If so, it might not be able to meet all of its needs just on this site. That could be tricky, because that might mean that the best solution would include addressing a situation on someone else's property, too, and you might not be able to do that.

All of this detective work is meant to help you decide how many individuals this site would naturally support. Of course, there could be other factors at work here, and you need to take those into account. If someone is feeding the wildlife you could encounter unbelievably high population levels, way beyond what the books and experience would suggest.

As you consider those questions, you may eliminate some control techniques because they're not practical, safe, or discreet enough for this situation. Legal restrictions may eliminate some options. The presence of free-roaming pets, children, or protected wildlife might also lead you to favor some approaches and avoid others. The preferences of your client will also be important.

Okay, you're ready to decide on a management strategy. We'll discuss each major strategy now, and then, in Step Three, will describe the common techniques used to achieve that method. In some cases, there are many different ways to achieve a particular strategy. For example, we'll describe more than a half-dozen ways to remove an animal from a site.

What are your options when confronted with a nuisance wildlife situation?

- Do nothing. Let the problem resolve itself.
- Make the environment less attractive.
- Scare the nuisance animal away.
- Remove the culprit.
- Reduce the local breeding population.
- Keep the animal out (exclusion).

Regardless of which strategy you favor, keep these few points in mind. First, a combination of techniques almost always works better than relying on only one method. Be as selective and discreet as you possibly can. Proper timing will increase your success rate (you wouldn't try to capture a nocturnal animal during the day, would you?). Some techniques are only effective in certain seasons; if used at the wrong time, a perfectly good exclusion technique might orphan young wildlife, for example. Take advantage of the species' vulnerabilities and habits, and watch out for risky conditions.

Option: Do nothing. Let the problem resolve itself.

Techniques: Education and persuasion.

Talk to your customers, so they understand why this problem developed and what's likely to happen next. The best thing to do may be—nothing.

What if a customer calls in mid-June, complaining about a fox den under the porch? There are five pups and she's afraid her yard is going to be overrun with foxes. Won't they attack her Irish wolfhound?

After asking a few questions, you determine that the foxes are behaving normally and look healthy. In fact, she's not even planning to use the porch until later in the season because it needs repair.

You know that in the spring, foxes sometimes den under a porch for a short time while they're raising their pups. But they generally use more than one den and may move their pups as many as four times. The foxes will stop using that den on their own, usually by the end of June.

They don't pose much of a threat to her dog, either. Foxes are actually smaller than many people realize. They only weigh about 10–12 pounds—about as much as a plush house cat. These smallish predators usually hunt mice and rarely attack dogs. (Even a coyote would be unlikely to physically threaten such a large dog.)

Can you persuade her to tolerate the situation?

We all have our moments. It's easy enough to work yourself into a bit of a frenzy over something that's not such a big deal, or not even true. For example, many people believe "drooling = rabid animal." The sight of a drooling opossum in their backyard might frighten some people into demanding the removal of the animal. Yet healthy opossums drool and shake as part of their defensive behavior, and this species doesn't often contract rabies. In such a case, sharing correct information may make a "problem" disappear.

American robins may attack windowpanes for a few weeks during the early spring as they establish their nesting territories. The birds attack their reflection in the glass because they mistake it for an intruder threatening their territory. But they'll stop soon. Some people will be satisfied to wait it out, once they realize the problem will go away on its own. Or, with just a little bit of effort, you can probably stop the situation by reducing the reflectivity of the window. Close the shades or hang Mylar ribbons above the window, where they'll move in the breeze. If the robin doesn't see the "intruder," which is just its own reflection, then it won't have a reason to defend its territory. (Quite a few problems with migratory birds take care of themselves within a few weeks.)

If someone doesn't know much about wildlife, how can they evaluate the situation sensibly? They lack information and the context for understanding it correctly, which you can provide. Perhaps you can handle this over the phone and save yourself the trip.

There are many times when this "Zen" approach is common sense. As soon as the snow melts, you'll probably get calls from people concerned about moles or voles ruining their lawns. There may be tunnels all over the place! But there's a good chance that this clue is all that's left. Both moles and voles change their habits seasonally. These tunnels are the sign of their winter activity. They may have left the area. It's much more sensible to wait a little while to see if there's evidence of new damage.

Sometimes, by the time you've arrived, the problem is fixed. Maybe a bat entered a room through a torn screen and has already escaped. You may only need to explain the cause of the problem and how to prevent repeats of this scenario. Consider creating a short tip sheet to explain the predictable problems. This could save time on the job. Instead of your long explanation, hit the high points and leave the customer with the written details.

Option: Make the environment less attractive to the culprit.

Techniques: Remove food sources; keep buildings in good repair; protect vulnerable areas with barriers; maintain a tidy landscape; move livestock to protected areas; switch to landscape plants that the nuisance animals find less tasty.

What do most wildlife seek when they invade our spaces? Food and shelter. Safe places to raise their young are very attractive; wildlife may also seek safe places to rest. They're also enticed by water.

If you can remove those enticements, or make them less appealing, you may be able to solve the problem for good. This approach is known as "habitat modification." Although it's sometimes more expensive up front, the long-term payoff is

FIGURE 5–10a *(Connecticut Nuisance Wildlife Control Operators Association, Inc.)*

FIGURE 5–10b *(Connecticut Nuisance Wildlife Control Operators Association, Inc.)*

FIGURE 5–10c *(Lynn A. Braband)*

FIGURES 5–10a–c People unintentionally feed wildlife. Spilled bird seed, pet food left in dishes outdoors, and garbage can become buffets for a variety of wildlife from rodents, bears, foxes, raccoons, to skunks. When people intentionally feed wildlife, they can cause or worsen other problems. The gulls in the parking lot may well mob people who try to feed them. Their droppings may foul cars.

usually good because this gets at the root of the problem. Sure, you can just trap and remove an animal. Year after year after year . . . that adds up.

There are scientists and WCOs who believe that some animals, such as raccoons, learn to recognize a good deal. "Once a nuisance, always a nuisance," they say. It's even possible that the young of a female who has such habits learn to take advantage of the riches we provide because of experiences with their mother. If true, then the quick-fix approach could be even more costly than we imagine.

Some of the most profound ways to modify a habitat may only involve persuading people to change their habits (Figures 5-10a, 5-10b, and 5-10c). For example, here's a no-brainer for customers annoyed by bears that raid the bird feeder during the late spring and summer: don't feed the birds during those seasons! Feed birds only during the winter, problem solved.

More examples: raccoons tipping over the garbage cans and making a mess? Store the cans in a protected area. Look for a sturdy garbage can with a high twist-on lid that can't be opened by the raccoons. Build a storage box for the garbage cans that has a hinged top that closes with a latch and snap hook. Or hang the cans on a hook above ground so they can't be tipped over (make sure the raccoons can't climb up the pole or wall to reach the garbage can). If possible, put the trash out in the morning. Has the compost heap turned into a restaurant for skunks, opossums, rats, and raccoons? Install an animal-proof compost bin. (For information, consult Cooperative Extension's "Master Gardener" program.) You'll find many suggestions in each of the species accounts at the end of this book.

Many wildlife species prefer to use the same path to food, shelter, water, and loafing areas. Can you block that route, or make it unpleasant? Canada geese, for example, love to forage on mowed grass. Let the grass bordering a pond grow longer to discourage the geese from using the site. Eliminate ponds from malls and business parks, and you can avoid many goose problems. This general idea works for many species. If a tree branch overhangs a roof, trim it 8–10 feet away, and the most acrobatic squirrel won't be able to make the leap anymore.

Other ways to change the conditions of the habitat may require efforts on a grander scale. Take a few tips from farmers, who combat pests by tilling, rotating crops, controlling water levels, and heating up the soil. Can you change the temperature, water level, or amount of light that reaches an area? Bats like warm, dark places. If they're roosting behind shutters, put a block between the shutter and the siding. This holds the shutters farther away from the wall and exposes the bats to drafts and more light. That might convince them to move on.

If wildlife dine on flowers and shrubs, consider switching to plant varieties that they don't find as tasty. For a list of plants that are more resistant to deer, see the Cornell fact sheet, *Resistance of Woody Ornamental Plants to Deer Damage* (See http://wildlifecontrol.info/ccewdmp/publications.html).

Option: Scare the nuisance animal away.

Techniques: Use of visual repellents (lasers; Mylar tape; balloons); use of noisemakers (distress and alarm calls; pyrotechnics; sirens); hazing (with predators or radio-controlled craft); use of guard animals. WCOs with a commercial pesticide applicator license can also use a variety of chemical repellents.

"Run 'em out of town." That's the idea of this method. But it's easier said than done. To make this work, think like the animal. What scares it? If you rely on something the animal doesn't experience in nature, the technique may not work because it's not meaningful to the animal. Many frightening devices fail for just this reason.

Repellents are objects, substances, or techniques that repel, or drive an animal away. They work by either scaring the wildlife or persuading them that the desired object is just nasty and better left alone.

Many repellents are meant to frighten the wildlife away from the site. These include scary-looking objects, such as Mylar tape, models of predators, and "scare-eye" balloons; noisemakers, such as propane cannons, bangers, clappers, crackers, and distress calls; devices that combine lights and noise to scare animals. The wildlife may be "hazed," or driven away from a site, when chased by predators (dogs, falcons) or vehicles (radio-controlled planes or boats). Guard animals (dogs, llamas) are sometimes used by landowners to repel wildlife, too.

FIGURE 5–11 See the mothballs to the left of the rocks? Not exactly repelling the skunks, are they? Repellents often work better when several are used together, in an unpredictable fashion. They're generally better at preventing new behavior than stopping a well-established habit. *(Connecticut Nuisance Wildlife Control Operators Association, Inc.)*

Devices such as spikes and wires or chemical irritants repel wildlife by making them physically uncomfortable. Some of these repellents make the desired object (food source, loafing area) smell awful or taste nasty. Others, such as sticky gels, spikes, porcupine wire, and bird coils, turn a cozy perch into an intolerably uncomfortable one.

Chemical repellents, which are classified as pesticides, may only be used by WCOs who have a commercial pesticide applicator license. If using these products, remember to check each one to make sure it's registered in your state for use on that species, in the way you'd like to apply it. Read the label carefully because the label is the law.

A variety of commercial chemical repellents are available, especially to reduce mammal browsing in orchards and vineyards, and on nursery stock and ornamental plants. Most products are registered for deer and rabbit control on nonfood crops and ornamentals. (For more information, contact your state pesticide regulatory agency.) Quite a few home remedies, such as soap and baking soda, have repellent properties. Likewise, concoctions made from plant extracts such as mint, citrus peel, or marigolds may act as repellents as well as pesticides. But home remedies are out-of-bounds for WCOs; they are not legally registered products.

Repellents can be a valuable addition to your strategy. This nonlethal technique saves you the trouble of capturing or handling wildlife—when it works. Your best chances are to try them when these three things are true: there's only light-to-moderate damage; the site is small; and a few applications should provide adequate control. In most cases, repellents provide short-term control. No repellent is 100% effective (Figure 5-11).

Sometimes, frightening techniques fail for a very basic reason: the animal can't really hear or see the thing that's supposed to scare it. Oops. This is why ultrasonic devices do not frighten birds. Ultrasonic devices usually aren't effective when used with mammals because they don't seem to associate the sound with real danger.

To avoid this mistake, think like the animal. Which senses does it rely on most: smell, sight, hearing, taste, or touch? How do individuals warn each other of danger? Beavers noisily slap their tails while deer flash theirs, a quiet but highly visible warning. Birds often rely on sight and sound to detect danger, while mammals rely more heavily on odor cues.

Another biological factor that influences the effectiveness of some repellents is the nervous system. Birds and mammals are "wired" a little differently so they perceive the world differently. Capsaicin, the active ingredient in hot peppers, burns the mouth, eyes, and nose of mammals. Hot hot hot! Not so for birds. They are insensitive to capsaicin, so they ignore it. On the other hand, birds find methyl anthranilate, a sweet grape flavoring, highly irritating.

Distress and alarm calls, hazing with live predators and objects that mimic the animal's real predators are often more effective than many other frightening devices because they make more sense to wildlife. Animals naturally fear predators, for good reason. Given time, many animals learn that other scary objects actually pose no danger to them, so they ignore them (this process is called "habituation"). That's why these devices are often useful for only a short time.

Option: Remove the culprit.

Nonlethal techniques: Direct capture; use of live traps, one-way doors, or nonchemical repellents. Lethal techniques: Use of lethal traps; shooting. WCOs with a commercial pesticide applicator license may also use a variety of pesticides (such as toxicants for rodents).

There are several methods for removing animals from areas in which they are unwanted. Nonlethal techniques include direct capture using a handheld device (Figure 5-12) or net, and the use of one-way doors or live traps. Lethal methods usually involve lethal traps, shooting, or rodenticides. Obviously, some techniques are better suited for some animals than others. With all of these methods, be aware of the possible presence of young so you don't create wildlife orphans.

Whether you've chosen a nonlethal or lethal removal technique, you'll be faced with the same question: What will you do once you've caught the animal(s)? In Chapter Two, we suggested that WCOs should consider the possible ecological consequences of each method. You have three basic choices once you've captured an animal: you can release it on-site, you can transfer it to a new place (in some cases, to a wildlife rehabilitator or licensed facility), or you can kill it.

Note: Once the animal is captured, it may be released; transferred to a wildlife rehabilitator (if it's injured or orphaned); or killed. In addition to the techniques mentioned above, there are other appropriate methods for killing wildlife, such as the use of a carbon dioxide chamber, which are described later in this chapter. These techniques and tools do not aid in the capture or removal of that animal from the site so they're not listed above.

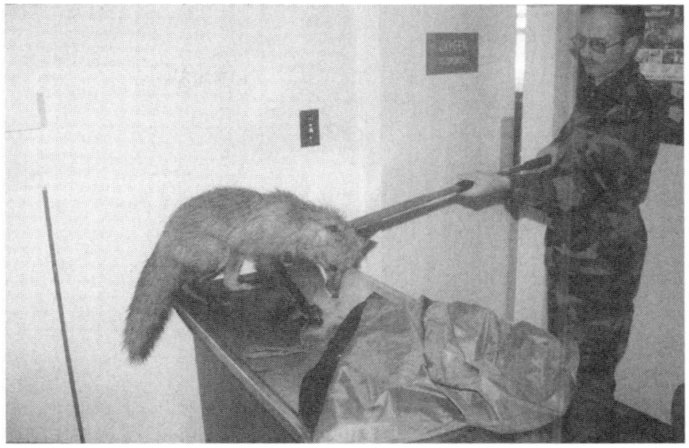

FIGURE 5–12 Foxes often calm down once they're restrained with a catchpole. Since this healthy fox had accidentally wandered in, it was released on site. *(Paul D. Curtis)*

On-site Release

There are two circumstances in which on-site release is almost a no-brainer: the accidental capture and the accidental entry. Say you're trying to catch a skunk but you find an opossum in your cage trap instead. Open the door and let it out. It's not causing any damage; it's not a nuisance; it's just in the wrong place at the wrong time. Likewise, animals sometimes end up inside buildings accidentally. Birds and squirrels sometimes fall down chimneys. Skunks fall into window wells. Bats will sometimes enter a building through an open window or door. If the animal poses little risk to nearby people, you should be able to capture it and release it outdoors. Help your customers understand how to prevent another animal from dropping in (maybe you'd offer to install a chimney cap, for example). Done. Everybody happy.

There are some situations in which the technique of on-site release would not be appropriate; for example, if the animal is obviously sick or may have exposed someone to rabies. In those cases, releasing the animal would be risky (and possibly illegal).

If your customer suffered damage, there are a few points to consider before releasing the culprit. What might happen to the animal, your customer, your customer's neighbors, and the local wildlife populations? If the customer opts for effective exclusion that will prevent the animal from re-entering the building, you should be able to release the animal on-site with reasonable assurance that your customer won't suffer the same problem again. But what about your customer's neighbors?

Consider which species caused the problem. Bats and snakes cannot create openings in buildings. Remove them, then repair the hole(s) that let them in, and they're not likely to trouble your customer again. They can't force their way into the neighbor's building, either. So on-site release is almost always appropriate with these species. In fact, it's actually preferred. Both bats and snakes congregate, which means that the individuals at that one site may represent a good percentage of the local population. What you do at this site could have broader effects than it would if you were dealing with a species whose population is more dispersed. (Please read the bat account for some additional factors to consider when dealing with large colonies.)

Now, let's apply the same logic to the great chewers, the rodents: mice, rats, and squirrels, and the strong and often destructive raccoons. They can most definitely chew or force their way back into the building. If the nuisance animal has developed the habit of using a building it may become a repeat offender, especially if den sites are rare, or the animal is ready to give birth and needs a den right away. Even though no one can predict the future, the ethics of this situation are less clear.

Look around the neighborhood. If the nearest building is miles away, then the risk of the animal troubling the neighbors could be low. If it's a condo complex of attached buildings that all share the same structural deficiency that allowed the animal to gain access, such as poorly designed louvers, it's much more of a concern (Figure 5-13). Is it ethical to potentially pass your problem on to the neighbors?

FIGURE 5–13 In this condominium complex, raccoons moved from one cupola to the next. Notice the bent louvers. *(Connecticut Nuisance Wildlife Control Operators Association, Inc.)*

(Or is this an opportunity to market preventive services? In the case of a large commercial property, that could be a lucrative account that allows you to demonstrate the effectiveness of inspections, proper maintenance, and exclusion in preventing wildlife damage.)

On-site release generally causes the least stress to the animal. It's a good example of balancing the needs of people and wildlife. With some modifications to their habits and sites, your customers will in most cases be able to protect their property and coexist with their local wildlife. Although denied their favorite site, the wild animals remain in their original territories, which is better for them than being moved or killed (and generally, much preferred by the public). Because these individuals are already part of the local wildlife population, they don't represent any additional threats to its survival. That can be a problem when an animal is relocated to a new habitat.

Relocation (a.k.a. "translocation" or "trap and transfer")

Moving the animal to a new place is a common but controversial technique. While solving some of the problems associated with releasing the animal on-site, it raises other concerns.

Let's try our six best practices questions from Chapter Two on this technique.

1. Is it safe?
   Reasonably so, for both the operator and wildlife. Relocation involves capturing and transporting the animal, which is stressful, so the animal may be quite agitated when you're ready to let it free.
2. What are the likely ecological consequences of this action?
   - The relocated animal can spread diseases to a healthy wildlife population.
   - An area can only support so many individuals of a particular species. Raising the level of competition at the release site for food, shelter, and den sites could harm more animals in the long run.
3. Is it practical?
   - At best, we can say "effectiveness unknown for nuisance species."
   - If the customer opted for capture and relocation without exclusion, chances are good a new animal will usually quickly replace the one that was removed.
   - The relocated animal may return to the site. Many animals have strong homing instincts. A male raccoon, for example, can travel as much as 5–10 miles each night. Bats routinely travel hundreds of miles to their hibernacula.
   - Is there a suitable release site close by, one that meets the habitat needs of this species?
   - Can you afford to make the time to take the animals to the site?
   - Do you even have a release site that you can legally use? In some states, it's illegal to release animals on public land, while in others you need the permission of the land manager. On private land, you'd need the permission of the land owner. (Ever try to convince someone they need more skunks?) Can you get permission in time? What if you're refused?
4. Is it humane?
   - How well does this species respond to transport? Will the animal be so stressed by the process that it's unlikely to survive? Studies have shown that the stress and trauma of capture causes some deer to die a few weeks later. We don't know if that holds true for smaller mammals or birds as well.
   - What are its chances of finding a new territory? The animals that are already established in the release site may drive off or attack this intruder. Between the capture, transport, and release in a foreign locale, this animal is probably already stressed. Does it stand a chance?
   - What are its chances of finding food? During the winter, when food is scarce for many of these species, placing an animal in an unfamiliar place can cause it to slowly starve to death. This is especially true if that animal would normally rely on stored food to make it through the winter, as tree squirrels do.
   - Add to the troubles: what if this is a mother with young? Check for signs of lactation before moving an animal during breeding season so you don't accidentally separate a mother from her offspring. If your plan is to move the entire family, realize that she might simply abandon or kill her young because of the stress of capture and translocation. Would she have time to find food, shelter, and a safe place to raise her young? Moving a family successfully takes more skill and time than relocating an individual adult. It's not as easy as releasing the family on-site. Don't be cavalier about it.
   - Some animals are killed by cars, dogs, or predators as they try to return to their original home ranges through unfamiliar territory.

5. Is it legal?
   - Is this a species that you are legally allowed to handle? If so, remember to check local laws, which might further restrict your activities.
   - WCOs may legally relocate wildlife (assuming, of course, that they have the permission of the owners of the release site), but in many states in our region, the general public cannot. Many people don't realize that.
   - Is the animal classified as a "threatened" or "endangered" species, either on a statewide or national basis? If so, stop! You cannot handle it. Do not harm it or its habitat.
   - Are you dealing with a migratory bird? Nearly all are protected by the Migratory Bird Treaty Act. You'd need federal and state permits, which would spell out approved techniques. Follow the instructions on your permits.
   - During times of disease epidemics, there may be restrictions on moving animals. You still have to follow your state wildlife agency and health department guidelines.
   - You are talking about wildlife, right? Because in many states in our region, your WCO license does not give you the authority to handle feral cats or dogs, or livestock, such as domestic ducks. Those species may be covered by other laws and agencies.

6. How would your actions play on the evening news?
   - This is hard to predict. So far, there haven't been enough scientific studies to answer some of the important ecological questions raised by this approach. We can't say whether this is likely to be a "happily ever after" story for the relocated animal, or a tear-jerker. How will people who live near the release site feel?

Is this technique all bad? Wildlife biologists relocate animals for restoration and research. This technique has been used to reintroduce wildlife into areas where they were once found, for example, or to stock certain areas with game species, such as wild turkey. But when it comes to nuisance wildlife, the usefulness of this technique is debatable. Your customers may request this option, so you'll need to think about it.

Should you decide to relocate an animal, there are several things you can do to give it the best possible chance of survival. Find an area with suitable habitat for that species, with good and varied sources of food and shelter. Release the animal during the time of day when it's naturally active; for example, release a nocturnal animal at dusk or in the evening, not in the middle of the day. Keep families together and release them into a covered area, such as a culvert or hollow log; this will increase the chances that the mother will stay with her young. Follow the guidelines for preventing orphaning (see Step Three). And don't overload an area.

If an animal is sick or injured, a WCO may be able to transfer the animal to a wildlife rehabilitator, which is a form of relocation. Even WCOs who are very careful to avoid orphaning young animals may sometimes find themselves with a litter that's been separated from its mother. Wildlife rehabilitators may be able to help. Of course, it's better to have a good relationship with your local rehabber before you need that person's services. To find licensed rehabbers who work in your area, contact your regional wildlife staff or your state's association of wildlife rehabilitators. Some WCOs donate equipment, money, or carcasses (to serve as food for captive predators) to rehabilitators, to help ensure that the service will be available in the future.

Killing the Captured Animal

You may choose to avoid relocating animals by refusing to offer that service, but as a WCO, if you handle mammals, you must be prepared to kill wildlife in case a rabies test is needed. You are legally required to follow the directions of the state wildlife agency and department of health. Most people understand and support measures taken to protect public health and safety. What about other circumstances? Like relocation, this technique is also controversial. Here's how it stacks up as a best practice.

1. Is it safe?
   - Care and skill are required when using a lethal technique, even one that's commonly used by the general public, such as setting a mouse trap. Consider the risks to people—including yourself—and those to other wildlife, pets, and livestock.
   - The risks vary depending on the technique. For example, a carbon dioxide chamber doesn't pose a risk to other species because you control which animal is placed inside it. Killing methods are described later in this chapter, with details about safety.
2. What are the likely ecological consequences of this action?
   - As we explained in Chapter Two, most of the species described in this book are thriving in our region. Their populations are rising, which is one of the reasons why they're more likely to come into conflict with people. In most cases, you would be removing a small number of animals and wouldn't significantly affect the local populations of these species. The only exceptions are bat colonies and snake hibernacula. Destroying either might hurt the local population of that species.
   - Could killing a nuisance animal affect non-target wildlife species? Yes. Some lethal tools, such as pesticides and lethal traps, may pose risks to other species, including people. That's why they must be used carefully and are not appropriate in all cases. For example, if you were attempting to control rodents outdoors, you should not use pesticides registered for controlling rats and mice indoors. If a coyote or fox ate a poisoned mouse or rat, it might also die. Consider the circumstances.
3. Is it practical?
   - If you've correctly diagnosed the situation, killing the nuisance animals will stop the problem—at least for a while. If the customer opted for removal without exclusion, chances are good that a new animal will take advantage of the situation.
   - On the other hand, if the customer opts for removal using lethal techniques and exclusion, then you've provided a long-term solution without the risk of passing the problem on.
4. Is it humane?
   - This is an important topic that will be discussed in much greater detail later in this chapter. In each species account, we've described preferred and acceptable killing methods for that species. Use a preferred method whenever you can.
   - You must match the technique to the species. For example, the use of a carbon dioxide chamber is not recommended for beaver, diving birds, or reptiles, because those animals can hold their breath for a long time, so that method wouldn't kill them quickly enough.
5. Is it legal?
   - Can you legally "take" (kill) that species as a WCO?
   - Check local laws. Some lethal techniques, such as shooting, for example, may be restricted in developed areas.
6. How would your actions play on the evening news?
   - Think about this question long and hard—especially when you're using lethal techniques. The best practices approach provides a solid framework for making decisions about wildlife damage situations. Don't just give it lip service. Think before you act. Imagine that a video camera is recording at all times.

- Be sensitive and savvy. Work in private. That's actually kinder to the animal, too, and there's no reason to add unnecessary stress to its last moments (Figure 5-14). Some of the killing methods described in this manual look gentler than others. If you have to kill an animal in front of observers—which you should avoid doing whenever possible—take a moment to explain what you're about to do, and why you've chosen this method for this situation.

FIGURE 5–14 Sometimes you cannot place your traps inside the building because you can't get to the attic or the rafters wouldn't support your weight. If you need to set a trap outdoors, especially in an area where it's visible to many people, cover the trap. If it's a live trap, this may help protect the animal from the weather and predators. People are less likely to interfere, too. (Some might try to free the animal, which, as you know, is not necessarily humane. Some might taunt the animal. Others might vandalize or steal a trap.) Discretion is also important when setting a lethal trap outdoors. *(Connecticut Nuisance Wildlife Control Operators Association, Inc.)*

Option: Reduce the local breeding population.

Techniques: Regulated hunting and trapping seasons; and special culling operations (permits may be needed)

With most of the species handled by WCOs, the focus is on an individual problem that generally affects a small area, such as one person's home. Some species, such as Canada geese and white-tailed deer, may

FIGURE 5–15a

FIGURES 5–15b

FIGURES 5–15a and 15b Goose roundups are an example of a special removal operation. They're most effective when the geese are molting and can't fly. If the birds are on the water, people in boats drive them onto the land. Then others walk behind the geese to direct them towards the pen. These temporary enclosures usually have a funnel-shaped entrance that guides the birds inside. *(Paul D. Curtis)*

cause problems for many parts of a community. In this case, one aspect of the solution may involve reducing a breeding population. This is often done through regulated hunting and trapping seasons or special culling operations of adults, eggs, or young (Figures 5-15a and 5-15b). Such efforts are usually planned by wildlife biologists but some WCOs participate in the project, so we'll give you an overview of the method.

These approaches can be biologically, socially, and legally complex. They require community support and government oversight, and are usually beyond the normal scope of WCO activity. The group in charge of this major effort may hire WCOs to provide specific services, such as puncturing or oiling eggs.

The bottom line for the average WCO: you may not even want to get involved in special culling operations, but you can encourage landowners to allow hunting and trapping on their land, especially if they're dealing with nuisance wildlife. This may contribute to solutions for problems caused by beavers, muskrats, raccoons, squirrels, and deer, for example.

Scientists are experimenting with fertility control techniques for wildlife. This hasn't yet proved generally effective for controlling populations, but contraceptive and sterilization methods are being tested in a wide number of species on a small scale. (To learn more, see Fagerstone, K.A., M.A. Coffey, P.D. Curtis, R.A. Dolbeer, G.J. Killian, L.A. Miller, and L.M. Wilmot. "Wildlife Fertility Control." *Wildlife Society Technical Review* 02-2 (2002): 29.)
Note: Some communities currently participate in research programs to test wildlife fertility control and sterilization techniques. These techniques are not yet proven, nor are they commercially available.

Option: Keep the animal out of the area (exclusion).

There's one strategy that should almost always be used as part of your solution: exclusion. In some cases, you may be able to solve the wildlife conflict by putting up a barrier that keeps the animal away from the vulnerable area, whether it's a garden, child's sandbox, orchard, or the foundation of a building (Figures 5-16a, 5-16b, 5-16c).

Exclusion is one of the best ways to prevent wildlife damage conflicts. It can stop repeats of the problem and may even prevent other conflicts from

Techniques: Repair buildings to seal entry holes; protect vents, louvers, and sewer pipe vents with animal-proof designs or add screens or shields; cap chimneys; fence outdoor areas; use netting, spikes, electric shock devices, or plastic strips to keep birds away from alcoves, ledges, and other perches; erect "post-and-wire" grids over large areas to discourage birds from landing there. For more information, see Step Four.

FIGURE 5–17 The smudge on this wall and the slight depression in the ground show where an animal has squeezed under the shed. It's important to locate the animal's travel routes and find the place where it enters the building. If you're going to remove the animal, those are among the best places to set a trap. In many cases, when you're done with the job, you'll block that entry hole to make sure animals can't get back into the building. *(Connecticut Nuisance Wildlife Control Operators Association, Inc.)*

happening. And it will probably increase the effectiveness of other strategies, such as habitat modification, the use of repellents, and the removal of the nuisance animals.

When you can, persuade your customers to take advantage of your wildlife expertise (Figure 5-17) and knowledge of building construction by hiring you to make their properties "animal-proof." Perhaps you only want to make recommendations, or maybe you're happy to do the actual repairs (this type of work can pay much better than trapping). In either case, you probably know more about the correct installation of exclusion products than the average building contractor.

FIGURE 5–16a

FIGURE 5–16b

FIGURE 5–16c

FIGURES 5–16a–c There are many exclusion products on the market. A chimney cover will keep many animals, such as birds, bats, squirrels, and raccoons, from getting into a chimney. This net is hung over a pond or field to discourage birds from landing in the area. Here's an example of an exclusion device that most WCOs custom-fit to the site. It's called a "rat wall." This L-shaped fence can be installed as a free-standing barrier or attached to a porch, shed, or stairs, as it is here. The bottom is bent into a shelf that keeps animals from digging underneath it. In this photo, you can see the bottom, but not the top, which is attached to the concrete. Once it's covered with soil, the rat wall won't be noticeable at all. *(Connecticut Nuisance Wildlife Control Operators Association, Inc.)*

To learn more...

- Survey your WCO friends to find out which management strategies they use most often. Are they happy, or do they wish they could change things?
- Find out if any nearby communities are actively trying to control white-tailed deer or Canada geese. What's happened so far? How is it working?
- Search the Web for fact sheets that explain why some temporary wildlife conflicts happen (such as robins pecking at windowpanes, or woodpeckers drumming on gutters). Create a file to help you respond to customer questions.

Review Questions

8. To make an area less attractive to nuisance wildlife, you might persuade your customers to:
  a) Keep the area around the bird feeder clean.
  b) Remove dog feces from the yard every day.
  c) Store the trash cans in the garage, then put them out right before the trash pick-up.
  d) All of the above.

9. To make a park less enticing to Canada geese, you advise the city officials to:
  a) Pave over the grass.
  b) Hold firework displays every night at 8 p.m.
  c) Let the grass around the pond grow taller.
  d) Use a trained dog.

10. In nearly every job, you'd expect to do this, or might advise your customers to take care of it. What is it?
  a) Clean up the area with antibacterial soap.
  b) Tell the neighbors they'll be next. But if they hire you now, there's a 10% discount.
  c) Get a federal trapping permit.
  d) Exclusion, to prevent problems.

11. Your customer tried a bird scare device, but it didn't work. He wants to know why. You explain (choose all possible reasons):

___ "Birds can't hear ultrasonic sounds."

___ "The birds have gotten used to seeing it there, so it's not scary anymore."

___ "You played the alarm call of a crow, but those are starlings roosting in your tree."

___ "Your border collie is a wuss. When I drove up, the Canada geese were chasing your dog!"

___ "Scare devices never work."

12. Your customer wants to consider every option for solving his squirrel problem. You say:
  a) "Well, there are several approaches. I could trap it, or try to scare it off, or put down some pesticides. That's about it. Trust me, let me trap the animal and take care of it for you. If we use pesticides, the squirrels could die in the walls, and that would stink so bad you'd be sorry you did it. You don't have to see a thing. But I don't have the time to talk all day."

b) "That's a trade secret. Don't you have to go to work, or something?"

c) "It mostly comes down to capturing the animal. We can live trap the animal, or use a lethal trap. If I see the animal, I might be able to catch it using my catchpole. I'd probably set this cage trap. Do you want to release the squirrels alive, or do you want me to take care of them? Because if you let them out, they might just get back in, so then it's important to fix all the holes to prevent that."

d) "Some wildlife problems stop after a short time, so you don't have to do anything. That's not true here. I suggest we use a combination of approaches. We make the place less cozy for the squirrels, while I work on removing them. You can try to scare them away. When we're done, we fix all the holes so they can't get back inside."

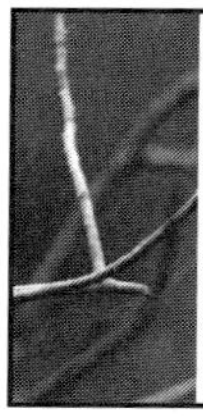

STEP THREE: Do It

LEARNING OBJECTIVES

5.12 Name five nonlethal and two lethal techniques for removing animals from an area.

5.13 Once you hit the road, you don't want to waste time driving back to the office. Describe five pieces of equipment you'd keep in your truck so you could respond to a wide variety of nuisance wildlife calls.

5.14 Which two animal removal techniques require additional training and other licenses or permits?

5.15 List six trapping tips that apply to the use of both live traps and kill traps.

5.16 You need to submit a raccoon for rabies testing. Which killing method should you use?

5.17 Describe how you'd clean an attic that had a family of raccoons living in it. What would you do if they'd been living in the chimney instead?

With your overall problem-solving strategy in mind, now it's time to choose the techniques that will help you accomplish your goals. If you decided to do nothing, well then—you're done. If you'd like to make the environment less attractive to the nuisance animal, you'll find more suggestions listed in the species accounts in Appendices B and C. This book won't go into any more detail about methods used to reduce local wildlife populations, but you can learn more about that from some of the resources listed in Appendix E. In this section, we'll explain some of the best practices for repelling, removing, excluding, and killing animals.

Remember to check your state and local laws, which may restrict the use of certain tools and techniques.

Avoid Creating Wildlife Orphans

Before you repel, remove, exclude, or kill an animal, take steps to prevent the orphaning of young wildlife. A lot of WCO work happens when wildlife are raising their young, so this is an important consideration. Sounds pretty reasonable. How do you do it?

Unfortunately, this is another hotly debated topic that hasn't been well studied. For raccoons and squirrels, there's some anecdotal information tracking the number of times a female will retrieve her young when certain removal techniques are used, but to our knowledge, no one has followed that all the way through to find out what happens later. How often does the female find a suitable den site in time? How easily does the family recover from the stress of the experience? Do a reasonable number of the young survive to adulthood?

Many people assume that den and nesting sites are plentiful and that females usually have several nearby alternatives. Perhaps this is true for some species, in some areas. It's certainly not true in all places.

The best way to prevent orphaning is to convince your customers to wait until the young are mobile before removing, repelling, or excluding the family from the site. Be sure they understand and are comfortable with the risks of allowing the animals to remain in the building (given the litigious nature of our society, offering such advice could increase your risk of being sued). Take measures to prevent the animals from entering the people's living spaces, which will reduce the risk of exposure to diseases.

Alternately, you can try to capture and remove both the female and all of her young and hope that she will retrieve them and continue to care for them. When is this most likely to work? Let's apply a little biological common sense. Older, more experienced females are probably better at finding resources than younger females—they're more likely to be "better" mothers and may be less likely to abandon their young. The more time the mother has had to bond with her young, the less likely she is to abandon them, so the age of the young may also be a good indicator. You can't change the ages of the animals you're dealing with, but if you think your chances of success are poor, you might choose a different option.

Some WCOs are trying to refine removal techniques to increase the chances that the female will retrieve her young. Here are their suggestions.

- Try this technique with raccoons, skunks, and squirrels.
- Time your activities to match the normal habits of that species. The sooner the female finds the young, the better. If the young are left alone too long, they may die of exposure. For a nocturnal species, start at dusk.
- Remove the female, preferably using a direct capture technique such as a catchpole (described later).
- Place the female and young in a release box. (We'll describe several variations on this theme.) Many WCOs use a simple cardboard box, others use a wooden nest box, such as a wood duck box, and some prefer plastic boxes. Match the size of the box and its entrance hole to the size of the species, using a smaller box with a smaller hole for squirrels, and a larger box with at least a 7-inch hole for raccoons. (One WCO recommends a 2 × 2 × 1 foot box.)

Make sure the mother cannot immediately get out of the box by covering the hole. Then move them to a quiet place outdoors. Unless they're likely to be disturbed, keep the box at ground level. Remove the cover so the female can get out of the box. Another option is to build a box with a sliding door. Leave the door open about an inch, to keep the heat inside but make it easy for the female to slide it fully open so she can retrieve her young.

Some WCOs prefer to use heated release boxes when it's cold outside. Make sure that the box doesn't get too hot. You may want to provide heat in just one area. Also, assume that if you put something in the box, the animals will chew on it. Don't give them access to anything that they shouldn't eat, such as wires. That means if you choose to use a household heating pad as the heat source, make sure the animals can't reach the wires. To avoid that problem, one WCO builds his boxes with a double floor, placing the heating pad in the space between the floors. Other options for heat sources include microwaveable heating pads and warm soapstones.

- If you can't catch the female, put the young in the release box and locate it as close to the entry site as possible. Check the next day to see if the young are still there. If so, they've probably been abandoned.
- Learn the typical birthing and rearing seasons of the species you handle. Know how to tell when a female is nursing and what young sound like.

How will you evict the animals? Here's an overview of the common techniques available to WCOs.

Nonlethal Techniques

Devices and techniques in this category include direct capture, the use of live traps, one-way doors, repellents, and other frightening techniques.

Direct Capture

Sometimes, you may be able to capture the animal immediately, using your hands (for immobile young) or a simple device, such as a catchpole (Figure 5-18). If you can do so safely, this is often appealing to the customer. It may also eliminate the need for repeated visits to the site. Remember all of the safety tips from Chapter Four. The right gloves are especially important if you're going to try to catch an animal by hand. Don't try to capture an adult animal by hand. It's just too risky. A squirrel, being "squirrelly," could pop out of your hand, shoot up your arm, run over your face and jump off the top of your head before you'd even had time to realize that you'd lost your grip on it. Other useful devices include:

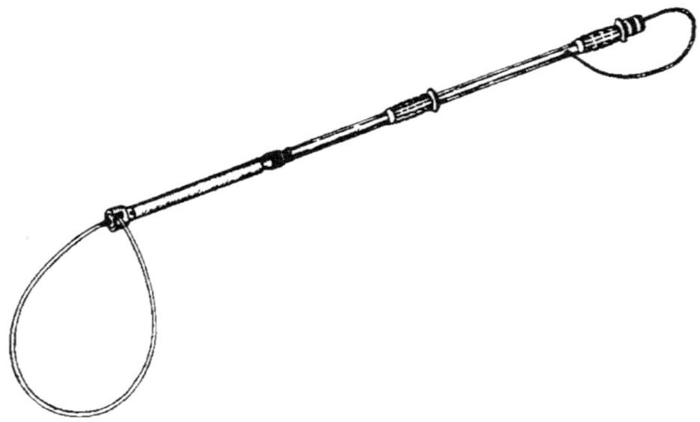

FIGURE 5–18 A catchpole. *(Michael Stickney)*

- **Catchpole,** or "snare pole" (similar to snake tongs, cat grasper). This is one of the most versatile tools used to capture and restrain animals. Basically, a catchpole is a long stick with a noose (cabled loop) on one end. For most species, place the loop over the animal's head and then tighten the cable to hold the animal. Bobcats and housecats can accidentally suffocate if the loop is only placed around their necks—it's better to place the loop over the cat's head and over one front leg. Minimize the amount of time an animal spends in this restraint.

  Some catchpoles swivel, allowing the animal to twist without being suffocated. Commercial catchpoles often lock once you've pulled the cable tight, and also have a quick release. (If you prefer to make your own, run a loop of plastic-coated cable through a piece of rigid aluminum pipe or conduit that's 3–4 feet long. You may still want to add a quick release mechanism.)

  Related hand-operated devices may substitute a tong-like closure for the noose on the end of the catchpole. Imagine the kind of pincer that some people use to grab cans off a high cabinet. This can be useful if you're trying to capture a small animal, such as a squirrel, or if you can only reach a part of the animal and would not be able to get a loop around it. Poles with this type of closure usually don't have the restraining power of catchpoles.

  Modified catchpoles are available for restraining snakes. They're often called snake "sticks," "tongs," or "hooks." These devices pin the snake's head to the ground. Use them carefully, because it's hard to tell how much pressure you're exerting. You could accidentally injure the snake's spine or even dislocate its head. Once the snake's head is restrained, grasp the snake just behind its jaws with your thumb and forefingers. This will give you control of its head. Support the snake's body (with your arm, a stick, or a pole) when you carry it. This will minimize its stress and prevent it from thrashing about.

What if you're confronted with a large nonvenomous snake, such as a boa constrictor, that's tightly wrapped around a person? If you can't work it loose with your hands, remember that snakes don't like cold temperatures. Have the person step into a cool shower. The snake will probably let go once the water hits it, and try to slither away.

Obviously, inexperienced people should not handle venomous snakes! WCOs with the proper snake-handling training should still take a few precautions when working with venomous snakes. Don't work alone. If you must, at least tell someone what you're doing. Call the local hospital before you go out on the job to see if they have the proper antivenin—and have them check that it's still good (they probably don't use it often, so it could be outdated). Bring along a garbage can with a lid, or a cage that you can put the snake in once you've captured it.

- **Buckets or small plastic containers.** This works well with bats and small snakes. Cover the animal with the container, and then slip a piece of stiff cardboard between the container and the wall or floor. (Or cut an "x" into the container's plastic lid so the bat can crawl into the container on its own.) You can then carry the animal outside and release it. (Remember: don't release bats if there was a possible rabies exposure.)

- **Nets.** Two designs are the throw net and "hoop" net. Throw nets are tossed over the target animal. Hoop nets are attached to the end of a long handle. They're used to scoop up the animal. The net should be deep enough to allow the hoop to be twisted to restrain the animal in the bottom of the net. A captured animal may climb out of a shallow net. The size of the mesh is important, too. If the mesh is too large, the animal may force its head through and injure itself or strangle. If you must use a shallow net, immediately place the frame against a flat surface to prevent escape. You can further restrict the animal's movement by carefully pressing on it with a stick. Whenever possible, encourage the animal to enter the net on its own. You might injure an animal while swinging a hoop net if you accidentally hit it with the frame.

- Customers may ask about the use of drugs to capture wildlife. **Immobilizing agents** (such as ketamine hydrochloride, tiletamine, alphachloralose, and metatomadine) are heavily regulated by federal and state agencies. Very few WCOs have access to these drugs.

Live Traps

This is probably the animal removal tool used most often by WCOs. A live trap is meant to capture an animal without killing it (Figure 5-19). Some types of live traps are cage traps (also known as "box traps"), multiple capture traps, foothold traps, nets, some specialized beaver traps, and a variety of bird traps.

Live trapping has some advantages. You can see what you've caught and prove your success to your customer. It also prevents animals from dying in inaccessible locations, one of the hazards of using pesticides. In addition to the foul odor caused by decay, the presence of a dead animal can attract other pests. In most cases, if you're using a live trap, you would be able to release an animal that had been caught accidentally (a "nontarget capture").

There are some disadvantages to live trapping as well. It's usually labor-intensive and you might capture the wrong animal. If the live trap is used improperly, an animal may die in it from lack of food or water, exposure to heat during the summer,

General trapping tips

- The trap design should be matched to the species and to the situation.
- Always choose trap types that minimize the risk of catching unintended wildlife species or pets.
- Leave a phone number so your customer can reach you in an emergency.
- Set traps (including cage traps) so that children and domestic animals can't reach them. Either may harass captured animals and possibly get bitten.
- Be discreet; keep traps out of view of the general public.
- Secure and label all traps with your name and address (required by law in most states).
- Take steps to protect captured animals from bad weather including summer heat, rain, snow, and cold. Covering a trap or setting it in a more protected location can help.
- For the most common nuisance species, you don't need to worry about wearing rubber or cotton gloves to avoid getting human scent on the traps, which is a concern for fur trappers. Consider where the nuisance animals are living and what they do—they're used to people.
- The proper use of baits and lures may increase your capture rate and help you avoid capturing animals you didn't want to trap.
- To prevent trapped animals from reaching through the cage and tearing up turf or nearby objects, place the trap on pavement, or put a large piece of hardware cloth, an old tarp, or a bit of scrap lumber underneath the trap.
- Take steps to avoid orphaning wildlife.
- Set traps where they're most likely to capture the target animal and not others. ("Location, location, location.")
- Clean the traps between uses to help prevent the spread of wildlife diseases and parasites.

weather extremes, or attacks by wildlife, pets, or people. Some animals might die from stress; perhaps they were already sick. Some animals might hurt themselves because of the stress of being restrained, or while trying to escape.

If you're not experienced with live traps, we strongly recommend that you seek hands-on training—especially before using foothold traps. Sessions on trapping techniques are often incorporated into seminars, conferences, and conventions sponsored by WCO associations and trapping associations (see Appendix E). There are opportunities for one-on-one instruction, too. Experienced trappers often advertise such services in trade journals. You'll also find many books, videos, and magazines about trapping.

■ **Cage traps** (a.k.a. "box traps") are the main stock-in-trade of most WCOs. These traps are often made of wire or plastic and come in many sizes and styles (Figure 5-20). They may open on one or both ends. Customers may call all cage traps "Havaharts" after a popular model, but there are many designs and manufacturers. For example, there are special, smaller versions of squirrel-sized cage traps that fit into woodstoves and chimney pipes.

FIGURE 5–19 Live traps are effective devices for capturing many species. Shown here are a raccoon, opossum, and woodchuck in the back of a WCO's truck. Notice the cardboard between the cage traps to keep the animals from seeing each other. This may reduce their stress during transport. *(Connecticut Nuisance Wildlife Control Operators Association, Inc.)*

FIGURE 5–20 Cage trap in set position. *(Michael Stickney)*

Tips for live trapping mammals on a roof using a cage trap

- Set the trap on the roof, as close to the entry hole as possible.
- Remember to protect the roof by placing a barrier underneath the trap that extends out far enough so the animal can't reach through the wire to damage the shingles. Raccoons in particular can do a lot of damage to roofs while restrained in cage traps.
- Avoid oily baits because they might stain the roof.
- With a roof set, you can check the trap from the outside of the building using a mirror on a long pole or binoculars.
- How will you keep the trap from sliding off the roof? If the owner agrees, drive screws part way into the roof, then wire the trap to the screws. Once the trap is secure, finish driving the screws into the roof then dab some roof sealant onto the heads of the screws.
- Check the trap early and often. Roofs get hot. The animal could die of heat stress if it's left in the trap too long.
- Use a rope to lower the animal in the cage to the ground. Don't descend the ladder carrying the trap. The animal might distract you and cause you to fall off the ladder.

An animal enters the cage, then steps on a treadle, which causes the door(s) at the end(s) of the trap to close. Cage traps are easy to set and do double-duty as a carrier. Little site preparation is usually needed. They are generally safe for children and pets. Most people think they're humane. But as one experienced trapper cautions, "a cage trap is not a pillow trap." Animals can be harmed when cage traps are used poorly.

Unfortunately, most cage traps are bulky, and because of their size and shape, they're hard to conceal. This makes them more vulnerable to theft and vandalism. They are relatively expensive and need frequent cleaning. Cage traps are not universally effective in capturing animals. Some species, such as foxes, are usually shy of these traps. Even individuals of species that are generally easy to catch in a cage trap, such as raccoons and gray squirrels, can be "trap-shy," especially if they have been captured before.

■ **Multiple capture traps** can catch more than one animal without having to be reset. Most multiple capture traps are designed for mice. Some brands (such as Ketch-All) will catch animals up to the size of chipmunks, but larger animals are likely to suffer harm. It's better to restrict their use to smaller animals. Some designs (such as Ketch-All and Kwick

Tips for live trapping mammals on the ground using a cage trap

- Set the trap near the animal's exit hole, in the animal's travel path, or in an area the animal uses often. (Look for sign.)
- Choose a bait that will appeal to the target animal but not attract unwanted animals. For example, pet food attracts raccoons—and skunks, cats, and opposums. If you want to trap a raccoon, a better option is a commercial sweet bait, which wouldn't appeal to cats. (Marshmallows are sometimes recommended as a bait for raccoons, but they quickly turn into a sticky mess in the heat. That's if the ants haven't devoured them entirely, leaving your trap unbaited.) Skunks love vanilla, which doesn't appeal to cats. Apples or raw vegetables are good woodchuck baits.
- Covering the bottom of the trap with soil or leaves may also convince an animal to enter.
- Raccoons may respond to a visual attractant such as a piece of aluminum foil suspended from the roof of the trap.
- Remember to protect the landscape by placing a barrier underneath the trap that extends out far enough so the animal can't reach through the wire to damage the grass or nearby plants.
- There are many commercially available scent lures. Just be aware that lures may attract unwanted species, too. WCOs tend to use lures when they're dealing with a trap-shy individual, or must set their trap away from the nuisance animal's favorite areas.
- After you have the trap set and baited, "dry fire" or trip the trap. Make sure the trap door(s) close smoothly and firmly.
- Check the trap early and often.

Katch) have a wind-up spring that powers a rotating mechanism. The mice are shuttled into a holding compartment. Other traps (such as the Victor Tin Cat) have one-way doors that allow mice to enter, but not leave. If you intend to use this as a live trap, you must ensure that it's checked often enough so that animals don't overheat and die in the trap.

■ **Foothold traps** restrain animals by holding the foot. Foothold traps can be used on land and in the water. Although they've gotten a lot of bad press over the years, when used properly, foothold traps generally result in a humane capture. Much recent research has gone into improving several trap designs. Most mammals have ample fur and cartilage padding their feet. Some people imagine what it would feel like to be caught in such a trap, but that's not what it feels like for wildlife. To treat an animal humanely, you have to consider what the situation is like for that animal, not what it would be like for a person.

Foothold traps are among the most efficient tools for catching coyotes and foxes and are also important for raccoon, beaver, and muskrat. There are several different designs, including the coilspring trap, which is probably the most commonly used foothold trap; the longspring trap; and several new designs (Lil' Grizz Get'rz, Duffer trap, EGG trap) meant specifically for raccoons (Figures 5-21a and 5-21b). These foot encapsulating traps reduce the chance of catching the wrong species and the risk of the raccoon injuring itself.

To set a coilspring trap, you must fully depress the spring levers. This compresses the springs. Then, place the dog across the nearest jaw until the tip of the dog fits into the notch of the pan. The dog holds the trap open in its set position. When an animal steps on the pan, it dislodges the dog from its notch and springs the trap, closing the jaws around its foot (Figures 5-22a and 5-22b).

There are several modifications to the standard coilspring trap design that will reduce the chance of injuring the captured animal; some may also increase the effectiveness of the trap. Padding or lamination creates a smooth side to the jaws of the trap. Wide-edge jaws, double jaws, and offset jaw designs, which are available commercially, also reduce the risk of injury.

Once the animal is captured, you want to keep it to one side of the trap to prevent it from circling around or moving its foot excessively, which might

FIGURE 5–21a *(Scott Smith)*

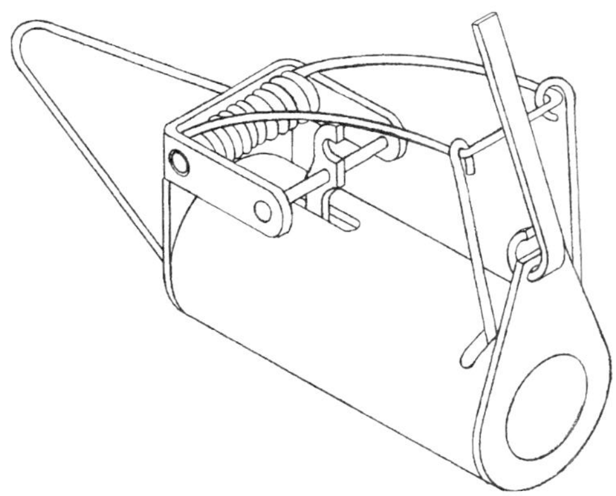

FIGURE 5–21b *(Jim Engel)*

FIGURES 5–21a and 21b This trap (Lil' Grizz Get'rz) takes advantage of the raccoon's dexterity. Many animals can stick their paws into a "hole" (the trap) but most can't grasp and manipulate an object inside it. Raccoons can. They'll try to pull out the trigger, which fires the trap.

FIGURE 5–22a A coilspring trap, which is a type of foothold trap, shown in its set position. *(Michael Stickney)*

FIGURE 5–22b The same live trap shown in its sprung position. *(Michael Stickney)*

cause it to injure itself. Place the chain to one side of the trap's base plate (off-center) so the animal would be pulling against the stake, which would help keep its foot from sliding around. A "stop loss," which is a secondary spring placed on the trap, will also force the animal to one side.

Many WCOs also add at least one swivel between the trap chain and the stake that anchors the trap to the ground. These swivels allow the captured animal to move around without binding the chain and twisting it around its leg. Most swivels placed within the chain are double swivels, but single swivels are usually used at the end of the chain. It's better to use multiple swivels because if the animal moves around a lot, and one swivel becomes gunked up with debris and stops working, there are others to keep the chain from binding.

Adjusting the pan tension on coilspring traps helps reduce the chance of capturing the wrong species (an animal that weighs less than the target species should be able to step on the pan without triggering the trap). There are several ways to adjust pan tension. With some trap designs, you can tighten or loosen the pan tension screw, which controls the amount of pressure needed to spring the trap. The tighter the screw, the heavier the animal must be to spring the trap.

When contending with snow or ice, which can disable foothold traps, consider switching to a cable restraint, a.k.a. "a snare," if the device is legal in your state, or to a four-coilspring trap with wider offset jaws.

- **Nets** may be used to capture birds, such as sparrows and starlings, inside malls or other large public spaces where shooting or trapping isn't appropriate. But even veteran users can become frustrated with this tool, especially if they're trying to capture a sparrow or two inside a large building. To use one type of fine-threaded net, called a "mist net," you'll need training and skill. Should you wish to use a mist net outdoors to capture wild songbirds, you'd also need federal and state permits. Purchasing mist nets may be difficult without the federal permit. Also, it takes skill to remove a bird from a mist net.

- **Bird live traps** are available in many designs. Most are designed to capture particular species. Usually such traps are used for unprotected exotic birds (e.g., house sparrows, European starlings, and feral pigeons). Some will capture one bird at a time, while others may capture many birds. To increase your chance of success, monitor the birds' daily movements and set the trap where they're most active. Give the birds time to get used to the cage trap. First, put out some bird seed, shelled corn for pigeons, or other appropriate bait to get the birds used to feeding in the area. Once the birds are feeding at the site, place the cage trap next to the bait and prop it open with wires. Place bait outside and inside the trap, to encourage the birds to enter the trap. Finally, when the birds are used to entering the trap, set it and place the bait only inside the trap. Leave one or two captured birds inside the trap as live decoys. Give them shelter, and daily water and food.

■ **Cable restraints (a.k.a. "relaxing snares")** are designed to capture larger mammals, including coyotes, foxes, bears, and beavers. Similar to a catchpole without the pole, cable restraints are highly efficient and, when used properly, have a low injury rate. Made of galvanized multi-strand cable, the device is suspended along the animal's travel path. When animals walk into the loop, they'll continue forward and try to push their way through it, as they would with a natural obstacle such as brush. Their movement causes the loop to tighten around the neck and often a front leg. The animal will then try to back out, again, as if it's a natural obstacle, but will stop as soon as they feel the constriction of the loop. The washer lock relaxes, securely restraining the animal.

Cable restraints are generally used with a free-floating reverse-bend washer lock, multiple swivels, loop stops, springs, and a breakaway device. The washer lock keeps the cable from over-tightening. The swivels keep the cable from twisting and binding. Loop stops and breakaway devices reduce the risk of capturing the wrong species. You use one loop stop to create the correctly sized noose for the target species, and another to make sure that the cable restraint can't become so small that it tightens down on a nontarget species. The breakaway device allows deer, livestock, or other large non-target animals to free themselves by pulling the loop apart.

For coyote sets, a "deer stop" is placed 9 inches from the end of the cable, to ensure a minimal loop of 2½ inches in diameter, which prevents the loop from closing down on a deer's leg. In most cases, the deer would be able to walk away. Of course this technique is not foolproof, so don't set cable restraints in known deer trails.

The maximum loop stop is set to a size that's appropriate for the target species. For foxes, try a maximum loop size of 6–8 inches, and set the cable 6–8 inches above the ground. For coyotes, use a 10- to 12-inch loop set 10–12 inches above the ground.

The cable should be anchored very securely in a spot where it cannot become tangled in plants or around rocks, fences, logs, or equipment. Look around in a full circle for anything that might get caught in the cable. Unlike foothold traps, animals captured in a cable restraint have the use of all four feet to pull against the anchor. Use strong earth anchors or double-stake with an 18- to 24-inch reinforcement rod. Commercial double stake devices are available.

In addition to properly anchoring the cable restraint, it must also be securely supported and stabilized (we suggest using 9-gauge wire). Bend the wire into an "S" loop before inserting it into the ground for added stability. Set the cable restraint lock on the support wire at either the 2 o'clock or 10 o'clock position to ensure that gravity will keep the loop open.

For coyotes and foxes, keep the cable restraints clean and free of human scent by placing each cable restraint in its own plastic zip bag. Wear clean cotton gloves when handling the cables. The cable restraint should be deodorized before use by simmering it in a mixture of two gallons of water and one pound of baking soda for an hour. This will turn the cable dull gray. You can then dye the cable to better camouflage with the surroundings.

Cable restraints can be modified with spring throwing arms for capturing bears (Aldrich foot snares) and coyotes (Collar-Um). Other similar models of active cable restraints are commercially available. Check the regulations in your state, as some states may require special permits to use cable restraints to capture nuisance wildlife.

■ **Beaver "suitcase" traps (Hancock or Bailey design)** are large rectangular steel cages constructed of chain-link fencing. They close like a suitcase over the beaver. Position the trap so that once it's sprung, the beaver is held about halfway out of the water and can breathe easily. The trap should be set in about 10–12 inches of water with the trigger about 4 inches below the surface so muskrats can swim over the trap without springing it. With care, this trap can be set in deeper water, using a deep stake so the top of the trap remains above the water's surface; hung off a dock; or staked so it flips the beaver onto land.

Bait the trap with willow, poplar, alder, or maple branches on the shore or use castoreum as a lure. You can direct the beaver to swim over the pan by positioning some branches or small logs in a "V" over the center of the trap. The opening of the "V" should be about 14–16 inches wide.

Check your set by splashing water over the trap. After the water has cleared, make sure that nothing is interfering with the trigger arms at the hinges. Splash water over everything that was handled and over the ground you walked on to help disguise your scent.

Take care while setting these large, strong traps. If the trap fired prematurely and hit you in the head, it could knock you unconscious. Manufacturers suggest that users wear a helmet. Hands-on training with an experienced trapper is also a wise precaution. The trap weighs over 30 pounds, which should be considered if it must be hauled long distances. If the area is a popular fishing ground, you may want to post a warning sign on a trail, alerting people that you've set a very strong trap nearby. This could invite vandalism, unfortunately.

One-Way Doors ("Checkvalves")

These devices allow an animal to leave but not re-enter a building. They come in a variety of designs, sizes, and materials, from plastic checkvalves or nets used for bat control to wire frames that are specially designed to fit certain cage traps (Figures 5-23a, 5-23b, 5-23c). Some WCOs build their own one-way doors. One of the advantages of using this tool is that you don't have to handle any animals. This is a relatively new tool in wildlife control, but its use is increasing.

Before installing a one-way door, conduct a thorough inspection to make sure there are no young present that are still immobile and dependent on their mother. One-way doors are only effective if the animal can find and use the exit but cannot find or force its way back through the door—or find another way into the building. Remember, if a mother has been separated from her young, she will be highly motivated to find another way inside to return to them and might cause additional damage in her attempt.

Here's how a one-way door is used in bat control. A commercial checkvalve or simple netting of ¼-inch mesh is installed over the bat's primary exit hole. Secure the top and two sides of the net, leaving the bottom loose and open. All the other holes into the structure are sealed. Bats exit at the bottom of the one-way door, but when they attempt to return, their sense of smell guides them back to the hole. They land on the mesh (or checkvalve) near the hole, and stay there, sniffing around. They just don't crawl down the mesh. After a suitable period of time, the one-way doors are removed, and the main entry sites are closed. Please note that the mesh size is important. Bats can get their heads stuck in a mesh of one-inch diameter and die, so we recommend a fine mesh of ¼ inch. This type of checkvalve should not be used when young, flightless bats are in the structure. (For more information, see the "house bats" account in Appendix B.)

FIGURE 5–23a

FIGURE 5–23b

FIGURE 5–23c

FIGURES 5–23a–c One-way doors may look very different but they all do the same thing: let the animal out and keep it from getting back in. Looking into the device, you can see its trap door in the set position. The same one-way door design is used to remove flying squirrels from an attic. If you opened the left-hand side of the device, they'd leave on their own. Mesh is often used to make a one-way door for bats; the result is called a "checkvalve." *(Connecticut Nuisance Wildlife Control Operators Association, Inc.)*

Frightening Techniques and Repellents

Repellents, you may remember from the earlier discussion, are objects, substances, or techniques that frighten an animal away. Or they make the desired object, such as a crop or nesting site, undesirable—the repellent may make that object smell or taste nasty, or feel bad to the touch.

There are several tools and techniques used to repel animals: scare devices; chemicals (which can only be used by WCOs with a commercial pesticide applicator license); hazing with dogs, hawks, falcons, or radio-controlled boats and cars; and the use of guard animals.

- **Objects that look scary,** such as Mylar tape, lasers, models of predators, and "scare-eye" balloons, are often used to drive off birds. Geese and crows can be easily dispersed from a night-time roost by pointing a spotlight, laser pointer, or commercial laser pistol (such as the Avian Dissauder) at them (be careful when shining lasers skyward, because they can cause difficulties for airplanes).

Visual repellents range dramatically in price, sophistication, and effectiveness. For example, everyone recognizes a scarecrow, but there are versions that actually move. These animated "human effigies" tend to work much better than the charming homemade kind, which are better considered as garden decorations. Whenever possible, choose a repellent that moves, such as by swaying in the wind (Figure 5-24). Visual repellents should be clearly visible.

■ **Objects that sound frightening,** include a variety of firecracker-like noisemakers and recordings of distress calls and alarm calls. Sometimes, even banging a pot or rattling tin plates will drive off birds. Avian distress and alarm calls are often most effective but may be species-specific. It's not clear if birds can hear ultrasound. Such devices have generally been ineffective for bird control. Before using any noisemakers, check local ordinances and consider the effects on the neighbors. Fireworks-based noisemakers (a.k.a. "pyrotechnic devices") include:

Bangers (a.k.a. "bird bombs"): Makes a loud bang. It's often launched from a handheld pistol launcher. Bangers can be used at medium range (50–100 feet).

Screamers: Makes a long, drawn out whistle. The shell flies out about 100 feet, screaming and whistling all the way.

Shell-crackers (or "crackers"): Makes a loud bang that sounds like a M-80 firecracker. These shells are fired from a twelve-gauge shotgun, exploding about 75 yards away.

Propane cannons: Unlike the other devices, propane cannons can be used with a timer. All of the other devices must be fired by a person.

There are a few devices that combine flashing lights and scary noises, for example, the Critter Gitter. Some of these can be very effective because they vary the pattern of the frightening element.

FIGURE 5–24 The spider-like devices on the top of this tank are called "Daddi Long Legs". The steel rods move in the breeze, scaring birds away from landing on the tank. Scare devices that move are generally more effective than stationary objects. *(Bird Barrier, Inc.)*

■ **Chemicals that make objects smell, taste, or feel bad.** You need a commercial pesticide applicator license to use any of these products in your WCO business. Contact your state pesticide regulatory agency for information.

■ **Hazing** describes a technique in which dogs, hawks, falcons, or radio-controlled aircraft and boats drive nuisance animals away from a site. Hazing Canada geese with border collies is one of the most effective ways to chase the birds away from golf courses, public parks, and similar locations. Hawks are sometimes used to chase other birds away from airport runways.

Don't haze birds in their nesting areas during the nesting season—that's a violation of the Migratory Bird Treaty Act. Be very careful if you want to haze Canada geese during their molt (usually

June 15–July 15 for most of our region) because the geese are vulnerable and can't fly.

■ **Guard animals** such as dogs and llamas are sometimes used to protect livestock, especially sheep, from predators. The livestock and the guard animal must be kept within a fenced area. Dogs contained within an invisible fencing system can sometimes protect orchards or vineyards from deer and turkey damage. This method may also be useful for protecting Christmas trees and nursery stock from deer browsing. Although WCOs don't use this technique commercially, some may want to recommend the option to certain customers.

You may encounter different definitions of "repellent" and "exclusion device." There is natural overlap between the two categories. Don't sweat it. We've classified products that are meant to be permanent installations, such as porcupine wire and post-and-grid systems, as exclusion devices. Although they repel birds, these devices are clearly barriers. Recordings of distress and alarm calls may have a long-lasting effect, but they are noisemakers.

Does this classification matter to anyone but an academic? Yes—because you need to remember that in general, repellents are a bit trickier to use. Although they may provide quick relief, it often doesn't last. Exclusion devices, on the other hand, often require an up-front investment because the materials and installation may be expensive. They tend to provide long-lasting results, and help to prevent problems. Exclusion can be a proactive approach, but repellents are generally a reactive strategy.

Lethal Techniques

Just as there is no magic pill solution to wildlife conflicts, there is no one killing technique that will work best for all species or in all situations.

General tips for the use of repellents and frightening techniques

- Before you use any repellents, double-check their legal status at all levels: federal, state, and local. Are you dealing with a protected species, such as a migratory bird? Any local ordinances that come into play? For example, noise ordinances would affect your ability to make use of pyrotechnic devices, which are purposefully loud. Local bird control ordinances might affect your choices, too.
- If you want to use chemical repellents, you must follow pesticide regulations.
- Often, a combination of techniques is more effective than using one technique by itself.
- Repellents tend to be more successful in preventing new behavior, such as roosting, nesting, or feeding, than in stopping a well-established behavior.
- Another thing to remember is that animals get used to both visual and sound repellents. To increase your chance of success, mix things up a little. Move the devices around. Change the patterns of the noise. Keep it unpredictable.
- Remove the devices after the problem has stopped (or if the animals have left the area for other reasons).

Some methods are not appropriate for young animals or diving or burrowing animals. Some work better with birds, or reptiles, or mammals. Some require additional permits, or a better than average level of skill, experience, or strength. They may be effective for some WCOs and not others. And if rabies is an issue, you must select your method carefully.

Sometimes, protecting people will be your highest priority. You may need to kill an animal as quickly as possible. Even in such cases, do the best you can to provide the most humane death under the circumstances. Most people agree that a "humane death" is one that is as painless as possible. The ideal is to be quick. Either kill the animal quickly, or cause rapid unconsciousness, then rapid death.

You need to know just a little bit about how animals experience pain to understand how this works. A sensation triggers an impulse that travels along the nerve endings to the brain. The brain interprets this as—ouch!—pain. The most important part of the brain involved in the sensation of pain is the cerebral cortex, which is in the front of the brain. Other parts of the brain are involved in the feelings of fear, anxiety, and discomfort. The brain must be working in order for the animal to experience pain. If you can break the circuit between the brain and the rest of the body, the message will not be delivered and understood; the animal will not feel pain.

Bottom line: No working brain = no pain.

That seems straightforward so far, doesn't it? Unfortunately, there is disagreement about which methods are appropriate to kill wildlife under field conditions. There hasn't been much research that's really relevant to the situations WCOs encounter. And to make things even more difficult, some of the methods that are well accepted for wildlife, such as shooting, cannot be legally used in all areas.

The following recommendations are meant for the species discussed in this book. In the species accounts in Appendices B and C, we've listed preferred and acceptable killing methods for each species. Whenever possible, use a preferred method.

Some of the lethal techniques we'll describe can also be used as removal methods, such as lethal trapping, shooting, and the use of barbiturates or pesticides. The others are strictly killing methods: the use of a carbon dioxide chamber, cervical dislocation, stunning, and chest compression.

This list reflects current scientific information and the judgment of experienced WCOs and wildlife biologists—people who care deeply about the treatment of wildlife and have actually used these methods in the field over many years. Many factors were considered when evaluating a method. Is it as quick and painless as possible? Is the technique or product generally safe for the WCO, the public, other animals, and the environment? Is it practical for use in the field? Does it require additional permits? Is it legal in most places? How reliable is the method? Can most WCOs be expected to master the technique, or does it require more than average skill or strength? Although we did consider public opinion, we did not rule out techniques that don't look pretty because in some situations, these methods may provide a more humane death than a technique that looks better.

Some of these methods also meet the guidelines described in the *2000 Report of the American Veterinary Medical Association Panel on Euthanasia*. Although their report focuses on domestic animals and the techniques used to kill them in a controlled indoor setting, you may still wish to read the document to better understand some of these techniques. The report is available at www.avma.org/resources/euthanasia.pdf.

This is the aspect of your job that might require the most discretion. In general, it's best to work in private. If a crowd has gathered, ask them to leave, or move the animal to a private setting before you kill it. This is safer for the people and kinder to the animal, which may be stressed by the presence of the crowd. If you must kill an animal in front of others, first explain what you are doing and why.

Carry a variety of tools so you can choose from the full range of methods, selecting the one that's most appropriate for the situation. In some cases, you may use one method to cause unconsciousness and another to kill the animal.

With the exception of lethal traps and the use of pesticides, which are set and then left in place, the WCO is able to monitor the animal to make sure it experiences as little distress as possible. Some lethal methods, such as the use of carbon dioxide chambers, may require adjustment during the process. It's also important to confirm that the animal is dead. An easy way to do this is to touch the animal's eye with a long stick. If it's dead, it won't blink.

Lethal Trapping

The types of lethal traps most frequently used by WCOs are mouse and rat snap traps, body-gripping

General tips for the use of lethal traps

- The trap's size and design should be matched to the species and to the situation.
- Use the smallest trap appropriate for the targeted species.
- Set the trap at the time of day when the nuisance species will be most active.
- Leave a phone number so your customer can reach you in an emergency.
- Set traps so that children and domestic animals can't reach them and get hurt.
- Be discreet. Keep traps out of view of the general public. If you can't switch to a less visible place, then hide the trap with a cardboard cover. There are commercial models that are enclosed within a tunnel, which partially conceal the trapped animal (designed for squirrels).
- Label all traps with your name and address (required by law in most states).
- Secure traps so predators can't remove them.
- At least one of the trap's jaws must be able to move freely.
- Take steps to avoid orphaning wildlife.
- The proper use of baits and lures may increase your capture rate and help you avoid capturing animals you didn't want to trap.
- Always choose trap designs and methods that minimize the risk of catching unintended wildlife or pets. This might include:
  - Using baits to selectively attract the target;
  - Placing the trap in a vertical "cubby" set, in a deep-notch box, or in a bucket with a restricted opening so other animals can't reach the trap;
  - Covering the burrow entrance and trap so the nuisance animal encounters the trap when it tries to exit through its hole, but other animals can't reach the trap;
  - Using a one-way trigger, which will only trip in one direction. When an animal exits its hole, the trap fires. This protects animals that encounter the trap from the outside;
  - Adjusting the trigger position to match the size and habits of the target species.

traps (such as the Conibear), tunnel traps, and mole traps. Most lethal traps kill an animal with a mechanical blow. When properly set, lethal traps are usually highly effective and selective. Although death usually comes quickly, it's not always instantaneous. Lethal traps may catch animals more quickly than cage traps, reducing the number of visits needed.

Lethal traps are appropriate for use with small to large mammals (mice to beaver size) and unprotected birds. Match the size of the trap to the size of the targeted animal, but use the smallest trap appropriate for the target. The lethal traps currently available in the East are generally not suitable or legal for use with deer, bears, coyotes, foxes, and most birds. Know which traps are legal in your state. Cable restraints, for example, can be set for lethal wildlife control in some Western states; such use is illegal in some Eastern states.

There are some disadvantages associated with the use of lethal traps. If you've caught the wrong animal, you can't release it unharmed. If a trap is not set properly, you may get a poor or nonfatal capture. It takes more skill, experience, and time to set a lethal trap (you might need more strength or setting tools). The site often needs to be prepared.

Again, if you are not an experienced trapper, we strongly recommend that you seek hands-on training—especially before using body-gripping traps. Sessions on trapping techniques are often incorporated into seminars, conferences, and conventions sponsored by WCO associations and trapping associations (see the resource list in Appendix E for state and national contacts). There are opportunities for one-on-one instruction, too. Experienced trappers often advertise such services in trade journals. You'll also find many books, videos, and magazines about trapping.

■ **Body-gripping traps** are spring-loaded lethal traps available in many sizes. They're usually square, but some specialized models are round. Body-gripping traps are often called "Conibears," which is the name of a popular model manufactured by Oneida-Victor, Inc., but there are many manufacturers. There are also new, smaller designs, including some with one-way triggers, which are more selective, and traps designed specifically for squirrels. Tunnel traps, which are metal tubes with a body-gripping trap inside, are commonly used by WCOs.

"Magnum" or "zero tolerance" versions are also available. The jaws of this model close more tightly than those of a standard body-gripping trap, so it often kills faster and more consistently than the standard trap. This may increase the chance of a proper strike with squirrels or raccoons, or other small, flexible animals that might pull back if there's a slight gap between the jaws. Magnum versions are very strong—a #220 Magnum body-gripping trap could break bones—so be very careful when setting the trap. Consider using setting tools and safety devices.

To set a conventional body-gripping trap, you must compress the spring until its tips nearly meet at the rotating point of the jaws (Figures 5-25a and 5-25b). Then hold both jaws open and fit the dog (a.k.a. the "trigger hook"), which is notched, into the notch located on the trigger. Once securely in place, the dog holds the jaws open. In its correct

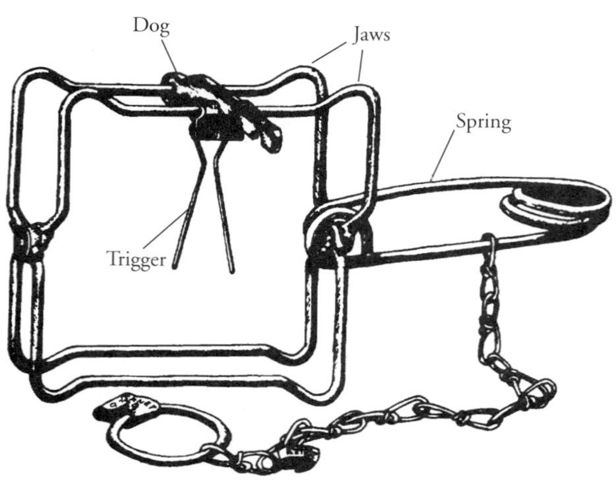

FIGURE 5–25a Body-gripping trap in set position. *(NYS DEC)*

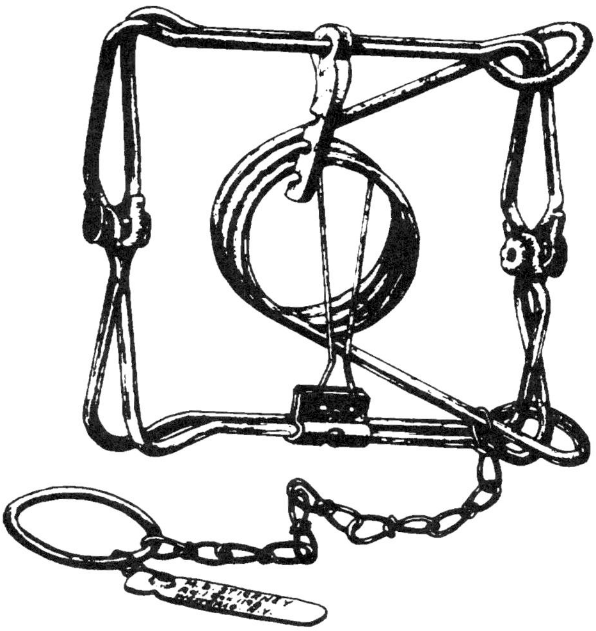

FIGURE 5–25b A body-gripping trap in its sprung position. *(NYS DEC)*

position, the jaws of the trap close on the top and bottom of the animal. Stabilize the trap to keep it in this top-to-bottom strike position, and to ensure that it can't be knocked over. Securely anchor the trap, as it is lightweight enough to be carried off by predators attracted by the captured animal.

Some models of body-gripping traps, usually those with two springs, have safety hooks that hold one spring while you're setting the other (Figure 5-26). If you use a safety hook, remember to release it when you're done setting the trap, so the trap can fire.

When an animal passes through the jaws of the trap, it moves the trigger, which dislodges the dog from its notch and springs the trap, closing the jaws around the animal's neck or chest. Ideally, this trap catches the animal directly behind the head, snapping the part of the spine that's in the upper third of the neck (called the "cervical spine" area). A proper hit provides a quick death.

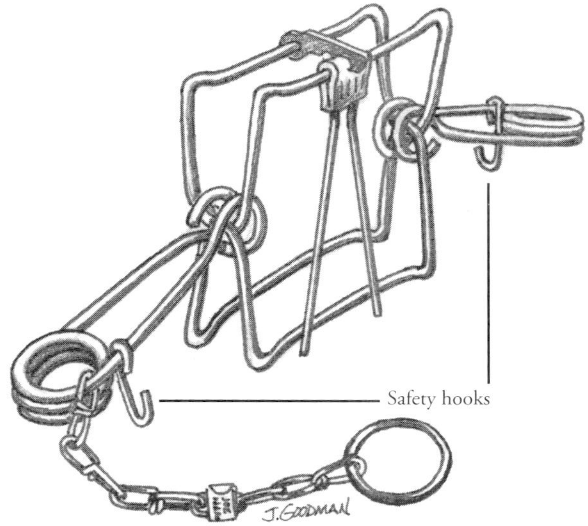

FIGURE 5–26 Notice the safety hooks on each spring. *(Joe Goodman)*

If you are trapping raccoons, skunks, squirrels, or woodchucks, modify the trigger to help to ensure a top-to-bottom strike (which is more humane) and prevent the animal from refusing to enter the trap. These species don't like to have anything brush against their eyes or whiskers, so separate the trigger and center it on the bottom of the trap.

As with all lethal techniques, care is needed to make sure that only the intended nuisance animals are caught. Body-gripping traps are often set in front of the animal's entrance hole because the animal must pass through the jaws of the trap to be captured (Figure 5-27). This way, only an animal entering or leaving the hole will be caught. If necessary, guide the animal into the trap. Use hardware cloth to reduce the size of the entry hole (shown in the photo) or to block escape routes.

This tool often provides the quickest way to remove a raccoon or squirrel from an attic. Another advantage is that nontarget animals are not nearly as likely to be on the roof as they are on the ground. But this trap is not appropriate for all settings. Use extra caution if you're setting a body-gripping trap

FIGURE 5–27 Match the size of body-gripping traps to the size of the target animal. In this case, the hole was larger than the trap. Hardware cloth was used to cover those gaps. *(Connecticut Nuisance Wildlife Control Operators Association, Inc.)*

on the ground, because of the risk to people, pets, and other wildlife. You can modify your technique and your equipment to minimize these risks. Here are some suggestions.

Cover the trap and burrow entrance. Let's say you've been unsuccessful trying to catch a woodchuck in a cage trap, or you need to trap many woodchucks. A body-gripping trap, if used cautiously, may be the right tool. Dig out the opening to the woodchuck's den a bit, so you can set the body-gripping trap in the burrow's entrance. The trap should be attached to a nonmovable object or a stake in the ground. Cover the burrow entrance and the trap with a large, loosely attached piece of hardware cloth, or with a piece of wax-coated corrugated cardboard. When the woodchuck leaves its den, it will spring the trap, but an animal poking its head into the burrow entrance wouldn't.

Place the trap in a container (a vertical cubby set, deep-notch box, or bucket with small hole). One way to lower the risk of catching an unintended species is to use a vertical "cubby" set, shown in the photographs (Figures 5-28a, 5-28b, 5-28c). This is a baited box that's open on one end, with the trap set well inside, usually held in place by friction between the coilsprings and the narrow notch in the box. You must use a model with two coilsprings, because one with a single spring won't stay in place. Make sure the open bottom of the box is no more than five inches off the ground. If a dog investigates, it may be hit on the nose, but it's unlikely that the trap would capture the dog. Cats cannot easily enter a vertical cubby set.

The deep-notch box (Figure 5-29) is an alternative to the vertical cubby set. It's set horizontally on the ground. If you're using a #220 body-gripping trap, the notches should be 8 inches deep. The trap is placed in the midpoint of the notches, which hold the trap in place. Make sure the trap is securely anchored. Center the trigger on the top of the trap. Bait should be placed deep inside the box, at least 6 inches behind the trap. The top of the deep-notch box is blocked with a piece of wood.

By restricting the size of the opening, you reduce the risk that a dog will spring the trap. The opening of a deep-notch box should be no more than 7 inches high.

Although the vertical cubby set and the deep-notch box are based on the same idea and are interchangeable, they have different strengths. The vertical cubby is less likely to attract cats (unless you've used the wrong bait) and is even more dog-proof than the deep-notch box. Sometimes, a raccoon will avoid a vertical cubby but investigate a deep-notch box. Some WCOs prefer one design over the other, as well.

Another option is to place the body-gripping trap inside a plastic bucket that has a restricted opening. The bucket can be round or square, as long as it has a lid (Figures 5-30a, 5-30b). An opening no larger than 7 inches is cut into the lid, slightly off-center. It works like a deep-notch box, only it's made from plastic instead of wood. Have some extra lids prepared, in case one is damaged or lost.

Use baits that selectively attract the nuisance animal. Commercial sweet baits will attract raccoons, but won't entice cats, so that's a safer bait

FIGURE 5–28a

FIGURE 5–28b

FIGURE 5–28c

FIGURES 5–28a–c Three views of a body-gripping trap placed in a vertical cubby set. The vertical cubby is attached to the tree at about five inches off the ground, to make sure that a dog can't reach up into the box and spring the trap. In this set, the trap was held in place by friction, so when it sprung the trap fell out of the box. You can anchor the trap so it will remain in the cubby after it's sprung. Bait is placed on a shelf near the top of the box. Notice the body-gripping trap in its set position. You can see the trigger and jaws within the box, and the spring sticking out to the left. *(Scott Smith)*

to choose if you must trap in an area where there are free-roaming cats. Mice will be attracted by many baits that are of varying appeal to other species, but you could avoid using bait entirely. Tie a cotton ball to the trigger instead. That's an attractive bit of nest material to a mouse, but cotton is of no interest to many other species. Whenever possible, use baits and lures that will attract only the nuisance animals.

Use a trap with a one-way trigger. The one-way trigger rests against the bottom jaws of the trap and only moves freely in one direction. In Figure 5-31, the animal would have to move from left to right to

FIGURE 5–29 A deep-notch box. *(Connecticut Nuisance Wildlife Control Operators Association, Inc.)*

FIGURE 5–30a A plastic bucket with a restricted opening. *(Angie Berchielli)*

FIGURE 5–30b The shape of the bucket doesn't matter, as long as it has a lid. *(Angie Berchielli)*

lift the trigger and spring the trap. An animal coming from the other direction might bump the trigger, but it's not likely to lift it high enough to spring the trap.

Compare the trap with the one-way trigger to the standard model shown below (Figure 5-32).

FIGURE 5–31 Setting a trap with a one-way trigger over an exit hole ensures that only animals leaving the den can spring the trap. Animals on the outside that might investigate won't be able to move the trigger enough to fire the trap. *(Joe Goodman)*

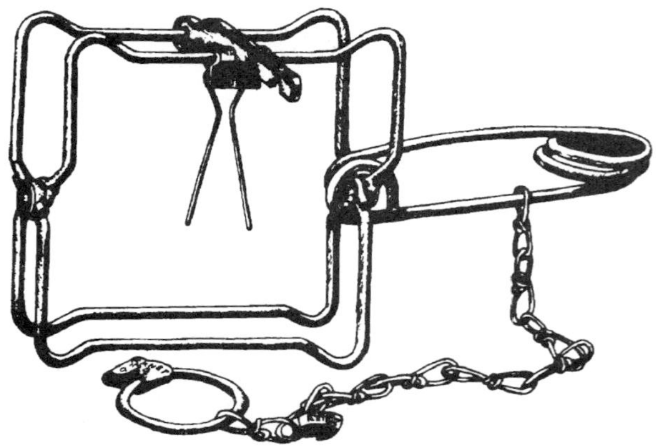

FIGURE 5–32 The standard trigger is much smaller and swings freely, which means the trap can be sprung from either direction. This could result in the capture of the wrong animal. *(NYS DEC)*

See how that trigger swings freely? It can spring the trap regardless of which way the animal approaches. That means it can catch an animal on its way out of its den, which is what you want—but it might also accidentally capture an animal that stopped to investigate. In some cases, that might be the wrong animal.

There are commercial traps with one-way triggers but you can modify the traps you currently own, too. For example, you can remove the standard trigger from a #120 body-gripping trap and replace it with the trigger from a #220 body-gripping trap, which is large enough to reach the bottom jaws of the trap. With the trigger resting against the bottom of the trap, it can only swing freely one way.

■ **Mouse and rat snap traps.** The familiar mouse trap is a form of body-gripping trap. And yes, people have "built a better mouse trap." Several, in fact.

Your trapping strategy will vary depending on whether you're trapping mice, rats, or other small mammals. See the species accounts for mice and Norway rats in Appendix B for information about the number of traps to use and how to place them effectively.

If you're trapping mice or rats, you'll need to use many traps, so you may want a model that's easy to set. There are many options besides the traditional mouse trap, which can be cumbersome. Snap traps with expanded triggers and the "clothespin" design (Figure 5-33) are much easier to set than the traditional mouse trap. The Quick Kill Mouse Trap made by Victor has a lid over the bait cup (Figure 5-34). Only animals that are motivated to seek the bait will lift that lid—and that's what triggers the trap.

FIGURE 5–33 A snapback trap with a clothespin design. *(Susan MacKay)*

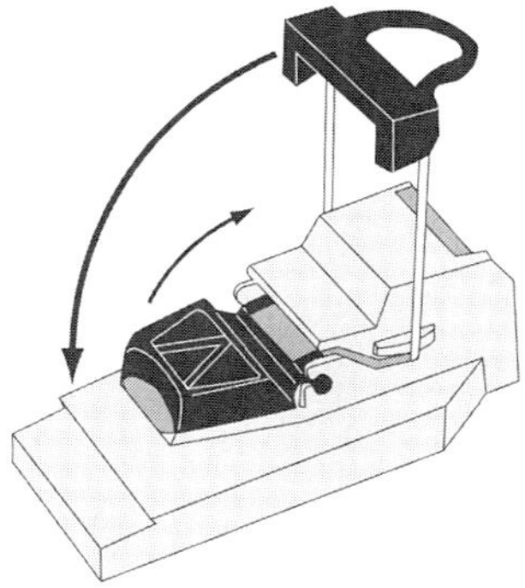

FIGURE 5–34 The Victor Quick Kill Mouse Trap. The bait cover is also the trap's trigger. When that lid is lifted, the strike bar lowers. *(Susan MacKay)*

This means that an animal can accidentally step on the lid without setting off the trap. There's another advantage to this design. The bait cup is located to position the mouse in the perfect strike position. So this trap is both more selective and more effective than the traditional mouse trap. It also has a safety catch and will not fire if it's picked up.

■ **Mole traps,** another form of body-gripping trap, come in several designs. Some spear the moles with a harpoon, others have scissor-like jaws. They are all set underground in the moles' tunnels. The key to successful mole trapping is to identify active tunnels. Look for dead grass or soft spots in the lawn. Prepare the site and set the trap according to the instructions given for the particular trap design. If there is no activity after a few days, move the trap. If moles are active near the trap but you're not catching them, either add more traps or switch to another type of trap.

■ **Foothold traps** can be used as lethal traps, as well. This is generally limited to a submersion set used to capture beaver, muskrat, or mink.

Advantages of lethal trapping:

- When they work right, lethal traps are one of the fastest and most humane killing methods
- Same piece of equipment can be used to capture and kill animal, so there's less handling involved
- Equipment is easy to get and reasonably priced

- It's easy to transport equipment
- Equipment is lightweight and compact, especially when compared to cage traps
- The disposal of the animal is easier and safer for the WCO
- Eliminates some of the concerns related to the release or relocation of live animals

Disadvantages:

- Can capture and kill nontarget animals
- Some people find these tools objectionable
- Can hurt people
- May require additional tools to set the trap
- Misfire could result in injury to animal
- Site often needs preparation, so it may take longer to set trap

Shooting

Firearms include pistols, shotguns, rifles, hand guns, and air rifles (high-end pellet guns). This technique is appropriate for use with medium to large mammals (squirrel-sized and larger), birds, and reptiles. Obviously, shooting requires training and skill. There are safety concerns and legal restrictions to consider, too. For proper training in the use of fire-arms, attend a hunter education course or a training course sponsored by the National Rifle Association (NRA).

General tips for the use of shooting

- Take the time you need to make the best shot.
- Practice!
- The type of firearm you choose and the ammunition should be matched to the size and species of the animal and your location. For example, BBs bounce more than lead, so they're riskier to use indoors.
- In most cases, small-caliber, low-energy projectiles are best. A .22 caliber rifle is adequate for most small mammals. Air rifles may be used on squirrels and birds. Among the most effective types of ammunition are hollow point bullets or low velocity .22 rimfire cartridges, such as shorts or CB caps, which are also quieter.
- Pay attention to the surface underneath, around, and behind the animal. Could a bullet ricochet? Bullets are less likely to ricochet off softer surfaces such as dirt or grass than off hard surfaces like concrete, asphalt, or rocks, or water.
- You must follow both state and local firearms regulations.
- Make sure the situation is safe. If a crowd has gathered, disperse them before you shoot the animal, or take the animal elsewhere.
- Some species continue to move after they've been shot, such as squirrels, birds, raccoons, opossums, and woodchucks. This is a reflex but can be difficult to explain to someone who's watching. This is a good reason to be discreet.
- Animals that are shot may retreat to inaccessible areas, such as dropping down into a hollow wall. Try to block such troublesome routes.

FIGURE 5–35a

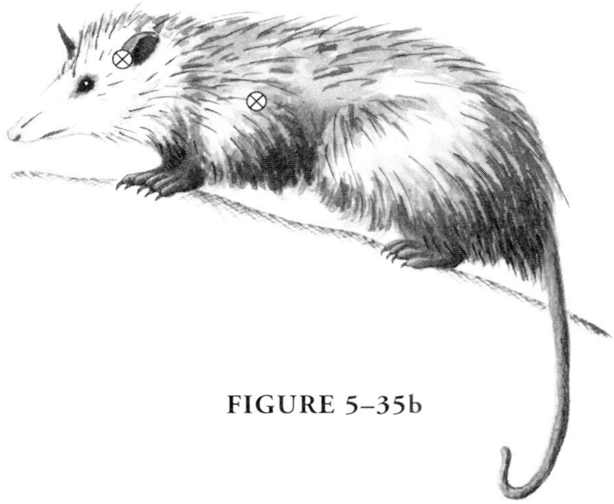

FIGURE 5–35b

FIGURES 5–35a and 35b Target areas for the brain or heart/lung shots shown on an opossum (not to scale). Opossums have very small brains housed in a reasonably large skull. This means the target area—the brain—is actually much smaller than you'd imagine, just from looking at the size of the animal's head. Their brain is about the size of a pea. Opossums also have a very big crest that runs down the center of their skulls, called the "sagittal crest." It's strong enough to deflect bullets. Unfortunately, if you try to aim slightly off to the left or the right to avoid hitting the crest, you could miss the brain entirely. A side target might be a little easier. Imagine a line drawn between the eye and the ear, and aim closer to the base of the ear. *(Karen English)*

FIGURE 5–36a

FIGURE 5–36b

FIGURES 5–36a and 36b Target areas for the brain or heart/lungs shots shown on a fox (not to scale). These target areas are the same for most mammals, with the exception of the opossum. *(Karen English)*

If an animal is restrained, shooting may be one of the fastest and most practical ways to humanely kill a wild animal.

Guidelines for shooting mammals: If you need to do a rabies test, don't shoot the animal in the head. You might destroy brain tissue that the lab needs to conduct the test. You could also spray potentially contaminated brain tissue into the air, which might expose you to the virus if the tissue came into contact with your eyes, mouth, or nose. (Rabies virus by itself is not airborne.) Instead, aim for the heart and lungs. The heart/lung target may also be a better option if dealing with a free-roaming animal, such as a deer.

What if you shot an animal in the head and later learned that a rabies test is needed? Don't panic. Submit the specimen. In many cases, the specimen will be adequate. What about that worst-case scenario, when it's not possible to do an accurate rabies test? Then, as a precaution, the exposed person or animal would receive the post-exposure rabies vaccinations. That's not fun, and there's a limited supply of one of the products that's given as part of this series, so don't be cavalier about the quality of the specimens you submit for rabies tests.

Although the head shot (aiming for the brain) is often considered faster and more humane than the heart/lung shot, this isn't necessarily true. Please note that few people would be able to ensure a proper head shot at anything except close range.

To properly target either the brain or the heart/lungs, you must think in three dimensions. Without proper aim, the bullet could deflect off the skull (Figures 5-35a, 5-35b). For the brain shot, the barrel of the firearm should be a few inches from the head. Ideally, aim so the bullet will travel through the brain and spine and lodge in the animal's body (Figures 5-36a and 5-36b). If the animal's head is turned so you don't have the right target, you may be able to distract it and get it to move its head by tossing a rock. Or wait patiently.

Guidelines for shooting birds (unprotected species): House sparrows, starlings, and pigeons can become trapped within large buildings such as malls, warehouses, and airport terminals. These structures often have food, water, and even nesting areas that allow the birds to survive. One of the most effective ways to remove a small number of birds from a large building is with an air rifle. Choose a higher-end model with a scope for improved accuracy and performance. Most often, you'll shoot the birds when the building is closed to the public. (Although pellet guns are not as powerful as gunpowder firearms, there are still risks to human safety and to property.) Learn where the birds roost, then choose a position that will provide you with good shots. The building management may have a lift that will provide you with an elevated platform.

Advantages of shooting:

- One of the fastest and most humane methods for killing wildlife
- Works with many different species under many conditions
- Minimizes the handling of a live animal

Disadvantages:

- May splatter brain tissue, saliva, or blood, which could expose someone to wildlife diseases
- May interfere with the collection of the tissue sample needed for the rabies test
- Cannot be done everywhere—there are legal restrictions and safety concerns that limit its use
- Requires skill and proper equipment
- Can be dangerous to people and other species

Carbon Dioxide Chamber

The technique is fairly simple: the animal is placed in an enclosed space into which carbon dioxide gas is added at a controlled rate (Figure 5-37). When the animal breathes this gas, it quickly loses consciousness and then dies. The entire process takes about five minutes.

How does it work?

Carbon dioxide (CO_2) is a colorless, odorless gas. It's called an "inhalant agent" because the animal must breathe it in. CO_2 affects the nervous system, the lungs, and the heart.

Can I use CO_2 on any species?

Carbon dioxide chambers work well to humanely kill birds, rodents, and most small mammals. But

FIGURE 5–37 This CO_2 chamber can hold large cage traps. That's good, because it means you don't have to handle the animal to place it in the chamber. Notice that the tubes feed the gas into the bottom of the chamber. CO_2 is heavier than air, so the chamber will fill from the bottom up. *(Connecticut Nuisance Wildlife Control Operators Association, Inc.)*

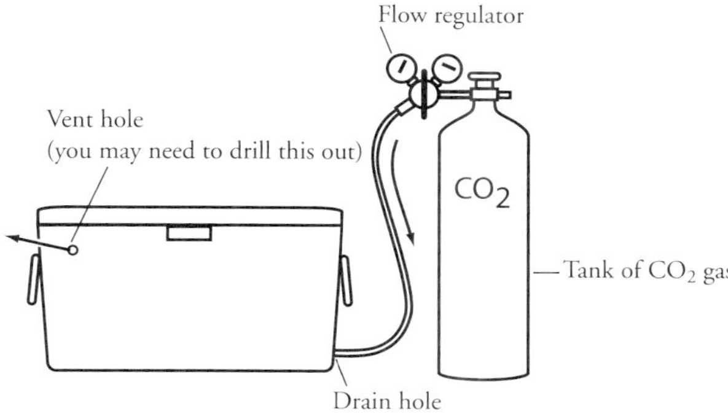

FIGURE 5–38 Many commercial coolers come with a drain hole at the bottom (if yours doesn't drill one). Attach the hose to the drain hole. To add, a window to your chamber, drill through the cooler using a jigsaw. Cut a piece of Plexiglas to fit, then attach it using epoxy. *(Karen English)*

you may encounter problems if you try this technique on animals that are old, very young (less than two weeks old), weak, or sick with a respiratory disease, because they're often resistant to the effects of carbon dioxide. If possible, choose another technique to humanely kill animals that breathe slowly (such as reptiles or amphibians) or are very good at holding their breath (such as beaver and diving birds). Under these circumstances, CO_2 might take too long to be considered a best practice.

Commercial carbon dioxide chambers are often made of metal and have windows for monitoring the animal, but the chamber can be made of many materials, such as plywood or plastic. Coolers, garbage cans, and other containers have been converted into chambers (Figure 5-38). The best chambers are easy to clean and transport and are clear or have windows so you can monitor the animal.

You may want to have several chambers of different sizes and designs, and a portable model. For example, bats can be more humanely killed in a small chamber that fills faster. A small, clear plastic container with two holes drilled into it (one for the tube feeding the gas, and a vent hole) would work well.

One WCO has rigged a garbage container on wheels to serve as both an enclosed trap carrier and carbon dioxide chamber. He sets a cage trap within the chamber, then, after capturing the animal, wheels the can to a place where he can safely and discreetly attach the carbon dioxide tank and kill the animal. This combination of a trap within the CO_2 chamber also offers a safety advantage, because there is no handling of the animal. Consider such an approach, especially when dealing with sick, rabies suspect, or highly aggressive animals.

There may be a market niche for those willing to provide on-site wildlife euthanasia services, especially in areas where other techniques that are generally accepted by the public, such as shooting, aren't legal or practical. With care, carbon dioxide chambers can be safely transported. Check local regulations.

Equipment needed

- Chamber that's twice as big as the animal
- Tank of carbon dioxide gas (These are available in several sizes, such as 6 lb., 20 lb., 50 lb., and 60 lb.

If you don't need to lug it around, a larger tank might be cheaper to refill.)

- Hose or rubber tubing
- Flow regulator
- Commercial carbon dioxide chambers, available in different designs
- Check with welding suppliers; you may be able to rent some of this equipment

Equipment tips:

- Choose a chamber that can hold a large cage trap so you don't have to transfer the animal from the trap into a holding cage. This is safer for you and less stressful for the animal.
- The animal should be able to sit and rest comfortably within the chamber.
- If the chamber is too tall, or if the vent hole is too low, the animal may be able to lift its head above the level of CO_2 gas, which would make the process take longer.
- *The chamber is not supposed to be airtight!* Air must be able to escape to leave room for the carbon dioxide. Carbon dioxide is heavier than air, so it fills from the bottom up (like filling a glass of water). A vent hole near the top of the chamber or a loosely fitted lid will let out the air but not the CO_2. The vent hole will also prevent pressure buildup.
- Have a spare tank of CO_2 ready.
- Don't mess around with dry ice, fire extinguishers, car exhaust, or antacids as sources of carbon dioxide. They will not work reliably and thus are not humane.

Technique tips:

- Always work in a well-ventilated space to minimize your exposure to the CO_2, which can be dangerous to people, too.
- Attach the hose from the CO_2 tank to the bottom of the chamber.
- Ideally, carbon dioxide should enter the chamber at a rate that displaces 20% of the oxygen each minute (more on this later).
- Should you fill the chamber with CO_2 before putting the animal in it? (This is called "pre-charging.") Some experts say "yes," others "no." Pre-charging will speed up the process, but many animals will react violently to high concentrations of CO_2 if they're awake. It's less stressful to expose the animal to the gas at a carefully controlled rate. Right now, we say "no," but watch for more research.
- Even if you have a commercial carbon dioxide chamber, you may want to consider building your own mobile unit.
- The chamber should be cleaned and aired out between uses. (A good reason to have more than one chamber.)

Using a CO2 Chamber, Step-by-Step:

1. Calculate the optimal CO_2 flow for your chamber. This depends on the chamber's size. Your math has to be done in the same units used by the flow regulator. Most are either calibrated in cubic feet/hour or liters/minute. If your flow regulator is calibrated to measure the amount of gas flow in "cubic feet/hour," here's how to convert your measurements to match:
   a. Measure the chamber's size (for this example, say it's $14 \times 14 \times 34$ inches)
   b. Convert this measurement into cubic feet

      First, divide each measurement by 12:
      $14 \div 12 = 1.2$
      $14 \div 12 = 1.2$
      $34 \div 12 = 2.8$
      Then multiply the three measurements:
      $1.2 \times 1.2 \times 2.8 = 4$ cubic feet
   c. Divide the chamber's size (in cubic feet) by the accepted amount of time (5 minutes, or .08 hours) to determine the flow rate: $4 \div 0.08 = 50$ cubic feet/hour. So in this example, the ideal

flow rate for this chamber is 50 cubic feet/hour.

2. Place the animal into the chamber. If it's in a trap, you can put the entire trap into the chamber.
3. Turn on the carbon dioxide gas until the flow regulator shows the flow rate you've just calculated.
4. Watch the animal. At first, the animal may move about or seem excitable. That's a normal reaction that often happens right before the animal loses consciousness. If you see signs of extreme distress (excessive vocalizations, head shaking, fierce sneezing), turn down the flow rate. Later on, when the animal's totally unconscious, it may still continue to move. That's a reflex, not a sign of pain. Some WCOs find that skunks react better to a slower flow rate.
5. Wait a few minutes after the animal has stopped moving and then turn off the gas. You can leave the animal in the chamber a little longer, if you'd like. It's important to confirm that the animal is dead because if it isn't, and you expose it to air, it could revive. How can you tell? Touch a long stick to its eyeball. If it's dead, it won't blink. Or hang a string in front of its nose (just keep your hands away from the animal's teeth). If it's dead, the string will stay still, because there won't be any breath to move it.
6. If the animal's not dead, or if you're not sure, put it back in the chamber and repeat the process, or use another method.

Advantages of carbon dioxide chambers:

- It's been well studied and has been proven highly reliable
- Is one of the most practical techniques for WCOs
- Does not interfere with rabies testing
- Equipment is easy to get and reasonably priced, or can be made easily
- Can be done on-site
- Poses few hazards to people
- Doesn't require special skills, strength, or additional permits
- Some of your customers may prefer this method, so there may be marketing advantages to its use

Disadvantages:

- May not work as well with burrowing or diving mammals; reptiles; amphibians; animals that are old or very young (less than two weeks old); weak animals; and those that are sick with a respiratory disease
- May be slower than some of the other humane killing methods
- May cause distress at first
- Equipment is a bit bulkier, and requires cleaning and airing between uses
- You have to do a little math

Cervical Dislocation

This method is commonly referred to as "breaking the neck" but would more accurately be described as "snapping the spine." The goal is to quickly separate the spinal cord from the brain to provide a fast and painless death. The separation must take place at the base of the brain or within the upper third of the neck (the cervical spine area).

This method requires skill, practice, and strength, especially with the larger animals, which have thicker, stronger necks. WCOs unfamiliar with this technique should receive training and practice on dead animals before attempting this on a live animal.

Cervical dislocation is used primarily for small- to medium-sized birds (duck sized or smaller) and small mammals, such as mice and rabbits. Be cautious if working with a mammal, especially a rabies vector species (it's safest to use a different method for them). This technique brings your hands into direct contact with the animal's head near its mouth. Remember that no glove can provide 100% guaranteed protection.

To snap the spine of a pigeon or duck-sized bird, grasp the base of the bird's skull in one hand and its body (usually at the base of the neck) in the other hand. Pull hard and fast—twist your hands in opposite directions.

Another cervical dislocation technique for birds uses pliers or vise grips. For smaller birds (up to 11 oz., about the weight of a pigeon): Hold the bird in one hand, and a pair of needle-nose pliers in the other. Place the open pliers over the bird's neck vertebrae (in the cervical spine area). Slide the pliers up the neck until they contact the head and are directly over the first and second vertebra in the top of the neck, which support the skull (the atlas and axis vertebra). Then close the pliers firmly and hold for 2–5 seconds.

For larger birds (12 oz–3 lb, the upper limit is about the weight of a gull): Hold the bird in one hand, and a pair of square-jawed vise grips in the other. Adjust the vise grips so its jaws will slide over the bird's neck but not over its head. Then slide the vise grips up to the base of the bird's head. With your other hand, pull the bird's body quickly, to snap its spine (separating the cervical vertebrae from the skull).

To snap the spine of a small mammal or larger bird, put it on a hard, flat surface. Hold a strong stick or metal rod firmly against the base of the animal's skull. Pull its body away from its head in a single, steady motion. Keep the stick in place then bend the body over the head.

Advantages of cervical dislocation:

- It's one of the fastest humane killing techniques
- Little equipment necessary
- Can be done discreetly in the field
- If done correctly, it's not bloody

Disadvantages:

- Brings your hands into direct contact with the animal's head, which, if dealing with mammals, increases your risk of exposure to rabies
- This technique doesn't look as good as others and may disturb onlookers. Animals often keep moving for several seconds or even minutes after death. That's not a sign of pain; it's a reflex.
- Requires skill, speed, and strength
- Best restricted to small mammals (excluding the rabies vector species) and birds

Stunning

Stunning may be used for two different purposes: it may be intended to make the animal unconscious so another killing method may be used safely, or it may be intended as a primary killing method, in which case it's usually referred to as a "lethal blow."

A lethal blow is a quick, very forceful blow to the head that is meant to kill the animal. It is appropriate for small animals, such as birds, rabbits, and small mammals (mice to squirrel sized). Work on a hard surface. The animal must be properly restrained so you can deliver the blow to the right location, which is the back of the head, unless using a special tool called a "penetrating captive bolt pistol."

The penetrating captive bolt pistol is a particularly helpful—although expensive—tool, which is often used on livestock. Powered by gunpowder charges or compressed air, this device forces a metal rod through the animal's skull into its brain. Proper placement of the bolt on the forehead is crucial, so the animal must be securely restrained. And make sure that no one is within the "firing range," which is the length of the bolt, because they could be hurt. (One thing you must watch out for: make sure you have a penetrating captive bolt pistol. There are captive bolt pistols that don't penetrate the brain. They just stun the animal.)

All of the cautions that apply to cervical dislocation also apply to the use of a lethal blow and to stunning. Although a lethal blow by itself may humanely kill the animal, it's best to use a second method to be absolutely sure. Sometimes, WCOs will stun an animal so they can more effectively use the killing method they've chosen. For example, a fox that's been captured using a catchpole may be stunned so it can be safely placed in a carbon dioxide chamber.

Stunning is a less forceful blow delivered to the center of the head. The animal must be properly restrained to ensure the correct positioning of the blow. Once the animal is stunned unconscious, you can choose among several techniques to kill it.

- Stunning and use of carbon dioxide chamber: If an animal has been captured using one of the direct capture techniques, such as with a catchpole, it may be stunned to allow the WCO to transfer it to the chamber safely.
- Stunning and shooting: If an animal is moving around too much to allow for a proper target, you can stun it into unconsciousness, and then orient it for the best shot. Or perhaps you need to move the animal to a place where it's safer to shoot it.
- Stunning and cervical dislocation: This combination may be used on duck-sized birds, to gain a little more time to properly snap the spine.
- Stunning and decapitation: "Decapitation" means the head is quickly cut from the body. This method works like cervical dislocation. The goal is to quickly separate the spinal cord from the brain to provide a fast and humane death. Decapitation is used primarily for birds that are too large for cervical dislocation, such as geese, and sometimes for snakes. A recently killed snake may bite by reflex, so don't handle its head. Use a heavy knife, ax, hatchet, or bolt cutters to cut the spine of a bird at the base of the head. A long-handled shovel or hoe can be used for a snake. The same uncontrolled movement that is seen after cervical dislocation will follow decapitation. A goose might flop around for up to two minutes.

The same safety precautions that apply to cervical dislocation should be considered when using stunning and decapitation. This method is messy and forceful. Blood and brain tissue could be splattered. If dealing with a mammal, the splattering of contaminated brain tissue or saliva could expose you to the rabies virus.

- Stunning and bleeding out (or "exsanguination"): The major blood vessels, usually those in the neck, are cut to rapidly drain blood from the body. This ensures death but is a messy and bloody technique.

Advantages of these combination methods:

- Stunning the animal first ensures that it is unconscious
- May make a situation safer for the WCO and other people
- May allow the WCO to use a preferred killing method, such as shooting or a CO_2 chamber, in a situation in which that method would otherwise be difficult to use
- When properly done, these are fast and humane techniques that quickly break the brain-body connection
- No specialized equipment necessary for some of the combined methods

Disadvantages of these combination methods:

- May bring your hands into direct contact with the animal's head, which, if dealing with mammals, increases your risk of exposure to rabies
- Some of these techniques may disturb onlookers. Some are bloody. Animals often keep twitching (or in the case of a goose, thrashing) for several seconds, or even minutes, after death. That's not a sign of pain; it's a reflex
- Will likely have more of a mess to clean up (consider working over a tarp to contain the mess)
- May require skill, speed, and strength

Chest Compression

The goal of this method is to quickly stop the heart. This technique may be applied to small- to medium-sized birds such as starlings or pigeons (but not diving birds) and small mammals (up to about 15 pounds which includes animals the size of a young raccoon or fox). Don't attempt this technique on a larger mammal because you'd be unlikely to accomplish it in a timely and humane fashion. This is not a preferred technique for mammals ranging closer to the 15-pound limit, but it is acceptable, and may be your best option in some cases.

For birds: Bring the thumb and forefinger of one hand under the bird's wing from the back, and hold them against the bird's ribs. Place the forefinger of your other hand against the bird's breastbone ("sternum") just below the spot where the "wishbone" forks. Squeeze your fingers together forcefully and hold the pressure to stop the heart.

For mammals: First, stun the animal between the eyes to make it unconscious. Then strike it again at the back of the head, which should be a lethal blow. Immobilize the animal by standing on its neck with one of your feet. Locate its heart, then compress the heart by standing on it.

Advantages of chest compression:

- When properly done, this is a fast technique
- No specialized equipment necessary
- May be one of the few practical options available in some situations

Disadvantages:

- May disturb onlookers
- Requires skill, speed, and strength

Barbiturates

This isn't a practical method for most WCOs because they don't have access to the drugs. Even state wildlife biologists and university researchers have difficulty securing licenses for obtaining barbiturates. But some of your customers may ask about their use, so you should be able to answer their questions and explain why you're not using them.

These drugs, such as sodium pentobarbital, sedate and can kill animals by interfering with the central nervous system. Barbiturates are the drugs veterinarians use to "put animals to sleep." And that's not a bad description of what it looks like, because they usually work smoothly. Animals must be well restrained for the injection.

These are dangerous drugs that can kill people, so their use is heavily regulated by the federal government. Only trained personnel registered with the U.S. Drug Enforcement Agency (such as veterinarians) have access to barbiturates. You'd need a license. There are strict security issues and lots and lots of paperwork.

There are two possible ways that a WCO might be able to use this method. First, you could transport the nuisance animal to a veterinary clinic, so the vet can inject it with barbiturates. But is that a great choice? Consider how the capture and transportation could stress the animal—would this cancel out the advantages of the use of barbiturates? Would your customers pay fairly for this service?

In rare cases, a WCO partners with a veterinarian who agrees to supervise and take responsibility for the WCO's use of barbiturates in the field. This is a significant liability risk for the vet, so few are willing or able to offer this assistance. There are many practical limitations, too, because the WCO must go to the vet's office before going out on each job that requires the use of the drugs. The vet would give you the amount needed to kill that animal. Could you fit this into your schedule in any reasonable way? You must also consider how you'd dispose of the carcass, because barbiturates can persist in the animal's body. A predator or scavenger that feeds on an animal that was killed by barbiturates might also die.

Advantages of barbiturates:

- One of the fastest and most reliable humane killing methods for small animals
- A gentler process that causes minimal distress

- Because barbiturates sedate animals, they may make the handling of a dangerous animal safer
- Although the unconscious animal may gasp right before death, the technique minimizes the thrashing sometimes seen with other killing methods

Disadvantages:

- These drugs are not available to most WCOs
- These drugs can kill people and other animals
- You'd need to get close enough to the animal to use a dart gun or hypodermic, and that may not be possible or safe
- Requires special training and additional permits
- The drugs may persist in the carcass and affect any animal that feeds upon it
- These are "street drugs," so you would need approved, secure drug storage for transport

Pesticides

You'd need another license and separate training in the safe and effective use of pesticides, so we won't even try to summarize it here. There are relatively few pesticides available for the control of wildlife. Most are rodenticides, targeting such rodents as mice and the Norway rat. Fumigants and gas cartridges may be registered in your state for use against woodchucks, chipmunks, voles, and moles in their burrows. There are products, such as Avitrol, registered for use on birds. Some details are shared in Appendix B. Remember, pesticide regulations and registrations change frequently, so always check for current information.

Clean-up and Disposal of Dead Animals and Contaminated Materials

Most WCOs do a certain amount of cleaning as part of their service. If the site presents a formidable mess, some WCOs contract for cleaning separately while others recommend a service. Whether you do a little or a lot of cleaning on site, you'll still need to clean your gear and your truck. Here's a quick review. For more details, see Chapter Four.

Clean and disinfect your equipment with a disinfectant or a 10% chlorine bleach solution—one part bleach to nine parts water. Mix up a new batch each day. Never mix bleach and ammonia—that forms a toxic gas! For the same reason, don't apply bleach to bird droppings, which also contain ammonia.

Remember, if you're working around bird, bat, or rodent droppings or nest materials, don't stir up dust. Wear the proper safety gear. Use disinfectant to wet down anything that might have been contaminated, including any dead animals. Wipe up with a damp towel or sponge, or use a commercial, heavy-duty vacuum.

Cleaning up after raccoons is a bit trickier. The eggs of the parasite that causes raccoon roundworm are resistant to disinfectants. Areas of soil or concrete that have been contaminated can be flamed thoroughly using a handheld propane torch (a weed burner). Turn the soil over, then flame it; repeat this process a few times. Metal traps can also be flamed, or you could opt to clean them with boiling water and bleach. To decontaminate a fireplace, woodstove, and chimney, build a roaring fire. Contaminated materials that can't withstand burning should be cleaned with boiling water and bleach. This is a good option for wooden decks, porches, and contaminated clothing.

If you're dealing with a mangy animal, clean anything that may have picked up mites, such as your clothing, equipment, or truck. The bleach solution is one option, or you can freeze objects, which will kill any mites. Don't use bleach on your skin.

To refresh your memory for safety protocols, here are the diseases you're more likely to encounter when working with:

Birds: Histoplasmosis

Bats: Histoplasmosis, rabies

Rodents: Hantavirus (squirrels often get mange)

Raccoons: Rabies, raccoon roundworm, distemper

Foxes and coyotes: Rabies, mange

Woodchucks: Rabies, mange

All of these species attract a variety of parasites, too. And although they don't contract it as often, woodchucks, squirrels, and birds may suffer from raccoon roundworm. Any mammal can be infected with rabies.

Carcasses and other potentially contaminated materials (gloves, protective clothing, nesting materials) must be disposed of properly, because they can also spread diseases. Your state wildlife agency and health department may give you specific disposal instructions for animal carcasses, which you must follow. Otherwise, the animals and other materials may be buried, burned, or sent to a landfill.

That takes care of health concerns, but most people won't consider a place clean if it still reeks. Some WCOs simply refer customers to commercial cleaning services. There are many commercial deodorizers that will eliminate wildlife odors. You'll find details in the skunk account.

One expensive machine, an ozone generator, has been marketed to WCOs as a way to clean or purify air. Here's what the EPA has to say about that: "Often the vendors of ozone generators make statements and distribute materials that lead the public to believe that these devices are always safe and effective in controlling indoor air pollution. For almost a century, health professionals have refuted these claims . . . Ozone can cause health problems at high concentrations." (For the full report, see: www.epa.gov/iedweb00/pubs/ozonegen.html.)

To learn more...

- Check catalogs or websites to compare different commercial models of carbon dioxide chambers. Consider buying one, or make your own.
- Attend your state agency's fur trapping course to learn more about trapping techniques.
- If you're unfamiliar with certain humane killing techniques but would like the option of using them in your WCO work, find an experienced wildlife professional who can teach you. Some techniques should be practiced on dead animals (cervical dislocation, decapitation, stunning).
- Attend your state agency's hunter education course or an NRA-sponsored firearms safety course, or the pesticide applicator course, if you'd like to use firearms or pesticides (including repellents) in your WCO work.
- Create the kind of stinky situations you might encounter on the job, then experiment with ways to control the odors. Try out various odor control products. Which work well?
- Read the *2000 Report of the American Veterinarian Medical Association Panel on Euthanasia* (available at: www.avma.org/resources/euthanasia.pdf). In addition to the added information in this report, it has an extensive bibliography that would lead you to other credible sources of information.
- Collect animal skulls. Use them to help employees understand the proper location of a head shot for each of the species you handle (if shooting is a preferred killing method for that species).

Review Questions

13. To use certain techniques, you need additional training and other licenses. This is true for which of the following? (Check all that apply.)

___ Trapping
___ Use of pesticides
___ Installing one-way doors
___ Use of barbiturates
___ Hazing with dogs
___ Use of chemical repellents

14a. If you must submit a specimen for a rabies test, certain techniques should not be used to kill the animal. Which ones could interfere with the test? (Check all that apply.)

___ Shooting (in the head)
___ Cervical dislocation
___ Carbon dioxide chamber
___ Decapitation
___ Lethal trap
___ Stunning and exsanguination

14b. Which of the remaining techniques would you avoid using in this case, because they might put you at risk? (Check all that apply.)

___ Shooting (in the head)
___ Cervical dislocation
___ Carbon dioxide chamber
___ Decapitation
___ Lethal trap
___ Stunning and exsanguination

15. Select the nonlethal techniques a WCO can use to remove an animal from an area. (Check all that apply.)

___ One-way door
___ Direct capture
___ Snare
___ Foothold trap
___ Hazing with radio-controlled car or boat
___ Chemical repellents
___ Cage trap
___ Hazing with dogs
___ Homemade repellents
___ Visual scare devices (scarecrows, Mylar tape)
___ Noisemakers (propane cannons, bangers)

16. Select the lethal techniques for removing an animal from an area that are available to (most) WCOs without additional licenses:

a) Lethal traps and CO_2 chamber.
b) Shooting and chemical repellents.
c) Rodenticides or other pesticides.
d) Barbiturates and CO_2 chamber.

17. If you could only pack five things to respond to a variety of wildlife problems, which items would you pick?

a) Catchpole; a few cage traps in different sizes; flashlight; a few body-gripping traps in different sizes; a few kinds of gloves.
b) A selection of firearms; bleach solution; rodent baits; a few foothold traps in different sizes; catchpole.
c) One-way door; nets; pyrotechnic devices; catchpole; a few cage traps in different sizes.
d) Snake tongs; buckets; respirators; glue boards; a shovel.

18. Which of the following techniques is recommended for cleaning up an area that's been contaminated by raccoons?

a) Your only option is to burn everything.
b) Burn what you can, then clean the rest with boiling water and bleach, or bury materials deeply.
c) Wet contaminated materials with a disinfectant, then double-bag them for disposal.
d) Vacuum the area.

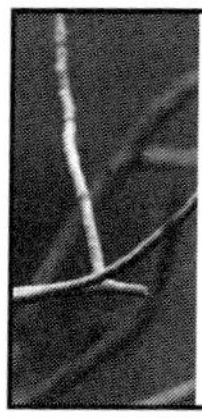

STEP FOUR: Prevent Future Problems

LEARNING OBJECTIVES

5.18 List a dozen tips you could share with your customers to help them reduce the amount of food and shelter available to nuisance wildlife.

5.19 Before you exclude an animal from an area, you should think about seven issues. Describe them.

5.20 You've been hired to bat-proof a house. Name three products you could use to seal small holes.

5.21 Explain the safety issue you need to consider when modifying vents.

5.22 Raccoons can remove a certain kind of chimney cover. Which one? Describe the type of chimney cover you'd choose to keep a raccoon out of a chimney.

5.23 Name two devices used to keep pigeons off ledges.

Most wildlife control work is in response to a problem that has already happened. Is there a way to turn that around, and to actually prevent problems? Or, can you prevent them from happening over and over again? Absolutely, and this is emerging as a more important part of wildlife control.

There are two major approaches to preventing wildlife conflicts: habitat modification and exclusion. We'll also discuss one strategy that's common in agricultural pest management, which we believe may become an important service offered by some WCOs: monitoring.

Habitat Modification

Animals look for food, water, and shelter. When practical, modifying the environment to reduce the amount of these resources will make the site less attractive to an animal.

Appendices B and C include a series of tip sheets for the animals that are most likely to cause nuisance problems in our region. Each account describes some of the basic biology you need to know to work with this species, and then lists control techniques for that animal.

The following two lists offer some general tips for modifying the habitat to make it less vulnerable to wildlife damage.

Remove Artificial Food Sources

- If anyone is feeding the nuisance animals, persuade them to stop. The wildlife might become dependent on the food source and learn to associate people with food, which could lead to other problems. Also, unnatural crowding could enhance the spread of wildlife diseases. In some areas in our region, there are statewide bans on feeding certain species such as deer and bear.

- "Animal-proof" the trash. In general, this means you can either keep garbage cans and dumpsters in protected locations, or use strong containers with secure twist-on lids. Obviously, a container that's strong enough to keep out mice may not even slow down a bear, so match your approach to the nuisance species. Attaching cans to posts will make them harder to tip over.
- Clean garbage cans, chutes, and dumpsters often. Check for cracks and holes. If you find any, repair them.
- Don't leave trash out all night for a morning pickup. Many of the nuisance species who rummage through trash are nocturnal. If you can, put the trash out in the morning before it's due to be collected.
- Enclose compost piles in a framed box using hardware cloth, or in a sturdy container. Don't compost meat products or cooked food.
- Feed birds during the fall and winter and gradually stop by April. Use sturdy poles for bird feeders. Keep the area underneath the feeder clean. Or use natural landscaping to provide good bird habitat instead. (See the National Wildlife Federation's backyard wildlife habitat program for information.)
- Feed pets indoors. Any food left outdoors should be removed at night. Bring the food bowls indoors.
- Promptly clean up spills of food, birdseed, grain, garbage.
- Remove and properly dispose of livestock carcasses immediately.
- Store food, birdseed, pet food, and grains in strong containers. Keep stored items off the floor and away from walls.
- Near buildings, rake up and remove fruits and nuts that fall off trees.
- Keep livestock feeding areas and grain storage areas as clean and secure as possible.
- Remove dog, cat, and horse droppings daily. (Feces are food to other animals.)
- Eliminate pools of standing water.
- Keep livestock in protected areas, especially when they're ready to have young.
- Switch to landscape plants that deer, voles, and other herbivores dislike.

Limit Their Shelter

- Maintain a foot-wide gravel border around the foundation that's free of plants (best) or at least keep foundation plantings well trimmed. Don't stack anything against the foundation.
- Remove brush piles, junk piles, and clutter. Keep woodpiles away from buildings.
- Keep a clean border around any vulnerable area (building, garden, field, orchard). Mow the grass often and trim shrubs.
- Mow openings through large patches of thick ground cover. Some animals don't like to cross areas where they can be easily seen. Canada geese, however, would make good use of such openings, so don't use this technique if geese are, or could become, a problem in the area.
- Trim or thin trees to reduce their appeal as roosts for starlings or crows.
- Cut trees that brush up against the building and limbs that overhang the roof.
- Wrap guards around trees to keep animals from climbing them. (Best done in late fall, when the wild animals have finished nesting in the tree. Keep the wrap loose so it doesn't girdle the tree.) This will only prove effective if the tree is isolated enough that animals cannot climb a nearby object and leap into the tree.
- Use netting, fencing, and other exclusion techniques to keep animals out of buildings, gardens, livestock areas, or any other vulnerable location.

Exclusion

Animal-proofing is the best way to prevent damage to buildings, gardens, livestock areas, and valuable agricultural fields. Before you start, there are a few things you need to consider.

Is the animal in or out? You don't want to trap animals inside because that can lead to worse problems. If you're not sure whether an entry site is active, monitor it for at least two days.

Place a "soft plug" over the hole, such as newspaper, cardboard, or duct tape. Or sprinkle flour on the floor. If you don't see any signs of an animal trying to force its way through the plug, or any tracks in the flour, then you can be reasonably certain there's no wildlife inside—except during the winter. Many animals are less active then, and may not go out on a daily basis.

Are there "hibernators" in there? In winter, many animals, such as bats, woodchucks, raccoons, chipmunks, skunks, and snakes are inactive for long periods. You may think that an entry hole is inactive only to be unpleasantly surprised in the spring, or during a warm spell.

Any young inside? During the spring and summer, the presence of young animals can complicate exclusion (Figure 5-39). Listen for their sounds, such as high-pitched squealing or chirping, in such places as walls and fireplaces. Another sign, if you can get close enough, is the condition of the female's teats: they're usually bigger and free of hair when she's nursing. Do your best to avoid creating wildlife orphans. Special care is needed to remove young from buildings.

FIGURE 5–39 The downspout was removed so the one-way door could be installed right in the animal's travel route. Great idea! But if there are young inside, that female will be very motivated to find another way back into the house. She might damage the building to return to her young. *(Connecticut Nuisance Wildlife Control Operators Association, Inc.)*

Does it work? The durability and effectiveness of an exclusion technique varies by species and situation. For example, bats usually can't chew or claw their way through most exclusion materials but they're very good at finding tiny, overlooked holes. Raccoons and rodents, on the other hand, are often able to chew or claw through a hasty repair job, or break in by creating a new hole. Be sure your methods are appropriate to the situation.

How does it look? Some customers will be concerned about choosing options that don't detract from the looks of the building (Figure 5-40). But don't sacrifice effectiveness just for the sake of attractiveness. For example, if something's chewed a huge hole through a piece of woodwork, just putting up fresh woodwork may not solve the problem. Covering the new wood with metal might be a better, though less attractive, choice (you can disguise the metal by painting it).

Can I do this job safely? Remember the issues discussed in Chapter Four? You might want to limit roof work during the winter, for example, or seek assistance from someone with needed skills or equipment.

Are the products and exclusion techniques legal? Building codes, fire codes, and other local ordinances are important to keep in mind when deciding how to exclude animals. For example, many homemade chimney covers do not meet legal safety requirements, and some communities ban fences or limit their heights. The legal fence height or style may not solve your customer's problem.

FIGURE 5–40 Can you see the bird exclusion product? The thin plastic spikes are barely visible along the curved ledge. Pigeon droppings would certainly detract from the looks of this historic building far more, and are acidic enough to damage sculptures. *(Connecticut Nuisance Wildlife Control Operators Association, Inc.)*

Will I need any special tools? Most exclusion work can be done with general carpentry tools such as hammers, staple guns, screwdrivers, caulking guns, pliers, and tin snips. In addition, you may want to have two battery-powered drills for those jobs that don't involve climbing a ladder. Why two? First, because you want a backup in case the battery runs low in one of them. But having two drills can make your work go faster, too. For example, if you need to attach sheet metal to a building, you could use one to drill the holes in the sheet metal and the other to drive the screws into the building. If you used only one drill, you'd have to change bits between these two tasks, which would take longer.

A foam gun (such as the Todol foam gun) is also recommended. This tool helps you spray expanding foam insulation into cracks and cavities quickly. It's particularly useful for bat exclusion. Spend some time practicing with the foam gun because even experienced users sometimes make a mess.

Will I need any special products? Yes! There are many materials available that can be used to repair holes in buildings, to screen vulnerable chimneys or vents, or to create barriers around yards and landscape plants. These products vary in their effectiveness, cost, durability, flexibility, and attractiveness. And, of course, some are more suitable for use with certain species. Consider an animal's size, habits, and abilities before you choose a product.

Bats, for example, would be the top bet in a limbo contest. They can squeeze through cracks that are ¼ inch wide by 1½ inches long, which is about the size of a stubby pencil. That means you'll have to search thoroughly to find all of the possible entry sites. Bats, however, would be a poor bet in a chewing contest. They don't chew holes in buildings and they aren't likely to chew through whatever you use to plug a hole, so you can use products such as caulk or expanding foam to bat-proof a building.

Now imagine you're trying to exclude mice from the same building. They're roughly the size of some of the small bats, but their abilities and habits are entirely different. Mice could probably chew through caulk so it would be better to use something more durable, such as hardware cloth. Raccoons are strong and can tear off chimney caps that deter squirrels and birds.

WCOs differ from many building contractors in their expertise in two areas: animal habits and wildlife exclusion devices. Even if you don't want to do the repair, help your customers understand which products to use and how to install them effectively, or recommend a knowledgeable contractor.

Here are some of the products available for excluding wildlife from buildings and landscapes.

To Seal Holes, Cracks, and Gaps

Galvanized sheet metal is durable and, when attached with screws, resistant to removal by raccoons and other animals. But it can be hard to bend and fit around corners.

Galvanized hardware cloth (or "metal mesh") is easier to shape than sheet metal and is reasonably durable. Hardware cloth is generally available in quarter-inch and half-inch mesh sizes. Half-inch hardware cloth is stronger but less flexible than quarter-inch. To keep smaller animals, such as bats or mice, out of an area, use quarter-inch hardware cloth. Hardware cloth is often used to create fences and build cages for pets such as rabbits.

Stainless steel or vinyl-coated hardware cloth is stronger than galvanized and less likely to rust. The disadvantages of stainless steel are that it's much more expensive and harder to cut and shape. The vinyl coating cracks when it's bent which can lead to rust.

Vinyl-coated, welded wire mesh is even stronger than hardware cloth. It lasts longer and is less prone to rusting, but it is more expensive than hardware cloth. Welded wire mesh is sold in rolls and is available in different heights, gauges, and mesh sizes. Some WCOs prefer to use welded wire mesh to create rat walls and for other installations that are intended to last a long time. Recommended size for larger animals is 1 × 1 inch mesh, while ½ × ½ inch mesh is suitable for most smaller animals.

Aluminum flashing is flexible and relatively easy to shape around corners, and is available in different gauges. It's best for bird and bat exclusion because raccoons and rodents may chew or claw through it, especially if it's loosely attached or they're highly motivated because their young are inside. Other exclusion materials include **caulk, sealant** (for movable joints), **copper mesh** (this resembles steel wool, but doesn't rust; typical brands include Stuf-Fit), and **expanding foam insulation.** These materials are great for sealing cracks and other small openings.

To Protect Vents

Animals frequently enter buildings through vents. Replace damaged and vulnerable vents with sturdier, more animal-proof designs. Some vents can be modified with homemade screens. For example, you could attach quarter-inch hardware cloth to screen a kitchen hood vent or protect an attic fan. Just be careful that you don't reduce the amount of air flow too much when you're modifying a vent, and check the requirements for each piece of equipment before you modify the vent. Don't attach screening to clothes-dryer vents because lint will accumulate on the screen, requiring routine removal. If the lint accumulated, it could damage the dryer and pose a fire hazard.

Roof vents (or louvers) should be made of either metal or heavy-duty plastic (Figure 5-41). The best models are totally enclosed to prevent birds and rodents from nesting inside them. There are also commercial stainless steel box screens that are secured over existing vents.

Ridgeline vents (Figure 5-42) come with end caps that frequently work themselves loose. This allows small animals, such as sparrows, mice, bats, and flying squirrels to easily get inside attics. Replace the caps to secure these vents.

A wide range of animals, from sparrows to raccoons, often find their way into a building through the ventilation openings in soffits that are located

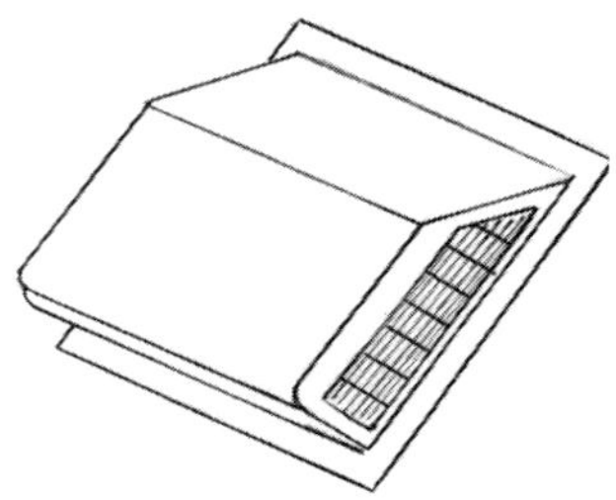

FIGURE 5–41 Lomanco 750 roof vent. *(Jim Engel)*

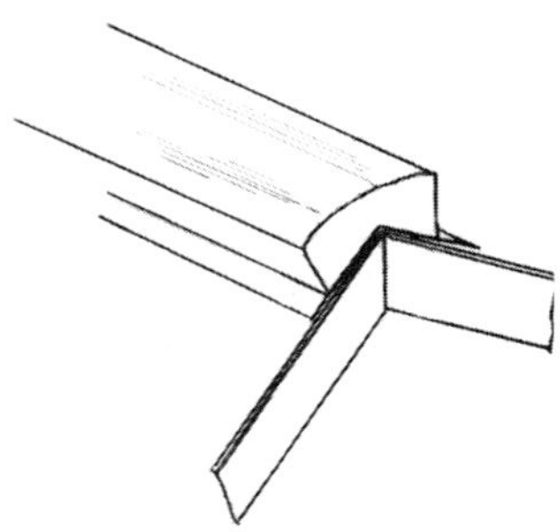

FIGURE 5–42 A ridgeline vent. *(Jim Engel)*

under the eaves (Figure 5-43). Securely attach metal louvers to the soffit to protect these openings, which are also called "soffit vents."

Plastic gable louvers on the sides of buildings ideally should be replaced with metal louvers. The gaps between individual louver slats should be narrow enough so birds can't nest in them. Screen the back (inside part) of the vent to keep bats and insects out of the attic. It's sometimes difficult to access the louvers because there's no attic or the space is very cramped. Overlapping siding and trim can also hamper your ability to replace louvers. In such cases, you can cover the outside of the louver with hardware cloth. This will prevent squirrels and other chewers from forcing their way in and is hardly noticeable.

Clothes-dryer vents are another popular route indoors used by small birds and mammals. The best option is to replace them with an animal-proof design.

Sewer vent pipes (Figure 5-44) can be covered with commercial shields to prevent rodents and birds from entering the building by slipping through gaps next to the pipes.

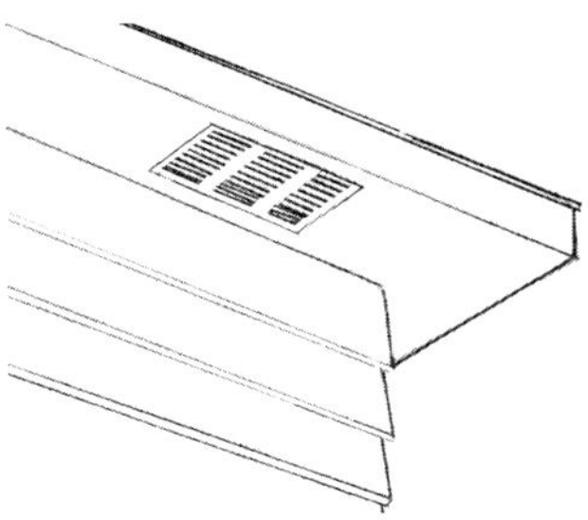

FIGURE 5–43 Soffit vent seen from below. *(Jim Engel)*

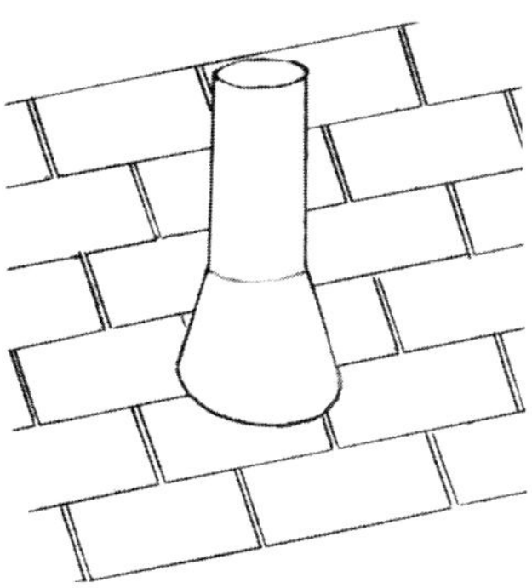

FIGURE 5–44 Sewer vent covered with a pipe shield. *(Jim Engel)*

To Protect Chimneys

Raccoons, squirrels, bats, many birds, or any animal that dens or nests in a cavity (such as a hole in a tree) will sometimes go down a chimney flue. You can prevent this by installing a chimney cover on the top of the chimney (Figure 5-45). Commercial models will meet fire codes. Most chimney covers are made of stainless steel or galvanized steel, but there are copper and aluminum models. Some work both as a cover and a damper (Figure 5-46).

Many chimney cover designs attach to a single tile flue liner. These generally bolt to the outside of the tile liner, or have legs that slip inside the flue. Covers that slip inside the tile liner keep squirrels and birds out, but raccoons can usually remove this kind of cover. If raccoons are a problem, choose a chimney cover that bolts to the side of the flue. Choose models with the smallest openings allowed by fire codes.

Other chimney covers attach to, or around, the top of the chimney. These covers are very helpful if there are several flues in each chimney, or if there are no tile liners extending through the top of the chimney.

There are commercial covers designed to fit metal chimneys. Be careful with gas vent flues because improper installation could restrict air flow or cause ice

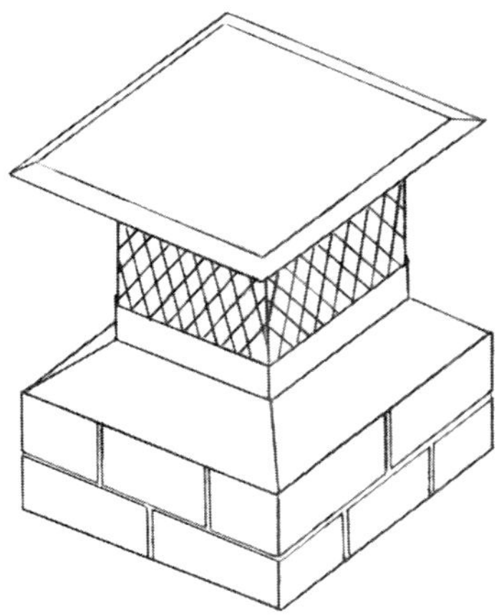

FIGURE 5–45 This chimney cover attaches to the outside of a tile liner. *(Jim Engel)*

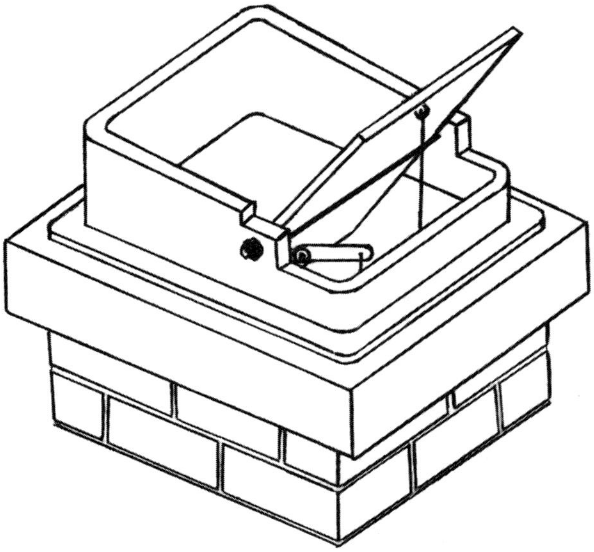

FIGURE 5–46 This model also functions as a damper. *(Jim Engel)*

to build up, which might result in carbon monoxide accumulating in the house—a potentially deadly problem. With care, you should be able to enclose the metal chimney cover with half-inch hardware cloth. Several chimney cover manufacturers are able to custom fit covers for unusual chimneys (for a price, of course). Call the manufacturer to find out which chimney measurements are needed.

To Protect Alcoves, Ledges, and Other Places That Attract Birds

Netting is often used to deny birds access to alcoves and other spaces. Bird netting is made from a variety of materials, including polyethylene twine and extruded polypropylene. It's available in different grid sizes and strand width, with specialized hardware to attach the netting to many kinds of materials.

Netting is often the most effective method to control bird damage. The cost varies considerably. Up-front costs may be quite high, because of the labor needed to install the netting (it must remain taut over time, which takes some doing), but it's often economical in the long term. The material tends to last three to ten years. In northern areas, consider ice and snow loads during winter.

By converting a flat perch into a sloped one using a piece of **wood** or a **Plexiglas panel,** you can deter birds from landing on ledges and ornamental architectural features.

Spikes (Bird Flite, Catclaw, Bird-B-Gone, ECOPIC, and Nixalite) help prevent birds from roosting on ledges, roof peaks, window sills, signs, and ornamental architectural features (Figure 5-47). There are different commercial products available, with some variety in the arrangement of the spikes in these products. Devices that feature sharp stainless steel prongs sticking out in many angles are sometimes called "porcupine wire" because the arrangement resembles a porcupine's spines.

Metal coils, which look like a Slinky, work the same way as metal spikes (Figure 5-48).

"Post-and-wire" grids discourage birds from landing in an area (Figure 5-49). The grid is made of stainless steel wire, thin cables, high-tech braided fishing line, or 80^{+}- pound test monofilament lines. Maintaining tension is essential, so steel wire is a better choice for a permanent installation because it will need less maintenance. Monofilament line stretches, degrades, and can break.

The cables are stretched tightly over the vulnerable area in a square pattern, as parallel lines, or just as a single line across a narrower area, such as a ledge. Birds react differently to this exclusion technique. It works best to discourage gulls, crows, and pigeons from such areas as rooftops, ledges, landfills, courtyards, and fish hatcheries. If using metal, consider the possibility that this installation could be a lightning hazard. Birds of prey might fly into overhead grid lines as they dive to attack their prey. If there are endangered or threatened birds of prey in the area, such as the bald eagle and peregrine falcon, don't use this exclusion device.

Electric shock devices (Avi-Away, Flock-Shock, Flyaway, VRS) deliver a deterrent shock, but they

FIGURE 5–47 Spikes turn ledges into uncomfortable roosts. *(Bird-Flite Spikes from Bird Barrier, Inc.)*

FIGURE 5–48 Sparrows can thwart spikes by filling them with nesting material, then they nest on top. In such a case, Bird Coils might be a better choice, because they form a physical barrier that can't be blocked as easily. *(Bird Coil from Bird Barrier, Inc.)*

don't kill the birds. They're used to keep birds off ledges. The cost of installing these systems is often high, but the systems generally have a long working life.

Commercial plastic strips can provide bird-proof barriers for doors to warehouses, grain storage areas, and other buildings. These strips can be hung from the top of the doorframe to ground, allowing people and equipment to easily pass through the door.

To Protect Outdoor Areas

Fences tend to provide the most effective exclusion. They can be made of many materials, such as woven wire, hardware cloth, electrified wire, rope, bird netting, or some combination of materials. Fences vary dramatically in design and cost. This device works on both a small scale (individual plant) and a large scale (garden, field, orchard, park). Some are permanent installations while others are temporary and portable. Permanent fences require maintenance. All fences need to be adequately secured.

FIGURE 5–49 The cables in this post-and-wire grid were stretched across this wide ledge, supported by four posts. This shows the parallel line installation. On a narrow ledge, one cable might be enough. *(Bird Barrier, Inc.)*

The most effective fences are designed with the particular abilities of the target animal in mind (Figure 5-50). For example, does it jump or burrow? Repellents will sometimes be used in combination

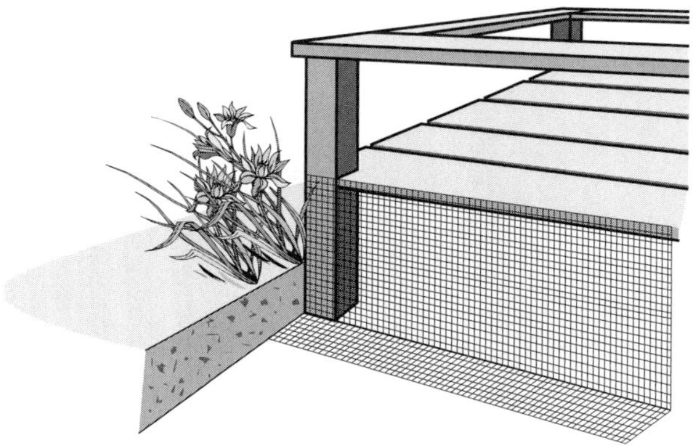

FIGURE 5–50 This fence design, called a "rat wall," is often attached to a foundation, deck, or porch, or installed as a free-standing barrier around a garden. Rat walls are effective against a variety of animals including skunks, woodchucks, raccoons, squirrels, and rats. Match the size of the mesh to the size of the animal you're trying to exclude.

The top of the fence is attached to a structure. The bottom is buried 6–12 inches deep. Notice that it's bent at a 90° angle, forming the letter "L." This shelf helps to stop animals from digging under the fence. The shelf should stick out 6–12 inches.

One-way doors can be "installed" in rat walls to release the animals. First, attach the rat wall. Leave one or two locations open and install one-way doors there. Make sure the animal can't dislodge or dig underneath the one-way doors. When there's been no sign of animal activity for several days, remove the one-way doors and finish the exclusion. (An easy way to test for animal activity is to put some nontoxic tracking powder or flour on the ground under the porch, in the animal's travel route. Check later for tracks.) *(Karen English)*

with a fence. A cloth dipped in repellent may be tied onto a rope fence, for example—enhancing a cheap and simple fence (but that does qualify as a pesticide).

Tree wrap, tree guards, and hardware cloth can be wrapped around trees and shrubs, or draped over individual plants to protect them from being chewed or girdled. Keep the mesh about an inch away from the plant so it has some room to grow. Don't staple the material to the tree because that could lead to rot. Later on, if someone wanted to cut that tree down, the staples could prove dangerous.

Monitoring

Here's a strategy that has a key role in modern agricultural pest management, which might be very useful for WCOs, especially those who service apartment complexes, large properties, or have corporate accounts. It may give you the chance to use your knowledge to promote a better approach to wildlife damage management. This can be good for your customers, for wildlife, and for your bottom line, all at the same time.

Monitoring refers to the routine inspection of a site to evaluate its current condition and look for vulnerabilities that could lead to wildlife conflicts later on. You gather the information your customer needs to make better decisions and then offer advice. Monitoring often helps people save money because you catch a problem when it's small and easier to manage, or better yet, you prevent it from happening altogether.

In agriculture, pests aren't controlled until monitoring reveals that their activity has reached a certain level—above the threshold, the point at which it pays to deal with the situation. In some cases, thresholds are very precise: "four flies on each leg of the cow counted during a 10-minute period," for example.

So how does this apply to wildlife control? Your customers are probably most concerned about three things: health and safety, economic damage, and quality of life issues. How bad does a situation have to be in order to justify control? You may be able to sit down with your customer and establish your own guidelines. Your customer might be willing to tolerate woodpeckers banging on a metal gutter but not squirrels in the attic, for example.

Or maybe you just skip the idea of thresholds altogether because your customer doesn't want any wildlife damage. Instead, you just agree on an inspection schedule. Your goal is to prevent problems. After a very thorough initial inspection, maybe you inspect a commercial site once a month. Homeowners might be willing to pay for an annual inspection.

Because monitoring isn't widely used in wildlife control work, you may need to explain the idea to convince your customers that this is a best practice. Pictures showing actual wildlife damage can help people understand the economic risks. Let them know how much the repairs cost for each of those situations. Compare that to the cost of an inspection. The best way to solve a wildlife conflict is to prevent it. Monitoring might be a way to achieve that goal.

Review Questions

19. Which type of chimney cover is best to use if you're trying to exclude a raccoon?

a) One that slips inside the tile liner.
b) One that bolts to the outside of the chimney.

20. Before you modify a vent to make it animal-proof, you should:

a) Clean it.
b) Check the ventilation requirements of the equipment to make sure your modification meets fire safety standards.
c) Paint it.
d) Spray it with a repellent.

21. Just as there are issues to consider before you choose a removal method, there are seven things to think about before you exclude an animal from an area. They are:

a) Practicality; legality; effectiveness of technique; cost; does the customer like the product; do I have to use a ladder; is it weather-proof?
b) Local laws; effectiveness of technique; safety; speed; cost; do I have time to do this now; how good does it look?
c) Could animals be hibernating inside; is the technique effective; what are the chances of trapping young inside; is this legal; am I sure I removed all of the animals; is this safe; how will it look when I'm done?

22. To keep pigeons off a ledge, you could use:

a) Plastic strips and ultrasonic devices.
b) Porcupine wire and electric shock devices.
c) Netting or parallel cables stretched tightly across the ledge.
d) Metal coils or pyrotechnics.
e) Answers "b" and "c" are correct.

23. To keep bats from crawling through small holes, seal the holes with:

a) Caulk, half-inch hardware cloth, expanding foam.
b) Quarter-inch hardware cloth, copper gauze, caulk.

c) Aluminum flashing, expanding foam, sealant.
d) Answers "b" and "c" are correct.

24. Select all of the tips you'd suggest to a customer to help them reduce the amount of food and shelter available to mice:

___ Keep area under the bird feeder clean
___ Maintain a foot-wide gravel border around the foundation
___ Enclose the compost pile
___ Rake up fruits and nuts that have fallen off trees
___ Clean up spills of food, grain, garbage
___ Mow the grass, and keep shrubs well-trimmed
___ Remove brush piles, junk piles, clutter
___ Store food in strong containers
___ Remove pet food once they're done eating
___ Wrap trees with tree guard
___ Don't stack things against the foundation
___ Store goods, especially grains, pet food, and bird seed, off the floor and away from the walls
___ Switch to landscape plants that mice don't find as tasty
___ Bring chickens into their coop at night
___ Keep livestock feeding areas and grain storage areas as clean as possible
___ Animal-proof the trash
___ Don't leave the trash out at night
___ Keep livestock in protected areas

STEP FIVE: Evaluate Success

LEARNING OBJECTIVES

5.24 Describe two benefits you might gain from evaluating your work.

5.25 List three different evaluation methods.

So you've finished the job. It's time to tip your hat and ride off into the sunset, right? Well, maybe not. Have you really been successful? Or is the "problem" just waiting to re-emerge, like some bad horror movie that spawns thirteen sequels? Should you care? Why spend time on follow-up when you're already so busy? After all, you're confident that you do good work, so you expect success. But it's not a bad idea to figure out how successful you've been (Figure 5-51). How happy is your customer with the work now?

Evaluations help you to improve your control techniques and your business planning, if that's relevant. For example, after studying his records, one WCO realized that it didn't make financial sense to go out on a call unless he'd earn $50. Otherwise, he couldn't justify the cost of the truck maintenance, the gas, the time, and all the rest.

There are many ways to evaluate your success. A low-key approach might simply be to give your customers your card, and ask them to call you if they have any problems. But keep in mind that many people don't return to a business if they're unhappy—they go to someone else. And they often tell friends and acquaintances about the raw deal that they believed they received. In other words, no news is not necessarily good news.

You could plan a follow-up visit to personally evaluate your success. This, of course, takes time, but you could view this as marketing effort. Many customers will really appreciate this extra effort, and that could lead to great word-of-mouth advertising. Also, you see the situation yourself, instead of relying on someone else's description. Be sure to leave an inspection report (preferably written) with the client.

FIGURE 5–51 Some jobs demand a little ingenuity and a little duct tape. The cage trap was too big to fit into the woodstove, so the WCO moved the stove and attached the trap directly to the chimney pipe. He's placed newspaper under the trap to protect the trunk from urine stains. Was the customer happy with this solution? Can you think of another way to solve this problem? *(Connecticut Nuisance Wildlife Control Operators Association, Inc.)*

If you can't invest the time to go back to all these sites, you could call or e-mail your customers, or leave them a brief evaluation form to fill out and return. Use a postcard with your address printed on it, and provide the postage. Then, all your customer has to do is fill out the form and drop it in the mail.

If you are performing nuisance control work to make money, you'll want to know how well you're doing. To figure out your profit margin, keep track of your costs for each job (such as materials and labor). Then add in the overhead (truck payments, gas, insurance, phone, taxes, marketing costs, and maintenance and replacement of equipment, for example). Your income, minus your costs of doing business, is your profit margin. Once you know your exact profit margin for a specific type of job, you can think about all sorts of questions that might affect how you run your business. For example, are you consistently making money on certain types of jobs while losing money on others? Then maybe you'd decide to specialize in the profitable work. Or increase the price for the less profitable jobs. On the other hand, some people consider the low-cost, less profitable jobs as a way to build your client base.

Evaluation helps you become a better WCO. No matter how many years of experience you have, you can still learn something new.

To learn more...

- Design a postcard you could use to ask your customers to evaluate your work.
- Write a few questions you might ask customers when calling to discuss their satisfaction.
- Start keeping track of the costs of each job. Figure out your profit margin.
- Find a small group of trusted WCOs who all want to improve their skills. Get together once a month to share tips and discuss any problems you encountered on the job.
- Join a professional organization. It's a great way to learn from your peers, and to find out about new developments in the industry.
- Learn more about housing construction. How does the unseen, internal structure of walls, roofs, and foundations affect your WCO work? Talk to some contractors. Ask them where buildings are most vulnerable, and therefore, more likely to be invaded by wildlife.

Review Questions

25. To evaluate your work, you might:
   a) Call your customer.
   b) Send a postcard or e-mail questionnaire.
   c) Visit the site again.
   d) All of the above.

26. What benefits might you gain from evaluating your work?
   a) You can compare the long-term effectiveness of different products and techniques.
   b) You further impress your customers, which is good advertising for your business.
   c) You can determine your profit margin, which may guide your business planning.
   d) All of the above.

Chapter 6
Professionalism

LEARNING OBJECTIVES

6.1 Wildlife control professionals must maintain certain technical and ethical standards. Describe them.

6.2 Professionalism is a self-improvement plan. List the four things that a professional is always trying to improve.

6.3 Explain three reasons why WCOs need to keep learning throughout their careers.

Okay, so technically speaking, not every job is a profession. The term usually applies to occupations that have established educational requirements, codes of ethics, and licenses, such as doctors and lawyers. But that's really nitpicky. The word "professional" means a lot more than that, and it can apply to any job. In fact, it doesn't matter if you work part-time or full-time. A part-timer might be more professional than someone who's running a full-time business. Whether you're a street cleaner or a rocket scientist, you can behave in a professional way (or not).

"A professional wildlife control operator is a person with demonstrated expertise in the art and science of applying the principles of wildlife damage management to the sound resolution of wildlife conflicts with humans."

—excerpted from the National Wildlife Control Operators Association (NWCOA) application for professional certification

Let's talk about what professionalism means for WCOs. Are you proud of your work? Do you have good reasons to be proud? Professionalism means that throughout your working life, you will improve your **knowledge**, your **skills**, your **wisdom**, and your **conduct**. Think about those four words.

Knowledge is not the same as wisdom. (Anyone with a teenager is probably painfully aware of that.) When you hear about a business scandal, the people involved may have had some of these traits, such as knowledge or skills, but probably not all four. It's the balance of the four, and the way they influence each other, that makes someone a professional—knowledge and skills, conduct that's tempered by wisdom.

Skilled, knowledgeable WCOs might have poor reputations if they communicate ineffectively with their customers. Successful communication is part of the conduct that establishes you as a professional. To achieve this, you must connect with your customers, establish your credibility, and then demonstrate your expertise.

Connecting with your customers is basic courtesy. This means listening to their concerns and responding in a respectful manner. Try to point out your shared values and goals; both you and your customer want to see the problem solved, for example. You both care about what happens to the wildlife. Many people will be more open to new information and a different point of view once they realize that both parties do agree on certain important issues.

In fact, many people won't care how much you know until they know *how much you care.* Remember those surveys we mentioned, which showed that the public cares deeply about how wildlife is treated? Reassure your customers that you care, too—and that's why you use best practices. Again, this should help your customers feel comfortable, which will allow them to be more receptive to your ideas.

After building rapport with your customers and establishing credibility, you can demonstrate your expertise. Provide accurate information. Rely on credible sources. Imagine you're the customer. Which would you trust more, information about a wildlife disease that was published by the Centers for Disease Control and Prevention or an anecdote from some WCO you don't know? Even though both might be accurate, the CDC is an internationally known and trusted authority.

Another way to establish your expertise is to be specific. There's nothing wrong with confirming a bit of information by looking it up in a book. That's why reference books are published. Over time, you'll know the most important facts by heart. Until then, if you have to double-check the size of a gray squirrel's tracks, don't sweat it. You can show the field guide entry to your customer to prove the point.

"I don't claim to be a mechanic because I can change the oil in my car, or a dentist because I can brush my teeth. So what defines the wildlife damage management professional? . . . technical and ethical standards.

First, the professional must have the highest possible technical standards. This includes performing to the best of your abilities and standing behind your work. It includes turning down a job or referring it to another technical expert whenever your skills are not appropriate or sufficient. It includes seeking out educational opportunities to continuously upgrade your skills.

In addition, wildlife damage management professionals must have high ethical standards. [They must be committed] to resolving damage complaints, [and to following] applicable laws and regulations. [They need] to respect—not necessarily agree—[with] varying viewpoints on the tools and strategies involved in wildlife damage management. These standards are in addition to traits like honesty, integrity, and sincerity, which all professionals should have."

—excerpted from Dr. Robert H. Schmidt's column, "The Professional Touch: The Wildlife Damage Management Professional" in the Feb./March 1993 issue of *Animal Damage Control* 1(1):25–26

Many professions create a code of ethics, defining their own standards for acceptable conduct and practices. Some people may choose to adopt the code of ethics as a way to demonstrate their support for these values; of course, those who don't aren't necessarily unethical. It just may not suit their styles. People who do adopt the code of ethics, however, are telling the public, their customers, and their peers that they will follow these guidelines for professional behavior.

In 1992, Dr. Schmidt proposed a voluntary code of ethics for wildlife damage management professionals, which has been discussed by many people who work with wildlife, including WCOs, wildlife biologists, and researchers. Some industry trade associations are also promoting standards, such as NWCOA's professional certification program, which judge applicants based on their expertise, education, experience, and ability to represent the profession as an ethical practitioner.

The code of ethics on the following page is based on those of NWCOA and Dr. Schmidt. (The two codes are similar.)

Does this code match your definition of professionalism for wildlife control operators? If not, how would you say it? For example, some WCOs might consider the promotion of long-term solutions to wildlife damage problems, instead of quick fixes, as an aspect of their professionalism. Do you? What about prevention?

If you were thinking about your customers as you read those suggestions, try re-reading them and consider how they apply to the other people you work with. Chapter One mentions the importance of a professional network. Some WCOs go a little bit further than just saying "thank you" when another professional refers a client or otherwise helps out. One WCO, for example, always donates some supplies or cash to a wildlife rehabilitator whenever he transfers an animal to that person's care. That's a nice, practical gesture that shows he values the rehabilitator's work and appreciates the challenge of offering such services for free. It also demonstrates one of the most important aspects of professionalism: it's got to be more than words. Professionalism must translate into action.

A voluntary code of ethics for WCOs

As a wildlife damage management professional, I will:

- Strictly follow all laws and regulations related to wildlife damage management.
- Behave in a professional manner, exemplifying the traits of honesty, sincerity, and dedication.
- Treat people, property, and wildlife with a great deal of respect.
- Be sensitive to different viewpoints on wildlife damage management.
- Encourage others to more fully understand and appreciate wildlife, biological diversity, and the different ways that people value and enjoy nature. At the same time, I'll also encourage people to understand the concerns of those who have suffered wildlife damage.
- Offer my expertise to all people upon request, within the limits of my experience, ability, and legal authority.
- Promote competence and present an image worthy of the profession by supporting high standards of education, employment, and performance. I will also encourage others to participate in state wildlife control associations and other professional organizations.
- Strive to broaden my knowledge, skills, and abilities to advance the practice of wildlife damage management.
- Attempt to resolve wildlife damage conflicts with the most humane, selective, practical, and effective management techniques available, whether they are traditional or new methods. I will encourage clients, co-workers, and others involved in the situation to do the same.
- Treat other practitioners and my customers in a courteous and honorable manner.
- Encourage, through word and through deed, all wildlife damage managers to adhere to this code.

A lifelong commitment to education is crucial, because wildlife control really is both an art and a science. Some people may have a knack for hand-capturing certain species, for example, and may prefer to use a catchpole instead of a cage trap. That's the art of wildlife control, the blend of personal skills and style. Of course, there's a lot of science involved, and you need to keep up with it.

Not only does our understanding of animal behavior change, but sometimes even the behaviors change. WCOs in the Northeast have noticed that more raccoons are breeding late in the season. If the only book on your shelf has a 1908 copyright, you might be unpleasantly surprised to find a raccoon still nursing her young in August. Techniques change, laws change, new and better tools hit the market, and new problems crop up. If you really care, you won't rely on the knowledge and skills you have today for your whole career.

Some people like to learn on their own, using books, videos, websites, trade journals, or other resources. Classes work well for many people, who find they learn a lot from the other students or from a hands-on experience. There are the natural scientists among us, people who have been tinkering as long as they've been alive. For them, experimentation is a great way to learn. Some people prefer to befriend experienced professionals and ask for their advice.

Others join professional organizations or attend conferences and meetings for inspiration. The method doesn't matter much, so pick the ones you like, because if you're having fun, you'll learn more. You may find some methods more appealing than others at certain times in your career.

It's easier to recognize professionalism than it is to define it. You know if someone's a professional or not. And you know if someone was a professional, but isn't anymore. How do you want people to think about you? The best practices strategy promoted in this book requires professionalism and encourages its development. Best practices represent the combined wisdom and experience of many people. Please contribute to the advancement of the field of wildlife damage management by sharing your knowledge with others.

To learn more...

- Get to know other WCOs. Figure out which ones are most deeply respected by their peers. Why?
- Contact a few wildlife damage management organizations. Find out about their goals and any standards of behavior promoted by their members.
- Check the websites of a few of the government agencies and animal welfare organizations involved in wildlife management. What ethical codes do they encourage?
- Express your views about professionalism by talking to your friends, peers, or neighbors. As you become more comfortable with this, you may want to speak to school groups or other organizations.

Review Questions

1. To maintain professionalism, you must:

a) Join professional trade associations.
b) Keep improving your skills, conduct, knowledge, and wisdom.
c) Go to the right parties.
d) Wear a uniform.

2. WCOs must meet ________ and ___________ standards. (Fill in the blanks.)

a) legal, trade
b) intelligence, athletic
c) height, weight
d) ethical, technical

3. You must accept every job you're offered. (Circle the correct answer.)

____ True ____ False

4. WCOs need to know about:

a) New wildlife disease outbreaks.
b) Changes in building codes, firearms regulations, and wildlife laws.
c) New solutions for wildlife conflicts.
d) All of the above.

Appendix A
Answers to Review Questions

APPENDIX A: Answers to Review Questions

Chapter One

1. d
2. False
3. a
4. d
5. c

Chapter Two

1. e
2. a
3. False
4. b (The best practices approach to solving wildlife problems takes the customer's needs and many other factors into account. It's a custom-tailored solution, not a "one-size-fits-all" approach.)
5. d
6. All of the answers are correct except "management by objective theory."
7. e
8. All are correct.

Chapter Three

1. House sparrow, upland game birds, European starling, and the pigeon.
2. b
3. True. (You must train your employees in the proper use of protective gear or equipment that could prove hazardous if misused.)
4. OSHA, FAA, USFWS, and the USEPA
5. False
6. d
7. True

Chapter Four

1. c
2. c (Apply the 4:1 ratio. $20 \div 4 = 5$)
3. False
4. e (The condition of the ladder and weather conditions affect your chances of falling. Your height and weight don't. Shoes play a minor role. Some types might provide better traction on some kinds of roofs, but it would be impractical to change your shoes throughout the day.)
5. a
6. c
7. d
8. f
9. a
10. False
11. d
12. e (Answers a, b, and d are correct.)
13. b
14. False
15. a
16. b
17. b
18. True
19. f
20. e
21. f
22. False. (Cats and wildlife use some of the same areas. Feral cats, strays, and pets that roam outdoors are also more likely to be infected than indoor house cats, because the former eat more wild animals.)
23. True
24. c
25. a
26. False
27. True
28. c
29. False. (Although West Nile virus is primarily spread by mosquito bites, it is possible to be exposed through handling infected birds, which WCOs may do, but most people don't.)
30. False
31. c
32. True

33a. Things to consider include tetanus, rabies, bites, raccoon roundworm, falls, heat stress, histoplasmosis, and hantavirus.
Tetanus: because of the chance of hurting yourself on a nail.
Rabies: raccoons are commonly infected, so you should approach all raccoons with caution.
Bites: especially if you use something other than a trap to capture the animal.
Raccoon roundworm: if you clean up raccoon droppings,

watch out for later hand-to-mouth contact.
Falls: if you need to use a ladder to get into the attic.
Heat stress: a concern in a confined space, especially during the summer.
Histoplasmosis and hantavirus: raccoons don't carry either of these diseases, but it's possible that an attic that's infested by raccoons is also home to mice, birds, or bats. Be cautious around any droppings, whether they're from rodents, birds, or bats.

33b. There's a certain degree of personal choice in this answer, unless you're covered by OSHA regulations. All of the listed items might be useful for some situations. There's no universal approach, so you may wish to have a variety of safety devices in your truck so you can choose the ones best suited for each job.
To protect yourself from falls: ladder brace, ropes or bungee cords, helmet (but you might not need any of these if there's a staircase to the attic).
To reduce the risk of exposure to wildlife diseases: gloves, respirator, disposable clothing, restraining device, trap.
To prevent heat stress: water bottle (remember to move to a safe area before removing your protective gear and taking your break).

34. Histoplasmosis, hantavirus, raccoon roundworm
Histoplasmosis: This is an airborne disease. The mold spores are microscopic and could be present in bird or bat droppings.
Hantavirus: This is another airborne disease that could be encountered if you disturb mouse nesting materials or droppings, especially under dry conditions.
Raccoon roundworm: The parasite's eggs, which are microscopic, are found in raccoon droppings.
Rabies: Transmission usually occurs through bites, or when handling the animal.
Mange: People usually pick up the mites by handling a mangy animal.
Tetanus: This is not a wildlife disease.
Toxoplasmosis: This is transmitted by the "dirty hand-to-mouth," "put dirty object in mouth," or "eat contaminated food" routes. This is the disease that most people catch through direct contact with cat litter or contaminated soil, or by eating raw or undercooked meat. You're not likely to encounter this parasite in an attic.
Distemper: This is probably an airborne disease.
Heat stress: This is not a wildlife disease.
Lyme disease: You're not likely to find the deer tick in an attic.
West Nile virus: You're not likely to be bitten by a mosquito in an attic.

35. Answers c, d, e, and g are correct. (Sounds like you're dealing with a stray dog that could have mange (fur loss) and rabies (running into an object). The kids are provoking the dog, which could cause even a healthy animal to attack. It's important to quickly restrain and isolate the animal. Many of the people listed could provide information about these wildlife diseases, but some are better equipped to respond to the actual emergency than others. For assistance, try calling the local animal control officer, the police, the health department, another WCO.)

36. b (All of the other answers are false. People sometimes walk right up to pigeons ("a"). Pigeons are diurnal (active during the day) ("c"). We know you wouldn't say "d" (even if you're thinking it). A deposit's not a bad idea. Some WCOs require that of all their customers. The health department wouldn't waste its time talking to people about rabies transmission from birds because it doesn't happen ("e").)

37. Financial risk: A bird strikes a plane and damages it.
Safety risks:
- A bird strike causes the pilot to lose control of plane, resulting in accident.
- Rodents chew wires, causing electrical failure, so the traffic control system fails.

Health risk:
- Workers are exposed to birds and rodents in planes or airport buildings. Severe infestations might expose workers to wildlife diseases.

Chapter Five

1. b (In addition to using a bigger trap to catch rats, you also need to pre-bait because rats are afraid of new objects. Mice are curious and will investigate traps.)
2. The five steps to best practices for nuisance wildlife control are: assess the situation; choose management options; do it (use those tools and techniques); prevent future problems; and evaluate your success.
3. e
4. c (These questions will help you identify which species is present. Some of the other questions address reasonable concerns, but in a tactless manner.)
5. a (Rodents and birds might be active in some of the same places. Most birds are diurnal. All of the ones most likely to cause problems in homes are. Rodents may be active at night, too.)
6. d
7. Most of these signs are made by rodents, except for livestock remains, broken egg shells, and crop damage, which you wouldn't expect to see in a kitchen. Rats do burrow, but usually outside.
8. d
9. c
10. d (Remember, most wildlife diseases that are common in the Northeast are not caused by bacteria, so an antibacterial soap isn't going to help that much. Choose a disinfectant, such as dilute bleach, instead.)
11. The only answer that's flat-out wrong is the one that suggests that scare devices and repellent techniques never work. It's unlikely that a border collie would be afraid of geese. This is one of the best breeds of dog to choose for hazing geese.
12. d
13. Pesticides, chemical repellents, barbiturates

14a. Stunning and shooting in the head would damage the brain. This makes it more difficult to work with the sample—but if you've used this technique and need to submit a sample, do it. They may still be able to get valid test results. With a small animal, you might also want to avoid using lethal traps, because it might misfire and hit the head. This isn't much of a problem with animals as big as raccoons or skunks, so for them, this could be acceptable.

14b. We've already ruled out shooting and stunning, and in some cases, lethal traps. To protect yourself from catching rabies, you need to avoid bites, scratches, and contact with saliva or brain tissue. It's best to minimize handling of the animal, and to avoid contact with the head. For those reasons, you might avoid the techniques of decapitation and cervical dislocation (which isn't recommended for use on the rabies vector species—raccoon, skunk, bats, and foxes). Both involve handling the animal, and close contact with the head. A CO_2 chamber is a much safer option for the operator.

15. The only things in that list that WCOs can't use are chemical repellents and homemade repellents (which are illegal even if you have a pesticide applicator license). Pyrotechnic devices may be prohibited in certain areas.
16. a (The use of barbiturates requires a federal license. Chemical repellents and rodenticides are pesticides and require a pesticide applicator license.)
17. a (All of the items are useful to solve certain wildlife problems, but some are more versatile than others. For example, answer "d" includes items that would be very useful if you specialized in snake removal, but not so good for skunks. Some of the items can be used only in certain situations, such as firearms, body-gripping traps, and pyrotechnic devices. Firearms and pesticides (rodent baits) also require additional licenses. Flashlights, respirators, and bleach solution would be good to have along, too.)
18. b (Raccoon roundworm eggs are resistant to disinfectants. Vacuuming may remove some eggs, but it's not practical outdoors.)
19. b
20. b
21. c (Answer "a" missed concerns about trapping animals inside, either because they're hibernating, young and immobile, or because you missed some while trapping. It also forgot to mention your safety. Answer "b"

missed concerns about hibernating animals and some that might have been missed during your capture.)

22. e (Answer "a" is wrong because birds can't hear ultrasonic devices, so they don't respond to them. Answer "d" is wrong because pyrotechnics won't deter pigeons.)
23. d (The mesh size of the hardware cloth in answer "a" is too big for bats.)
24. Almost all of these ideas apply to mice, except that mice don't attack livestock. They're also not as likely to damage trees or landscape plants, although they might do so. It's true that mice are not strictly nocturnal, but they are more active at night, so there may be some benefit to removing trash at night.
25. d
26. d

Chapter Six

1. b
2. d
3. False. (If you lack the technical skills, expertise, or equipment to successfully and safely solve a problem, it's better to refer the customer to someone who has the appropriate resources.)
4. d

Appendix B
Control Tips for Species that Commonly Cause Nuisances

APPENDIX B: Control Tips for Species that Commonly Cause Nuisances

Appendix B includes suggestions for the use of tools and techniques that are scientifically appropriate for each species. Legal restrictions vary, so please check your state and local regulations before attempting control.

MAMMALS

- Bats (little brown, big brown, Mexican free-tailed)
- Chipmunk, Eastern
- Coyote, Eastern
- Foxes (red and gray)
- Mice (house, deer, white-footed)
- Moles (hairy-tailed, star-nosed, Eastern)
- Opossum
- Rabbit, Eastern cottontail
- Raccoon
- Rat, Norway
- Skunks (striped, spotted)
- Squirrels, tree (gray, red, fox, northern flying, southern flying)
- Voles (meadow, pine, prairie)
- Woodchuck or Groundhog

REPTILES

- Snakes (common garter, eastern milk, black rat, and northern water, with some information about other common species)

BIRDS

- Pigeon
- Starling, European

BATS

Species' names

- Little brown bat, *Myotis lucifugus*
- Big brown bat, *Eptesicus fuscus*
- Mexican free-tailed bat, *Tadarida brasiliensis mexicana*

Little brown and big brown bats commonly roost in buildings in the Northeast. Farther south, Mexican free-tailed bats may also be found in structures.

Size

Little brown: 3–4 inches long, 1/16–1/2 ounce, 9-inch wingspan
Big brown: 4–5 inches long, 3/8–5/8 ounce, 12-inch wingspan
Mexican free-tailed: 3–4 inches long, 1/2 ounce, 11–13-inch wingspan

The Indiana bat, *Myotis sodalis*, an endangered species found in some parts of our region, can be confused for the little brown bat. Remember, an endangered species cannot be harassed, collected, harmed, or killed without a federal permit. Of course, an Indiana bat is not likely to be found in a house. It's not easy to tell them apart, but here's how:

Little brown bat	Indiana bat
Brownish nose	Pink nose
Long hairs on toes, hairs stick out beyond end of toes	Short, unremarkable toe hairs
Calcar* usually lacks keel, or has only a weak keel	Calcar has prominent Keel

* The calcar is the bone that juts back from the anklebone to support the tail membrane. It's "keeled" if there's a flat ridge of skin sticking off its side that looks like the keel of a boat. Some people consider this the most reliable feature for identification.

Signs of Their Presence

- At dawn or dusk, you may see bats entering or leaving the building. This is easiest to see outdoors at dusk because bats swarm and fly around the entrance hole a few times before leaving the site.
- Sounds: Bats can see, but they use ultrasonic pulses to guide their flight and locate insects (called "echolocation"). Roosting bats may squeak or scurry when disturbed.
- Scat: Piles of black, dry, guano usually found under the main exit hole in the attic. Can be found scattered throughout an area, particularly in roosts where bats enter and fly about, but also check specific areas where they roost, such as a wall void or the intersection of the beams and rafters. You may also see scat on the side of the house, usually below a hole or crack. Large piles of guano are usually found beneath areas that are used often or by large numbers of bats. Individual droppings for the little brown bat are about the size of a grain of rice. The scat of the big brown is about twice that size, and typically chunkier, due to the types of insects they eat, such as beetles. Bat droppings look like mouse droppings, but mouse scat isn't found in large piles, and is also a little smaller. Also, bat droppings will crumble into powdery dust, but mouse droppings won't. You may see pieces of insect wings, or their reflections, in bat scat.
- Crystallized urine: When bats have been present in a structure for quite some time, crystallized urine may coat the beams or walls below their roost. Large colonies may develop urine "stalactites" or "stalagmites" below their primary roosting area.
- Rub marks along the edges of exit holes: Slight brown discoloration that's a mix of body oils and dirt.
- Roosts: During the summer, they'll use different roosts during the day and night. The daytime roost is usually in an attic, barn, garage, soffit, cave, underneath shutters or roof shingles, in wall voids, or behind siding or chimneys. At night, they'll rest in a breezeway, under an awning, or in a garage or similar areas. In the winter, both species hibernate

in colonies in caves, mines, and deep rock crevices. Big brown bats are more likely to hibernate in buildings, often in the attic or in wall voids.

Diet

Entirely insects. A colony of 100 little brown bats can eat hundreds of thousands of insects each summer. They eat many insects that damage crops and ornamental plants, such as moths and flies.

Typical Activity Patterns

Social style: Females of these species are colonial, while the males are usually solitary or found in small bachelor groups. Females roost together to raise their young. Pregnant females look for a warm nook within the roost that will serve as an incubator. These maternity colonies are often found in attics, soffits, wall voids, behind chimneys or shutters, in barns and tree cavities.

Daily activity: Nocturnal, with peak feeding at dusk and dawn (crepuscular). After feeding in the early morning, they'll return to the daytime roost and hang out, in a slightly dormant state. After their evening feeding, they'll usually rest. Females return to the roost often to nurse their young.

Hibernator? Bats hibernate in colonies in caves, mines, and deep rock crevices. Big browns will also hibernate in buildings. Their body temperature drops to within several degrees of the "room temperature." Big brown bats can tolerate below-freezing temperatures for short periods, which is why they are often found roosting in buildings. Their heart rates also drop dramatically. The little brown's heart rate will fall below 20 beats/minute and the big brown's to 42–62 beats/minute. Compare that to their heart rates during the rest of the year: 250–450 beats/minute for bats when they're resting, 800+ beats/minute when they're flying.

Migrates? Both species will migrate locally. Their hibernacula are usually within several hundred miles of their summer roost, but some big brown bats will hibernate in their summer roost. Some of the other species of bats found in our region may migrate far to the south.

Where Found

Distribution: Big and little brown bats throughout the East. Mexican free-tailed bats are found in the western, central, and southeastern United States.

Habitat: Forests and forest edges, areas with lakes and ponds, parks, orchards, fields, suburbs, cities.

Territory and home range: The maternity colonies begin to disband shortly after the young are capable of flying, typically in late July. Depending on the weather, females will remain together through late September. From late July through early September, bats actively explore new roosts and often show up in locations where they weren't seen earlier in the year. During the summer, there are significant and as yet unexplained fluctuations in the population of maternity colonies.

Breeding Habits

Pair bonding style: Polygamous. Females raise young by themselves, although they share the roost with other females and their young.

Breeding dates: Bats have an unusual habit. They mate during the fall, but the females store the sperm in their bodies for months. Fertilization doesn't take place until the late winter or early spring, when the bats emerge from hibernation or return from the south. Gestation: 50–60 days.

Birthing period: Late May through early July (births are typically staggered over a 2–3 week period).

Litter size: Little brown: 1 pup; big brown: 2 pups; Mexican free-tailed: 1 pup.

Weaning dates: Young begin flying at 3–4 weeks old. In late July, they begin to leave the roost with their mothers.

Amount of time young remain with parents beyond weaning date: Not well studied yet. Some scientists believe that the young probably follow their mothers back to the hibernacula. Some of the young probably return to the roost in which they were born.

Common Nuisance Situations

Time of year: Peaks from the third week of July through the first week of August, although there may be calls any time of year.

What are they doing?

- Females may roost in colonies in buildings to raise their young. Their scat and urine can damage insulation and household goods and attract other pests.
- Sometimes a lone bat enters the house and flies around. This usually happens in July and August, when the young are learning to fly.
- During an extreme heat wave, several bats may enter the living quarters, seeking a cooler roost. This is when they'll show up in places they normally don't use.
- Big brown bats will hibernate in buildings (little browns don't seem to do this in the North, although they do in the South). In the winter, a big brown bat may leave its roost in the attic and fly around in the living spaces. This usually happens when the temperature of the attic roost changes dramatically, disturbing their hibernation—during a thaw, or during the very coldest part of the winter, if the attic is much colder than the rest of the house.
- Disease risks: Rabies, histoplasmosis (bats don't carry or transmit histoplasmosis, but their guano promotes the growth of the mold spores that cause the disease). Bats are a rabies vector species. Silver-haired bats (*Lasionycteris noctivagans*), which aren't found in buildings, are often associated with human rabies cases. How the person came into contact with the silver-haired bat often remains a mystery.

Debunking Myths about Bats

- There's an epidemic of rabies in bats. Wrong. The incidence of rabies in bats is very low and outbreaks in colonies appear to be rare. Each year, the CDC collects data on rabies cases from each state's health department and publishes a report. In 2004 (the most current data available), bats accounted for approximately 20% of all of the cases of rabid animals in the U.S., following raccoons (38%) and skunks (27%). But that number doesn't tell you how many bats actually had rabies because the sample is not representative. Only sick or otherwise rabies-suspect bats are tested. To determine the true incidence of a virus, a representative sample of the overall population, including healthy individuals, must be tested. So what's the chance that a particular bat has rabies? Let's take New York State as an example. Of the states, it reported the third-highest number of cases in bats (118) in 2004. The NYS Department of Health has been tracking the number of rabid bats since 1965. The total has remained remarkably constant at 3–5% of the sick bats that were tested. Yet the true statewide incidence is estimated at less than 1% of the bat population, adjusting for the sampling bias mentioned above.
- If one bat in a colony is rabid, they're all sick. Wrong again.
- Not all grounded bats are sick! Young pups sometimes become grounded when they're learning to fly.
- Overwhelming fear of contracting rabies from a bat. This is a tricky one. Although your chances of being exposed to a rabid bat are very low, it is true that most Americans who died from rabies in the past decade did have the bat strain. There are probably several explanations. Most people notice if they've tangled with a skunk, raccoon, or fox. Also, most people go right to the doctor if they think they've been potentially exposed to rabies, so they receive treatment and they don't die. But many people encounter bats at night, while they're drowsy and not thinking clearly, or even while they're sleeping. They may not realize they were

scratched or bitten, or may not remember the encounter by the time they wake. So they see no reason to go to the doctor. Unfortunately, that means that if they did get infected with the virus, it may not be detected in time for successful treatment. The bottom line: even though infection is unlikely, this is a fatal disease, so don't take chances. However, if someone's so terrified that it's causing problems, try to help that person understand this risk more clearly.

- Bats almost never attack people (they will bite in self-defense).
- Bats are not hair stylists. They don't aim for a person's hair. Bats are excellent fliers. They may swoop close to a person's face, but if you don't thrash about, they won't fly into you.
- Bats in the United States don't suck blood. The bats that live in our region all eat insects. The vampire bats found in the tropics and subtropics do feed on blood—from livestock and birds, mostly—but they will bite people.
- Bats don't chew holes in buildings.

Best Practices

- If there's a possibility that a person, pet, or livestock has been exposed to rabies, you are legally required to follow the directions of the health department, no matter what your customer requests.
- There's one way in which bats are quite different from many of the other nuisance species you'll deal with: for most of those animals, removing one nest site won't affect the local population. But bats have so few young that the destruction of one maternal colony might harm local bat populations. Please try to persuade your customers to accept an approach using nonlethal techniques. They are highly effective for bats. There are no toxicants registered by USEPA for bat control.

If one bat has accidentally flown into the house, and there have been no human or pet exposures:

- Remove any pets from the room, then seal the room. Shut the doors and place a towel under the door. Close any heat registers. Close all but one

Safety tips for contact with bats or their droppings

- Wear leather gloves, disposable Tyvek coveralls, goggles, and a proper respirator.
- Ventilate the area, if possible.
- Don't stir up dust. The dust could contain the spores of the fungus that causes the respiratory disease histoplasmosis; if you disturb the dust, you might inhale the spores. Don't sweep or vacuum, unless using a commercial vacuum intended for this purpose according to the manufacturer's instructions. Instead, thoroughly wet the materials with a household or commercial disinfectant. Wipe up with a damp sponge. Play it safe and don't use a bleach solution, because bat guano contains ammonia. Although the concentration of ammonia in bat droppings isn't as strong as it is in bird droppings, there could be a lot of guano. *Do not mix bleach with ammonia because the fumes are toxic.*
- Spray dead bats and their droppings with disinfectant, then double-bag for disposal.
- Be aware of potential pesticide residues in the attic from previous control attempts.
- For more information, refer to Chapter Four and the resource list.

window, which should be screened to let in air. Then turn on a dim light. You want just enough light to see—if you make the room too bright, you'll disturb the bat, and it will probably try to hide in the darkest, most remote place it can find. In a large, cluttered room, searching for a tiny bat is not easy.

- Remain nearby and wait quietly. The bat should calm down in 10–15 minutes. The air flow from the window should attract the bat, so it should land on the screen. When it lands, you can capture it by covering it with a plastic container and slipping a piece of cardboard underneath. If there's definitely no concern about possible rabies exposure, the bat can be taken outside and released. Or you could remove the screen from the window and let the bat fly out on its own. If the bat doesn't calm down quickly enough, leave the room sealed. Return the next day, at dusk. The bat should be calm. Repeat the dim light, window capture process.

The timing of bat-proofing (a.k.a. "exclusion") is critical. Do not separate females from their young!

- When there are probably young in the roost (mid-May through early August), *only limited bat-proofing is acceptable.* We'll describe that below.

Before and after that period, complete bat-proofing—with the use of a checkvalve to make sure that any bats accidentally trapped inside can escape—is okay. WCOs in cold climates will need to alter their activities once winter conditions make roof work unsafe.

- Don't worry about the bat flying at you, and don't chase it or swat at it because that will only cause the bat to panic. When indoors, bats normally fly around the room several times until they find an exit. They make steep, banking turns, flying up as they approach a wall, and swooping down as they cross the center of the room.
- Check the windows, especially if there's an air conditioner in the window. Gaps between the upper and lower window, or around a storm window, should be sealed.

How to Bat-proof a Building

1. **Find the bats' primary entry/exit hole (this is crucial; we'll explain why below).** Look around for piles of guano, which may accumulate beneath the primary exit hole, or stick to the wall near the hole. Rub marks are typically seen on the wall if there's guano stuck to the wall. Or stage a "bat watch." At dawn or dusk, stand outside the building and watch where the bats enter or leave.
2. **Inspect the building thoroughly to find all of the secondary holes and cracks.** There are probably many of them, because bats can wriggle through very small openings. Looks for cracks that are ¼ inch wide by 1½ inch long (about the size of a stubby pencil) and holes that are at least ½ inch by ½ inch. Inspect the house for signs of warped, shrunken, or loose materials, especially at joints. Pay special attention to roof drip edges, dormer tie-ins, roof corners, ridge caps, vents, and deteriorated walls, roofs, and eaves. In an old house that doesn't have soffit vents, you may be able to use light to your advantage. In the attic, cover the windows, then turn off the lights. Look for light peeping through gaps, cracks, and holes. Then use a flashlight to take a better look at these potential entry points. Modern houses typically have soffit and eave vents that let in enough light to make this

technique impractical. Some don't even have attics if the bats are roosting behind shutters, remove the shutters or install small blocks between the shutters and the wall. In either case, you're done.

3. **Seal potential points of entry into the people's living quarters.** Bats can't chew through exclusion materials the way rodents do, so you can use many different materials to seal holes and cracks, including caulk; expanding foam; Stuf-Fit, a knitted copper mesh that's like steel wool, only it doesn't rust; window screening; bird netting; or "flash-band," a self-adhesive aluminum-faced sealant. Some WCOs use the Todol foam gun to efficiently apply expanding foam, and then finish with a bead of silicone over that. The mesh size of screening or netting shouldn't be larger than ¼ by ¼ inch or the bats will be able to crawl through it. Although mature bats typically can't fit through a half-inch mesh, they can get their heads stuck in it, so the finer mesh is a better choice.

4a. ***There may be young in the roost* (mid-May through early August):** Leave the primary exit hole alone. It must remain open to allow the females to fly freely in and out of the building so they can hunt and care for their young in the roost. Seal all of the other entry holes and cracks on the outside of the building. Screen chimney flues and caps with mesh that's ¼ inch or smaller (check local fire codes). Seal any gaps in the flashing around the chimney, and where the chimney cap meets the chimney. Screen vent pipes with quarter-inch mesh. If the bats are roosting behind shutters, remove the shutters or install small blocks between the shutters and the wall. That's it for now. You'll finish the bat-proofing after the young have left the roost, usually by mid-August.

4b. ***No young in the roost* (late August through early May*):** Locate the primary exit hole. Seal all of the other holes and cracks. Screen chimney flues and caps with quarter-inch mesh (check local fire codes). Seal any gaps in the flashing around the chimney, and where the chimney cap meets the chimney. Screen vent pipes with quarter-inch mesh. Install a checkvalve (one-way door) over the primary exit hole, following the directions below. Leave the checkvalve in place for 5–7 days unless you're bat-proofing during the winter, when it's safest to leave the one-way door in place until it warms up in April. This way, if a few stray bats were over-wintering inside the attic, they would be able to leave. Once you're sure that all the bats have left, you can remove the one-way door and permanently seal the main hole.

*During the winter, WCOs in the North need to assess whether the roof is too slippery to walk on safely. Some will do bat-proofing jobs if there's no snow or ice on the roof, while others taper off such jobs beginning in December and schedule them again in April.

5. ***If you did limited bat-proofing to protect young in the roost:*** To finish the job after the young have left the roost, install a checkvalve over the main hole, following the directions below. Leave the checkvalve in place for 5–7 days. Once you're sure that all the bats have left, you can remove the one-way door and permanently seal the main hole. Now you're done.

Why the fuss about the main hole and checkvalves? Because the primary exit hole is the one that matters the most to the bats. If you accidentally trap bats inside or separate females from their young, the bats will frantically seek a way out of, or back into, the building. They may fly into the people's living spaces; that's a safety risk. To avoid this problem, do two things. First, do not separate mothers from their pups. That means you leave the main exit hole open during the period when bats are rearing their young.

At other times of the year, place a checkvalve over the main hole and leave it in place for 5–7 days. This way, if you accidentally trapped a bat inside, the bat has a way out. Once you're sure there are no bats inside, you can permanently seal the main hole.

How checkvalves work. One-way doors, or checkvalves, allow the bats to leave on their own, but they can't get back inside. You can use a commercial checkvalve or make your own, following this design developed by Dr. Stephen Frantz from the New York State Department of Health. Place quarter-inch polypropylene mesh or screening over the entrance holes, forming a long sleeve or tent. The screening should cover the hole and extend about 3 feet below the hole. It also should stick out about 3–5 inches from the wall, so the bats can crawl beneath the screen to leave. Secure the screening at the top and sides with duct tape or staples; leave the bottom open.

Bats exit at the bottom, but when they attempt to return, their sense of smell guides them back to the hole. They land on the mesh near the hole—and stay there, sniffing around. They just don't crawl down the mesh. Good thing, because if they did, they would be able to slip back underneath the mesh, crawl up the building, and find their hole. But don't worry about that. Bats trust their noses.

Trapping Strategies

(*Note:* Trapping or relocating bats is not legal in all areas of our region. Check your state regulations.)

Live traps:

- A resting bat can be captured with a plastic container or a butterfly net (wear leather gloves). If you can reach it, cover the bat with the container, then slip a piece of stiff cardboard between the container and the wall. (Or cut an "x" into the plastic lid, and hold the container so the bat can crawl into it on its own.) To release the bat, hold the container on its side on a secure place, such as against a tree, or on a ledge. Then remove the cardboard.
- One-way doors. See previous description.
- Multiple-capture traps. There are several different types and sizes of multiple-capture traps for bats. The Batrap consists of a tube that leads the bats through a one-way baffle into a chamber, in which they're confined. Like a one-way door, this trap allows the bats to leave the building but keeps them from re-entering at that spot. (Of course, you'd still need to bat-proof to make sure the bats can't enter the building in other places.) Multiple-capture traps are a bit trickier to use than a one-way door. If you're inexperienced with this tool, try to learn its proper use from a skilled WCO. For example, it's important to correctly estimate the number of bats in the roost. You need to match the size of the trap to the number of bats that may be captured, because if too many bats are confined in the trap, they might suffocate. Another danger is that if bats remain in the trap too long, they might overheat from exposure to the sun, and die. It's best to release bats in the early morning.

If your customers insist on seeing the bats, a multiple-capture trap will satisfy that need and allow you to then release or relocate the bats. (If the customer insists on having the bats killed, this is a safe way to capture them. But do try to talk them out of that.) One benefit of using a multiple-capture trap is that it may create a great "teachable moment." You can show the bats to the neighbors and talk to them about the benefits of having bats around, and debunk myths. You may be able to convince them to bat-proof.

Another option that allows you to prove that you got rid of the bats is to use a one-way door and a video camera. Place the camera near the exit holes, set it to record, and leave it running. Then you can show your customers the tape of the bats leaving the roost.

What to Do with the Bats

- Follow the health department's directions if there's been a potential rabies exposure, or if dealing with a sick bat, one's that grounded, or one that's behaving strangely.
- For healthy bats that can be released, on-site release (coupled with bat-proofing, of course) is preferred over relocation.
- Relocation is not recommended. Bats will fly great distances to return to a favored roost, so if the customer didn't bat-proof, the problem may happen again. You'd have to travel hundreds of miles to make sure the bats wouldn't be likely to return to that spot. If you did that, what would happen to the bats during the trip? If some became too stressed or overheated, they might be grounded after release. That might increase the disease exposure risk for people, pets, and wildlife in that area. Also, how fair is this to your customer's neighbors? Sure, they see you capture the bats and take them away, but you know the chances are good they'll return.

Special Concerns with Very Large Maternity Colonies

- If the site is really good, there could be thousands of bats, perhaps as many as 3,000. Obviously, the bats have been there for a long time. Can you convince your customer to choose a two-season solution, which might reduce long-term harm to the local bat populations? In the first year, install bat boxes in the early summer (a large maternity box will house 150–300 brown bats). Let the bats raise their young in the building, and give them time to check out those newly installed boxes. In the late fall, you can begin limited bat-proofing, but don't close the main entry hole. In early May of the following year, install a checkvalve over the main hole. Once you are certain all bats are out of the structure, you can finish permanent bat-proofing by sealing the primary entrance hole.
- Create an alternative roost nearby by installing a bat box in the yard (optional, but highly recommended, especially for larger colonies). Why? Many people find that they like having bats nearby, because of all the insects they eat. Another good reason to install a bat box is that it may prevent that colony from moving into a nearby building. Installing bat boxes in the northern sections of our region does require some careful planning. See *A Homeowner's Guide to Northeastern Bats and Bat Problems* from Penn State University or contact a bat specialist or Bat Conservation International (www.batcon.org) for details. It's best to locate the bat box where people are less likely to encounter the bats. If this isn't possible, the area immediately around the box can be fenced to prevent people from disturbing the colony and encountering a bat that's fallen out of the box. Once a year, remove the guano that's accumulated under the bat box to reduce the chance that it will promote the growth of the mold that causes histoplasmosis. Follow appropriate safety instructions for the clean-up (see Chapter Four).

Preferred Killing Methods

- CO_2 chamber (a smaller chamber is better)

Acceptable Killing Methods

- If there's been no rabies exposure, with care, can use stunning and chest compression. Be careful about contact.

Methods that Don't Work Well

- Ultrasonic devices are not effective.
- Customers may also ask about using mothballs or insect spray. These products are not registered for use in bat control, so that would be illegal. Mothballs can also be dangerous, especially to children. One of the active ingredients, napthalene, can be inhaled, eaten, or absorbed through the skin. That means that if you store clothing with mothballs, you can absorb the chemical the next time

you wear the clothes. Napthalene flakes may help to repel a bat colony if the site is confined and has little or no ventilation, but the recommended application rate is high, and vapors could enter the living spaces and endanger people (the WCO would also need a commercial pesticide applicator license).

- There are two ways to change an attic that may make it less attractive to the bats. Neither has been well studied. It's also possible that the bats will simply move into another area without leaving the building. But here they are: (1) Increase the ventilation and decrease the temperature of the roost area, which makes it less attractive as an incubator. In attics, install fans, windows, ridge or soffit vents, or insulation in the walls. (2) If you can't seal all of the holes, install a few floodlights or fluorescent bulbs in the roosting area. The lights should be left on all day and all night for several weeks. In later years, turn the lights on during the spring or summer occasionally, to discourage the bats from recolonizing the site. Leaving the lights on for so long could increase the risk of an electrical fire. Make sure that the wiring is installed properly and meets electrical codes.

EASTERN CHIPMUNK *(Tamias striatus)*

Size

- Body is 5–6 inches long. Tail is 3–4 inches. About 3 ounces.

Signs of Their Presence

- The animals themselves.
- Sounds: Chipmunks use a sharp repetitive chirp to alert others of danger. When startled, they'll often respond with a single chirp followed by a short burst (1–2 seconds) of chatter.
- Holes that are about 2 inches in diameter. Holes usually go straight down, with no dirt mound in front.
- Evidence of their feeding: Like squirrels, chipmunks leave gnawed nut shells.
- Garden and landscape damage: They dig up and eat flower bulbs and seeds. The digging and tunneling also cause damage.
- Occasionally, chipmunks will damage buildings, but not nearly as often as other rodents.

Diet

Opportunists, primarily herbivores. Their favorite foods are nuts, seeds, and fruits. Chipmunks eat nuts (acorns, hazel nuts, beechnuts, etc.), seeds (from many ornamental trees, wildflowers, clover, ragweed, and sunflowers, and bird seed), flower bulbs, berries (such as raspberries, strawberries, blackberries, and chokecherries), fruit (watermelon, apples, pears, peaches, cantaloupe, cherries), and wild mushrooms. They will occasionally eat corn, wheat, oats, grass seed, insects, worms, snails, slugs, bird eggs, nestlings, mice, moles, frogs, salamanders, small snakes, and carrion. Although they spend most of their time on the ground, they will climb trees to take nuts, fruits, and seeds. Chipmunks cache food in a storage chamber in their burrow. During the breeding season, they must drink up to a quarter of their body weight in water each day.

Typical Activity Patterns

Social style: Generally solitary, except for female with dependent young. They can be fairly aggressive to each other.

Daily activity: Diurnal.

Hibernator? No. They'll stay in their burrows in a sleepy state for days at a time during the winter. Chipmunks rely on their food caches during the winter. They may come out on warm days, often to travel to another food cache.

Migrates? No.

Where Found

Distribution: Across the Eastern U.S., except in its southern portions.

Habitat: They adapt to a variety of habitats, but are usually found in areas with at least a few mature trees. Common in rural, suburban, and urban areas (yards, gardens, campgrounds, parks, urban lots). Chipmunks often burrow under old stone walls bordering pastures or woods; under piles of brush, rocks, or garbage; among a tree's roots, or near buildings. The only places you're not likely to find them are marshy areas with very dense undergrowth.

Territory and home range: The home range varies from 1⁄10 to 3 acres, but most don't venture more than an acre. Males have larger home ranges than females. Densities may be as high as 10 chipmunks per acre. They will defend an area of about 50 feet around burrow entrance.

Breeding Habits

Pair bonding style: Polygamous. Females raise young alone.

Breeding dates: They mate twice a year, in April and July.

Birthing period: Two litters, in May and August. Gestation takes about 31 days.

Litter size: 2–7 young.

Weaning dates: 4–6 weeks.

Amount of time young remain with parents beyond weaning date: Not long. Young will leave the burrow at about 6–8 weeks old.

Common Nuisance Situations

Time of year: Spring through fall in the northern portions, while complaints may occur year-round in the southern parts of our region.

What are they doing?

- The most serious economic damage is caused by their burrowing under porches, stairs, patios, or into foundations. They may undermine the structure.
- They gnaw on the tubes used to collect maple syrup.
- Chipmunks raid bird feeders and stashes of pet food.
- They dig in flower beds and vegetable gardens, uprooting plants. They eat flower bulbs, seeds, seedlings.
- Their loud warning chirps in the morning may disturb some people's sleep.
- Disease risks: minimal. They are host to a variety of parasites, including fleas, lice, mites, worms, and botflies.

Best Practices

- Chipmunks rely on food caches to survive the winter, so if you move them too far away during that time, they'll probably starve to death. Limit relocation to times when food is readily available.
- Chipmunks raise their young in underground burrows. They don't raise their young in buildings (we can't say it would never happen but it's extremely unlikely). Very young chipmunks might enter a building on their own, leading someone to believe there's a "nest" in there. No. Those chipmunks are mobile.

Remove artificial food sources (bird seed, pet food):

- If anyone is feeding the chipmunks, persuade them to stop.
- Hang bird feeders on a rope between two pulleys. Ideally, feeders should be 15–30 feet away from the building so any seed that collects below doesn't lead the chipmunks right to the foundation.
- Squirrel-proof bird feeders that use the animal's weight to close the feeder won't stop chipmunks unless the feeder's set so it will close when a very light weight is applied. Unfortunately, at that setting, you'll also stop all but the smallest birds from using the feeder.
- Keep the area underneath the feeder clean, and feed birds only during the winter months.
- Feed pets indoors.
- Store food, bird seed, and pet food in metal, glass, ceramic, or heavy-duty plastic containers.

Block their travel routes:

- Screen gutter pipes, downspouts, and foundation drain pipes with quarter-inch hardware cloth.
- Chipmunks often gain access to houses through attached garages. Screen the eaves and overhangs of the garage with quarter-inch hardware cloth.
- Remove ground cover and hedges that are next to the foundation.
- Move firewood and piles of debris, rocks, and brush away from the foundation.
- Maintain a foot-wide gravel border around the foundation that's free of vegetation (best), or keep the foundation plantings well trimmed. Don't stack anything (such as firewood) against the foundation.
- Interrupt any routes that connect woods to the foundation in one continuous line by removing plants or a section of stone wall.

Protect vulnerable crops and ornamental plants: Plant bulbs within a cylinder of 1-inch poultry wire. Lay the wire in a trench then plant the bulbs in it. Add some dirt, finish wrapping the wire around the bulbs, then cover with soil.

- Another option for bulbs is to plant them and then lay a piece of half-inch hardware cloth over the soil surface to reduce the chipmunks' ability to dig up the bulbs. The hardware cloth should extend at least a foot around the plantings, and be covered with soil. Its mesh must be large enough for the stems to grow through, so you may need to experiment with different sizes for different plants.
- Establish a barrier around gardens and fields with wire mesh fences. Use quarter-inch galvanized hardware cloth or welded wire. The fence must be 30 inches high, buried 6–12 inches deep, with a foot-wide L-shaped shelf that sticks out to prevent the chipmunks from burrowing underneath it.
- If there aren't any mammals nesting in the tree, wrap two-foot-wide bands of sheet metal around fruit trees at 6–8 feet, to prevent chipmunks from climbing the tree. This will work only if the chipmunks can't leap from another tree or other object onto this tree. Attach the band loosely, so the tree has room to grow. Don't staple the band onto the tree because that can prove dangerous if someone needs to cut down the tree.

Prevent entry into building:

- Remove any chipmunks that are using the structure. Exclude them with a one-way door.
- Close the door! (Use screen doors.)
- Install door sweeps and weather-stripping (in garages, too).
- Seal openings with quarter-inch hardware cloth. Or poke steel wool, wire mesh, or flexible aluminum "gutter guard" into the hole, then caulk, or spray expanding foam (such as Great Stuff) over it to strengthen the barrier. Focus on holes at ground level, especially those in foundations. Chipmunks tend to stay on the ground, so you don't have to look for holes that are high up, as you would with squirrels.
- Check vents, especially clothes-dryer vents that are close to the ground. If the vent is damaged, replace it with an animal-proof design or screen it with quarter-inch hardware cloth (remember to

check building codes)—except for clothes-dryer vents, which shouldn't be screened because lint can build up, which is a fire hazard.

- Plug gaps around water, gas, and heating pipes with latex caulk. For large holes around pipes, use galvanized metal pipe chase covers, sheet metal plates, mortar, plaster of Paris, or cement.

Trapping Strategies

Live traps:

- Small cage traps (approx. 16 × 5 × 5 inches) or Sherman traps (approx. 10 × 3 × 3 inches).
- Bait with nuts, peanut butter, sunflower seeds, apple slices.
- Place the trap near the tunnel's entrance or along their travel route.

Lethal traps:

- Rat-sized snap traps. There are now models that have built-in safety catches. The bait's under a cover, which must be lifted before the trap will fire. This means that an animal that's just investigating won't set off the trap. The design also helps ensure proper positioning, which is more humane. Place the trap near entrances to active burrows. Cover the trap as described below.
- If using a traditional snap trap, place it within a cage trap, a box, a coffee can with both ends cut out, or in PVC pipe, to prevent the capture of songbirds. If the trap's next to the foundation, you could lean a board over it.

Other lethal techniques, for WCOs with a commercial pesticide applicator license:

- There are registered fumigants for chipmunk burrows (phosphine gas, carbon monoxide gas cartridges), and zinc phosphide bait. The fumigants could pose a fire hazard.

Preferred Killing Methods

- CO_2 chamber
- Lethal trap
- Shooting, using an air rifle or a .22 caliber rifle with bird shot
- Cervical dislocation
- Stunning and chest compression

Acceptable Killing Methods

- Fumigants and rodenticides

Control Strategies that Don't Work Particularly Well

- There are brands of sunflower seed and suet that are treated with a repellent. The active ingredient is capsaicin, the chemical that makes hot peppers taste hot. Although capsaicin is often effective against mammals, it doesn't bother chipmunks because they have fur-lined cheek pouches.

EASTERN COYOTE (*Canis latrans* var.)

Size

30–45 pounds, 4–5 feet long (including tail)

Signs of Their Presence

- Coyotes run with their tails down, while wolves and dogs hold their tails straight out. The tails of domestic dogs often curl; coyotes have straight tails, with a black tip.
- Sounds: May yip, yelp, howl, growl, bark, or woof. Family units (the adult pair and their pups of that year) often yip-howl when they reunite.
- Tracks: More compact, linear, and forward-directed than the dogs' often splayed, sloppy-looking track (Figures B-1 and B-2).
- Scats: Are twisted, often containing hair or berries (Figure B-3). Often found on rocks, logs, or at trail intersections. Their scat has a mild, musky odor, unlike that of the domestic dog.
- Large carcasses: Coyotes tend to eat the organs first. They'll pick bones clean, unlike dogs. Dogs start at the rear of a larger carcass and eat their way towards the head.

Diet

Opportunist. About 80% of their diet consists of small rodents and rabbits. Coyotes will also eat insects (especially grasshoppers); fruits (berries, watermelon); fish and crayfish; frogs; snakes; bird eggs; larger mammals, such as raccoons, opossums, muskrats, and deer; garbage; carrion; dog or cat food; bird seed; and even doughnuts. They can kill house cats and dogs. Some individual coyotes will kill livestock, especially chickens, ducks, and lambs.

Typical Activity Patterns

Social style: Most often, will see mated pairs (usually, life-long mates) but coyotes may live in family groups or alone, and may switch from one lifestyle to the other. Sometimes, "teenaged" coyotes form a loose group, often much larger than the family group, perhaps with as many as 15 individuals. Such groups

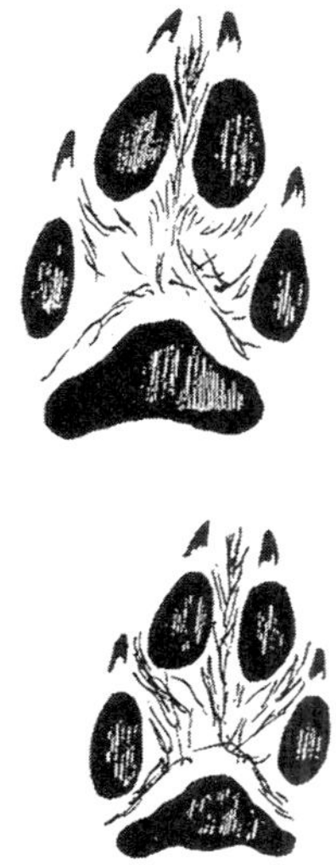

FIGURE B–1 Coyote tracks, close-up. *(Wayne Trimm)*

FIGURE B–2 Coyote track pattern. *(Wayne Trimm)*

FIGURE B–3 Coyote scat. *(Wayne Trimm)*

may behave more aggressively towards people and dogs. This subadult group is probably just a temporary arrangement.

Daily activity: They're usually nocturnal, especially during hot weather. Coyotes may be active throughout the day when they're rearing their pups, and in areas where they're left alone by people. In suburban areas, coyotes may alter their activity patterns, especially if there are daytime food sources available, such as dog food that's left outdoors.

Hibernator? No.

Migrates? No.

Where found

Distribution: Rural and suburban areas throughout the East.

Habitat: They prefer forested areas, shrubby fields, and marshy areas but are highly adaptable and will live in residential areas, including cities.

Territory and home range: Home ranges cover 5–25 square miles. Coyotes will scent-mark their territory. Home ranges may overlap with other coyote groups, especially when they're not raising their pups.

Breeding Habits

Pair bonding style: Usually monogamous. Coyotes are thought to mate for life. Both parents cooperate in rearing the young, sometimes with the help of an older pup. Sometimes, a male may mate with several females and they may share the den, raising more than one litter there.

Dens: Coyotes seek sheltered areas for rest or protection from severe weather, but use dens to raise their young. Coyotes commonly move their pups from one den to another. Most often, coyotes will renovate a burrow that was built and then abandoned by another animal, such as a woodchuck, but they can dig their own dens. Their dens are often found on rocky ledges or steep or brush-covered slopes. In urban areas, they may den in storm drains, under sheds, or in holes in parks, golf courses, and vacant lots. Coyotes may reuse dens from year to year.

Breeding dates: February, continuing into March in southern parts of our region. Females are receptive for only a few days. Gestation takes about nine weeks.

Litter size: 5–7. The better the food supply, the larger the litter.

Birthing period: Late April, continuing into May in southern areas.

Weaning dates: Pups begin exploring when they're three weeks old and are usually weaned by six weeks old.

Amount of time young remain with parents beyond weaning date: Pups are usually driven out of their parents' territory between September and March, with peak dispersal in December through January.

Common Nuisance Situations

Time of year: Peaks in late winter (February–March) when the coyotes are establishing their territories, then again in early spring and summer, when they need more food to raise their pups. During the winter peak (March–April), coyotes aggressively defend the area around their den site. This is when they often come into conflict with dogs, which they view as a threat to their pups. This is especially true if the coyotes are trying to move into the dog's turf (a yard).

Coyotes are opportunistic. In the early spring and summer, coyotes seek easy prey to keep up with the food demands of their pups. "Easy prey" may include cats and small dogs in suburban areas, and young livestock (lambs, chicks) in rural areas. When coyotes move between natural areas within their home range, they also encounter and may take domestic pets. There may also be complaints during the fall, as young coyotes try to establish their own territories, because that can be a noisy process. But they're fussing among themselves, and tend not to wrangle with dogs then.

What are they doing?

- Their yipping and howling may disturb some people.

- Their mere presence may frighten some people. Many people aren't used to seeing coyotes and may fear them.
- They can kill housecats or small dogs. Large- and medium-sized dogs (over 35 pounds) are rarely physically threatened, because the coyote recognizes that it's outmatched. They'll usually work out the territorial dispute (loudly) without either being hurt. Small dogs are at risk. The coyote expects to be dominant and will discipline the dog until it offers the correct submissive behavior. If the dog doesn't submit easily, it could be badly injured or killed. Very small dogs and cats are seen as easy prey. Free-roaming pets should be brought inside to keep them safe from cars as well as predators.
- Some coyotes kill livestock, usually free-ranging chickens, ducks, and sheep.
- They'll eat some vegetables and fruits, especially melons.
- Coyotes (and foxes) will chew holes in irrigation pipes in fields and orchards.
- Nationwide, a few people have been attacked, mostly in the western United States. Usually coyotes don't bother people. However, some coyotes may become bold and aggressive. If you see individuals showing these behaviors, take action. The potential does exist for coyote attacks in the East. People and coyotes can usually coexist if the coyotes maintain their natural fear of people (more on this later).
- Coyotes may travel along an airport's runways, causing delays and hazards to aviation.
- Disease risks: Distemper, hepatitis, parvovirus, rabies, mange, and tularemia.

Debunking Myths about Coyotes

- Coyotes are often mistaken for "coydogs." A cross between a coyote and dog is possible but highly unlikely because of differences in their breeding habits. There are enough coyotes; they can easily find mates.
- A coyote that's active during the day is not necessarily rabid. Most likely, it's a healthy animal that's feeding more often than usual, because of the demands of its young.
- Although many people fear coyotes and believe they are much more dangerous than domestic dogs, that's not true. More people are hurt by pet dogs each year in the U.S. than by coyotes. However, coyotes are strong enough to cause serious injuries to people, especially to children, and coyotes can kill pets, so they should be respected.

Best Practices

Remove artificial food sources (garbage, compost, pet food, pets):

- If anyone is feeding the coyotes, persuade them to stop.
- "Coyote-proof" garbage cans or dumpsters with a tight-fitting lid. Secure garbage cans with heavy-duty straps or bungee cords, or attach them to a post, or keep them out of reach in the garage (close garage doors at night), or place the cans in a covered and secure bin.
- Enclose compost piles in a framed box using hardware cloth; in a sturdy container, such as a 55-gallon drum; or in a commercial compost container.
- Feed pets indoors. Any food left outdoors should be removed at night. Bring pet food dishes inside, too.
- Keep the area around bird feeders clean, because coyotes will eat spilled seed. They'll also eat animals that are attracted to the bird feeder, such as birds, mice, and squirrels, so in some cases, you may want to remove the feeder during the summer months.
- Do not allow pets to roam freely.

Protect children and pets and keep coyotes out of yards:

Some coyotes in suburbia have lost their fear of people. A coyote that does not fear people should be considered potentially dangerous.

The foods that coyotes find in residential areas (garbage, pet food, pets) are full of human odors, so these coyotes quickly learn to associate people with food. That's bad. Many people become frightened when they see coyotes and run into their homes. That's even worse because to a coyote, that person has just behaved like prey. Running triggers their attack response.

In short, food smells like people and people behave like prey. Add to the mix people intentionally feeding coyotes, and the potential for a coyote attack becomes very real. Certain changes in coyote behavior seem to indicate a growing risk that coyotes will become aggressive toward people (based on studies of coyote-human conflicts in California). The following warning signs are shown in the order they may happen.

1. During the night, coyotes kill more pets than they did before.
2. During the night, coyotes are seen on streets and in yards more often than before.
3. Coyotes are now seen in those areas during the day, especially in the early morning and late afternoon.
4. During the day, coyotes chase or kill pets (previously, only a night-time activity).
5. During the day, coyotes kill pets that are on leash (previously, they took only free-roaming pets), or chase joggers and bikers.
6. At midday, coyotes are seen near children's play areas.

Coyotes are generally nocturnal, so increasing daytime activity is often a sign that those animals have become used to people. Such boldness should be taken seriously. The California studies suggest that if the situation is addressed before pet attacks are common, further problems can probably be avoided.

What can you do?

- Act tough. Be very aggressive. Yell, make loud noises, wave your arms, spray the coyote with water from a hose, and throw sticks or stones near the coyote.
- Teach children that coyotes are not dogs and they are not pets. Watch coyotes from a distance.
- Small children are most vulnerable, so don't leave kids in the yard unattended.
- Don't let pets roam freely. And don't feed them outdoors!
- Keep the yard clean and well mowed to remove cover for the coyotes (trim shrubs at ground level). Remove brush and junk piles that attract rabbits and other prey that might entice coyotes into the yard.
- Hunting and trapping helps to keep coyotes from losing their fear of people.
- Fences will help a great deal, but coyotes can climb over, burrow under, or squeeze through fences. So build them right and keep the fences well-maintained.
- Construct net wire fences (using 4-inch mesh) that are 5½ feet high. To prevent coyotes from digging under the fence, either add barbed wire at ground level or bury the fence 6 inches deep and bend a foot-wide section outward into an L shape. To discourage coyotes from climbing over the fence, add an electric wire at the top or create an overhang of regular wires.
- Electric fences: A fence of 13 strands of high-tensile electric wire is proven to protect sheep from coyotes. Other designs using fewer wires, including electrified netting, may work in some areas.
- Combination woven-wire and electric-wire fence: If there's an existing woven-wire fence, add an electric strand that's placed 6–8 inches above ground and 8–10 inches outside the fence, using an offset bracket. If the coyotes are climbing over the fence, add a top wire that's also outside the fence.
- Invisible fences may be an attractive solution for keeping pets within an area but they will not keep wildlife out. To protect pets from predators, switch to a traditional fence design.
- Ask neighbors to follow these same steps.

Protect vulnerable livestock

In some cases, you may be able to successfully scare off coyotes and solve the problem. This is most likely to work when you have access to the coyotes' den. It's reasonably easy to harass a coyote (or fox) enough to convince them to move their pups. These ideas haven't been well studied, but we believe they're worth trying.

- Act tough, as described previously.
- Use slingshots, an air rifle, or a shotgun loaded with rubber buckshot to frighten a coyote that's farther away (30 feet). You don't have to hit it to scare it off, but if you can, that will more effectively train the coyote to stay away (this refers to the use of the less lethal projectiles mentioned above).
- Experiment with noisemakers, such as bangers and screamers.
- One scare device, the Critter Gitter, combines a siren and flashing lights. It's triggered by a motion detector. The device switches patterns, so it should be effective longer than a scare device that doesn't vary.
- As with many scare tactics, some of the ideas above are labor intensive. You might have to sit and wait a long time before the coyote shows up. (That may not concern some customers.)
- In some situations, such as when the customer really doesn't want the coyote killed, you may be able to train the coyote to become wary of the capture location. Some biologists believe that if a coyote is trapped and released, it will avoid the area in which it was caught. If the customer's property is large enough, and you don't think the coyote will move on to trouble a neighbor, you may want to trap it and release it on site. This is a riskier technique because if it doesn't work, your job has just become much harder. Coyotes easily become trap-shy. Consider this a last-resort technique.

Protect vulnerable livestock (poultry, sheep):

- It is much easier to prevent a problem than it is to stop one.
- Close and tightly fasten the doors to poultry houses.
- Fence areas (see previous section). Another fencing option is to create a temporary pen for confining livestock at night. Portable electric fences can be installed quickly and easily. They're usually made of thin electric wires running through polyethylene twine or ribbon, called "polywire" or "polytape."
- Herd livestock into pens and corrals at night and during foggy or rainy days, when coyotes are more likely to hunt. Also, bring livestock into sheds or paddocks when they're ready to have young. If a particular pasture seems to be more enticing to predators, move the livestock into a less vulnerable area.
- Keep pastures open. Clear away brush and weeds that provide cover for the coyotes. Remove brush and junk piles that attract rabbits and other prey that might entice coyotes into the pasture.
- Sheep or goats kept near cattle are less likely to be preyed on by coyotes.
- Keep herds and flocks healthy. Coyotes often prey on smaller, weaker lambs and calves.
- If possible, change to a fall lambing and calving season. This doesn't coincide with the period when coyotes are raising their pups (late spring through September), which is when they are most likely to prey on young livestock.
- Concentrate the lambing and calving period (using synchronized or group breeding) to reduce the amount of time that young, vulnerable animals are present in the herd or flock.
- Well-trained guard dogs (usually Great Pyrenees, Komondor, Anatolian Shepherd, or Akbash breeds) and donkeys and llamas help prevent predation on sheep. This technique works even better if you keep the livestock and guard animals within proper fences.
- Add lights to corrals. Strobe lights may work best to discourage coyotes.
- Propane exploders and sirens may give temporary relief.

- The USDA's National Wildlife Research Center developed a device for protecting sheep in fenced pastures. The "Electronic Guard" combines a strobe light and siren. (Or try the Critter Gitter, described in the scare tactics section.)
- Parking a car or truck in a pasture may frighten away coyotes for a short time. If the coyotes are hunting during the day, put a dummy in the car.
- Remove and properly dispose of livestock carcasses. Do not allow coyotes to develop a taste for livestock, live or dead! Carrion will attract coyotes to an area.
- Contact your regional wildlife agency staff or USDA APHIS–Wildlife Services for more information.

Trapping Strategies

The Association of Fish and Wildlife Agencies has published *Best Management Practices: Trapping Coyotes in the Eastern United States* (see: http://www.fishwildlife.org/Furbearer/Aug%202006%20BMPs/EasternCoyote_rev1.pdf).

- It takes experience and skill to trap coyotes effectively. Don't wing it! Get help before you start. Coyotes easily become trap-shy, so a botched trapping attempt will leave the customer with a worse problem. There are excellent courses, videos, and books that explain how to trap coyotes.
- Good trap locations: Look for coyote tracks and scat! Favorite spots include high hills; near isolated land features or isolated bales of hay (which they may use as scent posts); trail and stream crossings; fences; roads through pastures; deer trails; dry or shallow creek beds; near the border of fields, or tree groves, or pond dams; and near brush piles.
- Set traps to one side of the trail, not directly in the trail. Look for a place where coyotes would naturally stop, such as near a gate, on a hilltop, or where the cover changes.
- Place traps in an open flat area upwind of the path, to carry the scent of the bait (or lure, or both) across the coyote's normal trail. Bury foothold traps.
- In the fall and winter, a dirt-hole set may work best. Flat sets and urine post sets are generally more effective during the summer.
- The trap should be clean and not smell of people. Minimize human scent around the trap, too. Wear clean gloves and rubber footwear.
- Traps should be dyed and waxed, or cold-dipped in a commercial product.
- Coyote urine is a good lure for year-round use.
- Use baits and lures that have mild scents during the warmer months, because a strong-smelling bait or lure might prompt the coyote to roll on the trap. The trap might fire without capturing the coyote, which might become trap-shy as a result.
- Coyotes are very sensitive to irregularities in the ground, so make sure that foothold traps are firmly packed into the soil so they don't move. Otherwise, the coyote will probably realize the trap is there and avoid it.
- Block the back of the trap with sticks to guide the animal to the trap.
- Put lure on a stick to create an attractive scent post.
- Attach survey tape to a branch or bush using a clothespin to mark the spot where you set your lure or trap.
- Coyotes are powerful. Use strong, well-anchored, and well-balanced traps. But don't overpower your traps, because that could cause injuries.
- Options for anchoring traps: two stakes inserted crosswise, or disposable earth anchors with aircraft cable. "In-line" shock springs on the anchors will help prevent the coyote from lunging and pulling out of the trap.
- Many trap manufacturers and suppliers now offer versions of traps with the modifications suggested below, or will modify their models upon request. Or you may choose to modify your own traps.
- Nuisance coyotes should not be relocated, especially if the problem concerned a threat to human

safety. They might cause a problem in the new location. Once caught, the coyote is extremely likely to be trap-shy.

- If using a cable restraint, set the maximum loop size to 10–12 inches. The cable should be 10–12 inches above the ground. The cable restraint must be placed in an area where it cannot become tangled in plants, rocks, fences, logs, or equipment. Support and stabilize the cable restraint with 9-gauge wire and anchor it securely with earth anchors or a double-staked 30-inch metal rod.

Live traps that meet the BMP performance criteria:

- #1¾ coilspring trap: standard model; one with offset, laminated jaws; or one with wider, offset jaws
- #2 coilspring: standard model; or a 4-coiled model with offset, laminated jaws
- #3 coilspring: padded coilspring, 4 coils; or an offset double laminated coilspring, 4-coils

Lethal techniques:

- Hunting coyotes with a pack of trained dogs (usually larger breeds, such as greyhounds and some of the larger fox hounds). This is a more specialized and costly technique.
- Shooting, with a shotgun or accurate rifle, depending on conditions. May use light at night. The shooter(s) will need effective cover (camouflage, blinds) and must stay as still and quiet as possible.

  Two approaches:

  1. If the coyote's keeping to a regular schedule, sit and wait.
  2. Attract the coyote within shooting range using a predator call (howls or distress calls). Call when you're downwind of the coyote.
     - Howls, barks, and yips: you can try to mimic these sounds yourself, but they're harder to master than distress calls. There are recordings and commercial mouth-blown devices that will do the trick. Once you know what they mean, you may be able to use these sounds (perhaps in combination with a distress call) to locate the coyote or its den, or to draw a coyote into shooting range.
     - Distress calls (usually of prey, but some people use distress calls of competing predators or coyote pups): Most people use recordings or commercial mouth-blown devices. Coyotes respond because they hear an injured animal. They're coming in to eat it (coyotes are nearly always hungry).

Preferred Killing Methods

- Shooting, using a shotgun with BB-sized shot or larger, or a .22 caliber rifle, or a center fire rifle (target the head, if no rabies testing is required, or the heart/lungs)
- Lethal injection of barbiturate, if possible

Acceptable Killing Methods

- Stunning and CO_2 chamber (You can probably fit a coyote into a cooler-sized chamber. The animal must be restrained or unconscious.)
- Stunning and shooting
- Stunning and exsanguination (take care to reduce your exposure to body fluids)

Control Strategies that Don't Work Particularly Well

- Bells placed on sheep, and radios blaring in pastures, don't work well.
- Many repellents have been tested to keep coyotes from preying on sheep and cattle, but none have worked well. Coyotes rely on visual cues when hunting. Taste and smell are not as important, so even when the repellent offended the coyotes, it didn't keep them from killing the animals—just from eating what they'd killed.

- An interesting idea that hasn't panned out yet was to train the coyotes to avoid eating sheep by offering a bad-tasting bait that would sicken, but not kill, the coyote. The theory was that coyotes are territorial, so if you could train "your" coyote that livestock taste bad, it wouldn't kill any more livestock and would also keep other coyotes away. Inconclusive results.

For More Information

Association of Fish and Wildlife Agencies' publication: *Best Management Practices: Trapping Coyotes in the Eastern United States*, http://www.fishwildlife.org/Furbearer/Aug%202006%20BMPs/EasternCoyote_rev1.pdf

Baker, R. O., and Timm, R. M. 1998. Management of conflicts between urban coyotes and humans in Southern California, pages 299–312 in *Proceedings, 18th Vertebrate Pest Conference* (R. O. Baker and A. C. Crab, eds.). U. California at Davis.

Timm, R. M., Baker R. O., Bennett J. R., and Coolahan, C. C. 2004. Coyote Attacks: An Increasing Suburban Problem, pages 47–57 in *Proceedings, 21st Vertebrate Pest Conference* (R. A. Timm and W. P. Gorenzel, eds.). U. California at Davis.

FOXES

Species' Names

- Red fox, *Vulpes vulpes*
- Gray fox, *Urocyon cinereoargenteus*

Size

10–15 pounds. Body is 20–25 inches long. Gray fox is smaller.

Signs of Their Presence

- Sounds: May scream, yap, growl, or bark.
- Scats: Twisted, often contain hair or berries, deposited on rocks or logs (Figures B-4 and B-5).
- Odor: Their scent markings have distinctive odors. Red fox urine smells skunk-like, while gray fox urine smells musky.
- Tracks (Figures B-6, B-7, B-8, and B-9).

Diet

Opportunist. Foxes eat mostly mice, voles, bird eggs, rabbits, insects, and native fruits such as blueberries, blackberries, chokecherries, grapes, and black cherries. They'll also eat poultry, lambs, piglets, carrion, and cats. Both species will bury food for later use, especially around the den site, when they're raising their pups.

FIGURE B–4 Gray fox scat. *(Wayne Trimm)*

FIGURE B–5 Red fox scat. *(Wayne Trimm)*

FIGURE B–6 Gray fox tracks (not to scale). *(Wayne Trimm)*

FIGURE B–7 Gray fox track pattern (not to scale). *(Wayne Trimm)*

Typical Activity Patterns

Social style: Solitary except during the breeding season, when mated pairs and their young live as a family unit.

Daily activity: Both day and night, but generally most active at dawn and dusk.

Hibernator? No.

Migrates? No.

Where Found

Distribution: In rural and suburban areas and parks throughout the East.

Habitat: The red fox prefers woodland edges along open fields, meadows, and lawns, while the gray fox tends more toward forested and brushy areas. Typical den sites: red foxes may dig their own den or use an abandoned woodchuck burrow. Gray foxes often den in wood or brush piles, rocky outcrops, or hollow trees. Foxes will often reuse these dens from year to year.

FIGURE B–8 Red fox tracks (not to scale). *(Wayne Trimm)*

FIGURE B–9 Red fox track pattern (not to scale). *(Wayne Trimm)*

Territory and home range: Varies. Sometimes the red fox is aggressively territorial; other times it has overlapping home ranges. Perhaps this varies according to season, breeding cycle, habitat, or population density.

Breeding Habits

Pair bonding style: Polygamous and promiscuous. Among red foxes, males rarely help females raise the young, but male gray foxes may help raise the pups.

Sometimes, females will raise their young alone, and sometimes one male and several females den together, with more than one female raising young. Both species will cache food around the den for the pups.

Breeding dates: December through March, peaking in late January. Gestation takes about 52 days.

Litter size: Red fox: 4–9. Gray fox: 3–7. Most average 4–6 pups.

Birthing period: March through April. Most red fox pups are born in mid-March, with the gray foxes following a few weeks later.

Weaning dates: 2–3 months.

Amount of time young remain with parents beyond weaning date: Most disperse in the fall, between September and December.

Common Nuisance Situations

Time of year: Any time of year.

What are they doing?

- Their mere presence may frighten some people.
- Getting into the chicken (or turkey, duck, or goose) coop or yard. Foxes may take piglets, lambs, and small pets.
- In the spring, they may den underneath a porch or in a yard for a while, while they're raising their pups. Foxes generally use more than one den to raise their pups and may move them as many as two to four times, so this may be a short-term situation. They'll usually leave by the end of June at the latest. These dens aren't used during other seasons.
- Foxes (and coyotes) will chew holes in irrigation pipes in fields and orchards.
- Disease risks: Rabies, distemper. Red foxes, but not gray, get mange.

Debunking Myths about Foxes

- Foxes are much smaller than many people think. They're about the weight of a house cat (10–15 pounds).
- They rarely attack dogs or people (unless the fox is rabid). These are relatively small predators that usually hunt mice.
- Pups that are alone during the day have not necessarily been abandoned. Their parents are probably out hunting for food for them. (This is also true of coyotes.)
- A fox that's active during the day is not necessarily rabid. Most likely, it's a healthy animal that's feeding more often than usual, because of the demands of its young.

Best Practices

Do nothing: In some cases, the mere presence of a fox will be perceived as a nuisance. As mentioned above, foxes will sometimes den for a short time under a porch. They will leave on their own, usually by the end of June. Can you educate your customers and convince them to tolerate the situation? If you can't persuade your customers to wait until July, then nonlethal controls will likely work (e.g., harassment to scare the foxes away from the den site).

Scare them away: This is most likely to work when you have access to their den. It's easy to harass red foxes enough to convince them to move their pups (this also works with gray foxes and coyotes, although not as easily).

- Make noise near their den site. Run motors, shout, increase your activity in that area.
- Use slingshots or a shotgun loaded with rubber buckshot to frighten a fox that's farther away (45 feet). You don't have to hit it to scare it off, but if you can, that will more effectively train the fox to stay away (this refers to the use of the less lethal projectiles mentioned above).

- Experiment with noisemakers, such as bangers and screamers. You may be able to use predator calls (such as the Coyote Howler) to frighten foxes away from their dens.
- One scare device, the Critter Gitter, combines a siren and flashing lights. It's triggered by a motion detector. The device switches patterns, so it should be effective longer than a scare device that doesn't vary.
- WCOs with a commercial pesticide applicator license: Use a repellent containing coyote urine around the den site. Then cause a disturbance: cut back the brush, flatten the dirt mound in front of the den.

Remove artificial food sources (garbage, compost, pet food):

- If anyone is feeding the foxes, persuade them to stop.
- "Fox-proof" garbage cans or dumpsters with a tight-fitting lid. Secure garbage cans with heavy-duty straps or bungee cords, or attach them to a post, or keep them out of reach in the garage (close garage doors at night), or place the cans in a covered and locked bin.
- Enclose compost piles in a framed box using hardware cloth or welded wire; in a sturdy container, such as a 55-gallon drum; or in a commercial compost container.
- Feed pets indoors. Any food left outdoors should be removed at night. Bring pet food dishes inside, too.
- Remove and properly dispose of livestock carcasses.

Keep foxes out of yards, prevent them from denning under porches and decks, and protect vulnerable livestock:

- Close and tightly fasten the doors to poultry houses.
- Bring livestock into sheds or paddocks when they're ready to have their young.
- Fences will help a great deal, but foxes will try to climb over, burrow under, or squeeze through fences. So build them right and keep the fences well maintained. You can fence the whole yard, or just the area around the vulnerable spot.
- Use a 2-inch net-wire fence, hardware cloth, welded wire, or galvanized sheet metal. (You may wish to recommend the use of smaller mesh because this will also prevent other animals, such as skunks and woodchucks, from gaining access.) Create a "rat wall." Attach the hardware cloth to the bottom of the deck. Then bury the bottom of your wall 6–12 inches, with a foot-wide shelf that sticks out, to prevent animals from digging underneath the barrier (this will look like the letter L).
- Three-wire electric fences, with wires spaced at 6, 12, and 18 inches, can repel foxes.
- Combination woven-wire and electric wire fence: If there's an existing woven-wire fence, add an electric strand that's placed 6–8 inches above ground and 8–10 inches outside the fence, using an offset bracket. If the foxes are climbing over the fence, add a top wire that's also outside the fence.
- Invisible fences may be an attractive solution for keeping pets within an area but they will not keep wildlife out. To protect pets from predators, switch to a traditional fence design.

If young are present, remove the entire family before blocking the entrance to their den:

- See fur or feathers outside the den during the early summer? That shows the adult fox is bringing food to young pups. Can you wait it out? They'll leave on their own once their pups are old enough (usually 12–14 weeks old).
- Cage traps may be effective in capturing young foxes. Place the trap near the den's entrance.
- A device called a "mechanical wire ferret" may help you chase the young out of the den. It's a long piece of steel wire that has a spring and wooden plug on one end, and a handle on the other. Twist the spring through the den to flush out the foxes so you can capture them directly

using a dip net, net bag, or cage trap as they leave the den. The mechanical wire ferret can become entangled in a young fox's fur, so use it carefully.

- If young are separated from their parents, you can place the pup in a different litter. As long as that female is nursing, she'll probably adopt the pup.

Trapping Strategies

Live capture:

Cage traps

- You may be able to use a cage trap to capture a young red fox in an urban area, or to remove young from a den, but don't expect to catch adult red foxes in cage traps.

Foothold traps

- Trapping foxes requires great skill. The amateur is likely to simply educate the fox, thus making it difficult for even experienced trappers to succeed later.
- Foxes are very sensitive to irregularities in the ground, so make sure that foothold traps are firmly packed into the soil so they don't move. Otherwise, the fox will probably realize the trap is there and avoid it.
- Block the back of the trap with sticks to guide the animal to the trap.
- Lure smeared on a stick creates an attractive scent post.
- Attach survey tape to a branch or bush using a clothespin to mark the spot where you set your lure or trap.
- Set foothold traps along trails or at entrances to fields.
- In the fall and winter, a dirt-hole set may work best. Flat sets and urine post sets are generally more effective during the summer.
- Foothold traps should be dyed and waxed, or cold-dipped in a commercial product.
- For adult foxes, use foothold traps, #1½ or #1¾ coilspring. For pups, use a #1 coilspring. Offset, padded, or laminated jaws may reduce injuries.

Cable restraints

- Set the maximum loop size to 6–8 inches. The cable should be 6–8 inches above the ground. The cable restraint must be placed in an area where it cannot become tangled in plants, rocks, fences, logs, or equipment. Support and stabilize the cable restraint with 9-gauge wire and anchor it securely with earth anchors or double-staked 30-inch metal rod.

Note: Cable restraints are not legal in all states, so check your local regulations.

Preferred Killing Methods

- CO_2 chamber (if caught in a cage trap, simply place the trap in the chamber; if caught directly, using a catchpole, for example, stun the fox, then transfer it into the chamber)
- Shooting, using a shotgun with #6 shot or larger, or a .22 caliber rifle (target the head if no rabies testing is needed, or the heart/lungs). Especially with gray foxes, may be able to successfully call the fox into the area with a predator call.
- Lethal injection of barbiturate, if possible

Acceptable Killing Methods

- Stunning and chest compression (rabies risk: handle them carefully)

Control Strategies that Don't Work Particularly Well

- Noisemakers, such as propane cannons, radios, and tape recordings, may convince the foxes to move to a new den, but these techniques aren't generally effective in preventing livestock depredation.
- The same holds true for flashing lights such as strobe lights or beacons. They're most effective in an enclosure or small area, when used intermittently with other techniques.

MICE

Species' Names

- House mouse, *Mus musculus* (an exotic species)
- Deer mouse, *Peromyscus maniculatus* (native)
- White-footed mouse, *Peromyscus leucopus* (native)

Size

Under one ounce. Body is 2–3½ inches long, excluding the 3- to 4-inch-long tail. The best way to tell these mice apart is to capture them. The house mouse has grayish brown fur and a nearly naked tail. Both wild mice are two-toned, with reddish-brown backs and white bellies. They have furry tails (also two-toned).

Signs of Their Presence

- Sounds: Squeaking, scuttling, or gnawing inside the walls, ceilings, or between floors of buildings.
- Scat: ⅛–¼ inches long, slender. Found in kitchen cabinets, drawers, and corners; on counters, under sinks, and near food. Urine stains on the woodwork can be seen with an ultraviolet light. If you notice a musty odor, the urine was from a house mouse.
- Trails: Mice use the same route over and over. Eventually, a faint, dark "trail" of body oil and dirt may be noticeable on the baseboards. Look as well for smooth, worn paths in insulation. Their burrow holes, often seen in baseboards, corners, walls, and foundations, are usually about an inch in diameter. They can fit through a dime-sized hole.
- Nests: A loosely woven ball of shredded fibers, such as clothing or paper, about 5 inches in diameter. Deer and white-footed mice line their nests with fur, feathers, cloth, and other fine materials. Droppings, seeds, and pet food are often found near the nest. If possible, mice will choose to nest near food.
- Damage to household goods and buildings: tooth marks on packaged goods, electric cables, baseboards, door and window casings, and cabinets. (Their teeth grow constantly, so they gnaw to keep them worn down.) Piles of wood chips or shavings.

Diet

Mice are primarily herbivorous, although during the summer they may eat mostly insects. The house mouse will sample everything in your larder, including your lard—and perfumed bar soap, too. They prefer seeds, grains, and grain products, like bread, but are happy with cheese, peanut butter, bird seed, potatoes, and pet food. They'll eat chocolate, bacon, and other foods that are high in sugar, protein, or fat. Deer and white-footed mice are mostly seed-eaters. They're more likely to stash acorns, the scales of pine and spruce cones, and beech nuts between wall studs.

Typical Activity Patterns

Social style: House mice are generally solitary, except for females with young. Mated pairs of deer and white-footed mice usually stay together during the breeding season.

Daily activity: Nocturnal, with peaks at dawn and dusk. Daytime activity is seen.

Hibernator? No, but mice living outdoors may den for a few days in very cold weather.

Migrates? No.

Where Found

Distribution: Throughout the East.

Habitat: The house mouse prefers a house, cabin, garage, barn, attic, shed, office, warehouse, or similar locale. Deer and white-footed mice prefer forested or brushy areas. White-footed mice spend a lot of time in trees, and will take over abandoned bird or squirrel nests. Other preferred nest sites include small pre-existing burrows, brush piles, the knotholes of trees, bird boxes, attics, and spaces under rocks or logs.

Territory and home range: House mice tend to stay close to their food, ranging up to 10–30 feet in diameter. In contrast, the wild mice tend to forage outside, covering from ⅓ to 4 acres, but will nest indoors when shelter is available. Deer and white-footed mice are probably territorial around their nests, and avoid contact with each other within their home ranges, although they nest together in family groups during the winter. They may establish a dominance hierarchy. Males' home ranges may overlap; females' apparently don't. Females in mixed habitats tend to have smaller but better-quality home ranges, and those females can be larger than males.

Breeding Habits

Pair bonding style: Mice are polygamous. Females raise the young alone.

Breeding dates: Indoor residents may breed year-round, but wild mice tend to breed in the spring and fall. They may have 4–10 litters per year. Gestation takes about 23 days. Females may breed again within a day or two of giving birth. Mice are sexually mature at 6–10 weeks old.

Litter size: 3–5. May see as few as 2, or as many as 8. A female house mouse can produce over 40 young in a year.

Weaning dates: Between 2–4 weeks of age.

Amount of time young remain with parents beyond weaning date: Not long!

Common Nuisance Situations

Time of year: Any time of year.

What are they doing?

- Mice can damage buildings and household goods as they seek food and nest sites. They'll gnaw through or foul woodwork, aluminum siding, plasterboard (Sheetrock), insulation, plastic food containers (including garbage cans), papers, packaged goods, clothing, mattresses, furniture, and even lead or copper pipes. Indoors, they nest in walls, kitchen cabinets, attics, basements, garages, sheds, barns, and under appliances. Outdoors, they'll nest in thick vegetation, wood or rock piles, and junk.
- Their nests might block a vent, causing a fire hazard.
- They also chew on wires, which in addition to creating a fire hazard could also short-circuit electrical systems, causing failures of alarm systems or refrigeration.
- Mice eat many human foods but prefer seeds and cereal grains. In barns and outbuildings, they get into stored grains, corn, feeds, and seeds. They'll raid bird feeders and pet dishes. They damage much more than they eat because they tend to nibble and discard foods.
- Scurry about at night. The noise may drive you and your pets to distraction.
- May dig up recently planted seeds in home gardens, agricultural fields, and areas that were supposed to be reforested.
- They contaminate stored foods, especially grains. Mice ruin a good portion of the world's food supply.
- Foul items stored in warehouses, museums, libraries, and other sites.
- Disease risks—among the possibilities:
    - for all three species: Lyme disease, salmonellosis (food poisoning), leptospirosis (Weil's disease), typhus, rat bite fever, ringworm, tapeworms
    - for deer and white-footed mice: hantavirus pulmonary syndrome
    - for house mouse: lymphocytic choriomeningitis (LCM) virus, rickettsial pox

Best Practices

The best way to deal with a mouse infestation is to clean up, get rid of the mice, and keep them from finding a way back in. Keep three words in mind: *sanitation, eviction, exile.*

Safety tips for contact with mice, their droppings, urine, or nest materials

- Wear rubber gloves and a proper respirator.
- Ventilate the area, if possible.
- Don't stir up dust by vacuuming or sweeping. The dust could contain hantaviruses, LCM virus, or other airborne disease organisms.
- Instead, thoroughly wet the materials with a 10% bleach solution (one part bleach in nine parts water) or household disinfectant. Wipe up with a damp sponge.
- Spray dead mice and their nests with disinfectant, then double-bag for disposal.
- Disinfect toys, silverware, or other items that may have been fouled. Throw away any contaminated foods, drinks, napkins, paper plates, or cups.
- For more information, refer to Chapter Four and the resource list.

Remove artificial food sources (garbage, compost, bird seed, pet food):

- Store food, bird seed, pet food, garbage, compost, and recyclables in metal, glass, or ceramic containers.
- Promptly clean up spills and crumbs.
- Feed pets at scheduled times, and then put unfinished food in the refrigerator. (Relax—with a proper trapping and exclusion program, you won't have to do this for too long.)
- Keep the area around bird feeders clean, especially of spills underneath the feeder. Feed birds only during the winter months.
- Clean garbage cans, dumpsters, and garbage chutes regularly. Screen dumpster drainage holes with hardware cloth.
- Elevate compost heaps or enclose with quarter-inch hardware cloth or welded-wire mesh.

Remove their nesting sites:

- Keep stored items off the floor and away from walls. In a warehouse, paint a 12-inch white band on the floor all the way around the room to make inspections easier, and to remind people to keep items away from the walls.
- Reduce clutter and remove cardboard boxes.
- Move firewood, debris piles, and garbage cans away from the house.
- Maintain a foot-wide gravel border around the foundation that's free of vegetation (best), or keep the foundation plantings well trimmed. Don't stack anything (such as firewood) against the foundation.

Prevent entry into building:

- Close the door! (Use screen doors.)
- Install door sweeps and weather-stripping (in garages, too).
- Repair holes in walls and screens. Poke steel wool, wire mesh, or flexible aluminum "gutter guard" into holes and then caulk, or spray expanding foam (such as Great Stuff) over it to strengthen the barrier.
- Use latex caulk to plug gaps around water, gas, and heating pipes, heat registers, air ducts, electrical chases, and false ceilings.
- For large holes around pipes, use galvanized metal pipe chase covers, sheet metal plates, mortar, plaster of Paris, or cement.
- Seal openings beneath and behind sinks, stoves, and dishwashers with latex caulk.
- Check vents (sewer, stove, clothes-dryers, roof, ridge-line, soffit, furnace ducts, and attic fans).

If the vent is damaged, replace it with an animal-proof design, or screen it with quarter-inch hardware cloth. End caps on ridge vents may loosen, providing access to the attic. Soffit vents are best protected with metal louvers.

- White-footed and deer mice are excellent climbers. They will often enter buildings through holes in the roof, even on two- or three-story buildings. Inspect thoroughly.

Protect valuable trees and shrubs:

- Commercial tree wrap or hardware cloth will protect young trees.
- Plant seedlings instead of seed.

Trapping Strategies

The house mouse is an exotic species, so please do not release any into the wild (Chapter Two explains why). Deer and white-footed mice, which are native in the East, can be live-trapped and released. If their entry points have been sealed they shouldn't be able to get back inside the building.

- To increase your success, trap intensively for several days. More is better.
- Place the traps in their runways, in corners, near food sources, nests, or holes—wherever the mice are most active. You may be able to lift some ceiling tiles to place traps in a dropped ceiling. Set traps at night, when mice are most active. Check the traps in the morning.

Live traps:

- Various models of small cage and box traps, such as Sherman traps.
- Multiple-capture traps, which hold the mice in a compartment.
- Set live traps parallel to the wall.

Lethal traps:

- The familiar mouse trap is a type of body-gripping trap called a "snap trap." There are many designs.
- The Quick Kill Mouse Trap made by Victor has a lid over the bait cup (Figure B-10). Only animals that are motivated to seek the bait will lift that lid—and that's what fires the trap (the "lid" is actually the trigger). This means that an animal can accidentally step on the lid without setting off the trap. There's another advantage to this design. The bait cup is located to position the mouse in the perfect strike position. So this trap is both more selective and more effective than the traditional mouse trap. It also has a safety catch and will not fire if it's picked up.
- Other snap traps have an expanded "trigger" (properly called the "pan") or a clothespin design (Figure B-11). These features make the traps easier to set than traditional mouse traps.

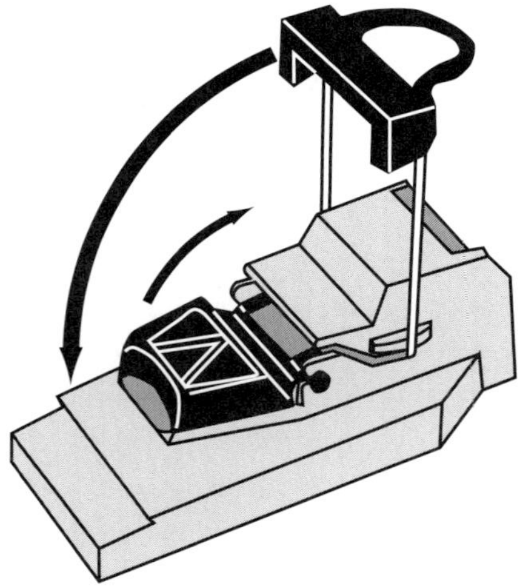

FIGURE B–10 Victor Quick-Kill Mouse Trap. *(Susan MacKay)*

FIGURE B–11 A snap-back trap featuring the clothe-spin design. *(Susan MacKay)*

- Place traps right against the wall, every 5–6 feet.
- Set snap traps in pairs. This is much more effective. Two sets work well:
  - Side-by-side, perpendicular to the wall, with the trigger snapping towards the wall
  - Parallel to the wall, with the trigger snapping to the outside (not into the center)
- Traps may be attached to rafters with nails, or to pipes with wire or Velcro strips.
- Bait with peanut butter, bacon, dried fruit, nutmeat, chocolate candy. Or tie a cotton ball to the trigger (it's enticing nesting material).
- To protect young children, place lethal traps inside a cage with 1-inch mesh, or in a coffee can with both ends cut out, or in PVC pipe. (Make sure the trap will still fire.) Wildlife rehabilitators may appreciate donations of mice, which are used to feed some snakes, birds of prey, and other animals. If you're going to do this, don't spray the mice with disinfectant and be sure that no poisons (rodenticides, poisonous tracking powder) have been used in the building during previous control efforts. You can double-bag the mice and freeze them.
- The use of glue boards, which the authors do not consider a best practice, may be warranted in the case of a *severe* infestation. A glue board is a layer of long-lasting adhesive spread over a surface, usually cardboard or plastic. The mice get stuck in the adhesive (glue boards are used to capture mice, rats, and snakes). Although some call the glue board a live trap, it's not often used that way. In fact, some biologists believe that you cannot remove the animal from the trap unharmed because the oil that is used to loosen the glue may harm the animal. In practice, mice are frequently left to die on glue boards. Snap traps are often as effective as glue boards and are more humane, although setting them does take more effort. If using glue boards, set them in protected areas, such as within a dropped ceiling. Check them frequently (at least every 12 hours) and kill any captured mice by stunning them. Do not leave dead mice on glue boards because the carcasses will stink and likely attract other pests.

Other lethal techniques, for WCOs with a commercial pesticide applicator license:

- Toxicants (in various forms, such as baits, fumigants, and tracking powder) can be effective and may be warranted in some situations. Rodenticides can be hazardous to children, pets, and animals that eat the poisoned mice. The mice may die in walls and stink, while also providing a breeding place for flies. Trapping is often a better solution.
- There's a fair amount to consider when using rodenticides. You'll learn the details during your pesticide applicator training.

Preferred Killing Methods

- Lethal trap
- Cervical dislocation
- Stunning, and if necessary, chest compression
- Carbon dioxide chamber (use a small chamber)

Acceptable Killing Methods

- Pesticides (for those with proper license)

Control Strategies that Don't Work Well

- Ultrasonic devices don't work against mice. Loud or unusual noises will frighten them and may drive them off for a short time.
- Electromagnetic devices don't work, either.
- Mothballs and ammonia don't do much, either, and are not EPA-approved for mouse control.
- Cats may kill some mice, but a single female house mouse can bear up to 40 young in a year. The mice also travel in places that cats can't reach, so don't expect success.
- Glue boards will capture mice but they can cause other pest infestations if used improperly. There are more humane options that are effective.

MOLES

Species' Names

- Hairy-tailed mole, *Parascalops breweri*
- Star-nosed mole, *Condylura cristata*
- Eastern mole, *Scalopus aquaticus*

Size

Depending on the species, 1–5 ounces. The hairy-tailed and star-nosed moles are about 5–5½ inches long, including the short tail, while the eastern mole is about 3¼–8¾ inches long. The snout of a star-nosed mole is ringed with 22 small, pink, fleshy projections that make it look like it has a sea anemone on the tip of its nose.

Signs of Their Presence

- Tunnels, or "runs," in the soil or lawn. (Runs of star-nosed moles are usually deeper and less noticeable than those of hairy-tailed moles, except, at times, in well-watered golf courses.) These tunnels are seen most often in the spring and fall, when the soil is moist, soft, and easy to dig. Moles make two kinds: feeder and travel tunnels. Feeder runs look like a long, squiggly, rounded ridge that's about two inches wide. Feeder tunnels tend to be short and very crooked, because if a mole finds an area that's full of food, it tends to dig all around, feeding. They'll abandon these runs when there's not much food left in them. Dead grass over the run is usually a sign of an old, abandoned run. (Moles don't eat grass, but they may loosen the roots from the soil, which can kill it.) Their travel tunnels are usually long and straight and often follow an edge, such as a driveway, fence, or foundation. Look for travel tunnels that continue into wooded areas because these will be the best spots in which to set traps.
- Molehills (also called boils or mounds): small, cone-shaped piles of soil that are usually just a few inches high and anywhere from a few inches to a foot wide. They vary in size. Often seen in the late fall, as the moles prepare for winter by digging deeper tunnels that are under the frost line. At that depth, the moles can't toss up the soil as they go, which is what they do when they're near the surface. So they'll usually dig forward for a while, then stop, and carry the soil up to the surface where they dump it, creating the molehill. Moles also dig deep tunnels in the summer when the soil is dry. Then, they're seeking earthworms, one of their favorite foods.
- It's unlikely you'll see or hear moles, or find scat or tracks because they spend their time underground. Although they feed and travel in the shallow, surface tunnels described above, they find shelter and raise their young in deeper tunnels that could be 6–24 inches below ground.
- Moles are often accused of crop damage that was actually caused by voles. Although moles rarely eat roots, their tunnels may damage them. So how do you tell a mole from a vole?

Moles have:	Voles have:
Very small eyes	Small eyes
No external ears	Small, but definitely noticeable ears
A naked, pointy snout	Furry noses
Large front feet that are turned sideways, and big claws (excellent shovels)	Small, mouse-like feet

Diet

Mostly insects. Grubs, beetle larvae, earthworms, and some carrion. Occasionally, frogs and mice. Star-nosed moles may catch minnows. They must eat 70–100% of their body weight each day to have enough energy to burrow. Occasionally, they'll eat seeds, roots, or bulbs.

Typical Activity Patterns

Social style: Hairy-tailed moles are solitary, except briefly while mating. Star-nosed moles are thought to live in colonies.

Daily activity: Moles are most likely active throughout the day and night. They need to eat a lot to keep up their energy levels.

Hibernator? No. They simply move deeper into the soil, tunneling below the frost line.

Migrates? No.

Where Found

Distribution: Throughout the region.

Habitat: Lawns, meadows, orchards, and woods with moist, loose soil. Hairy-tailed moles prefer loamy, sandy soils well covered with plants and avoid wet, dry, or heavy clay soils. Star-nosed moles prefer swamps, bogs, and low, wet meadows (they've even been seen swimming under ice in the winter) but can manage in somewhat drier locales.

Territory and home range: Territorial. Two moles usually fight when they meet, except during the mating season. The home ranges of male and female moles overlap, but the home ranges of the females do not seem to overlap with those of other females. Some tunnels overlap territories and are used like highways by two or more moles. The males range over about two acres, females over a half-acre.

Breeding Habits

Pair bonding style: Polygamous. Female raises the young alone in an underground nest chamber lined with leaves and grasses. The nest chamber is usually found in a deeper tunnel, perhaps as far as two feet underground.

Breeding dates: Late February to March. Gestation takes about 42 days.

Birthing period: April to May.

Litter size: 3–7.

Weaning dates: Between 4–5 weeks of age.

Common Nuisance Situations

Time of year: Spring (April–May) and fall (September–November), when surface soil is moist and easy to dig, and grubs and worms are nearest the surface. You may receive a few calls as soon as the snow melts, which reveals old damage, but should wait to see if there are still moles present. After the snow melts, the runways seen above ground in turf are often the result of meadow vole activity, not moles.

What are they doing?

- While helping rid lawns, gardens, and golf courses of grubs, moles create unsightly runs. Their tunnels disfigure lawns and can wreak havoc in a garden.
- Disease risks: Almost none.

Debunking Myths about Moles

- Moles are often mistaken for voles, mice, and shrews. If in doubt, check your field guides.
- Many people believe that there's a mole in every tunnel they see. The good news is that even though you may see dozens of tunnels, there are probably only a few moles in the yard. Possibly only one or two. Really! Moles dig fast: about 18 feet/hour. They may be able to tunnel 100 feet a day or more, depending on soil conditions. You may think your lawn is full of moles, when it's just the home of a few, very busy little guys.

Best Practices

First, decide whether this is really a problem or not. Moles eat a huge number of grubs that damage lawns and gardens. Is the sight of the tunnels tolerable? If not, trapping is currently considered most effective, but repellents, exclusion, and habitat modification techniques may also contribute to an effective strategy, and may be preferred by some of your customers.

Protect vulnerable plants and lawns:

- Small areas can be fenced with hardware cloth or sheet metal. The fence should be two feet high, buried a foot deep, with the bottom edge bent outward into an L-shaped shelf that sticks out a foot. This should form a 90° angle. This keeps the moles from burrowing under the fence.

Make the area less attractive for moles: Moles prefer wet, low areas that are rich in grubs. Moles follow their food sources, so if there are fewer grubs, the moles may move on—just remember that moles eat worms and other foods, too.

- Don't overwater your lawns.
- Improve soil drainage and try to eliminate low spots.

Trapping Strategies

Lethal traps:

- There are several effective lethal traps for moles, including harpoon-shaped or scissor-jawed traps. A newer model, the NoMol trap, doesn't contain a spear or heavy springs so you may find it easier to use.
- Trap in the spring or fall, when the soil is moist and the moles are closer to the surface.
- If the lawn is so dug up that you can't tell the feeder tunnels from the travel tunnels, roll it flat (if it's small, walk it flat). Flag the area so you can find it easily, then watch for a few days. If the flattened area is raised again, you're looking at an active run.
- Watering a dry lawn will entice worms and moles closer to the surface, where the moles will be easier to catch.
- Set multiple traps. If you can't choose between locations, set traps in both.
- Set locations: Best: an active travel tunnel that extends into a wooded area. Good: any active travel tunnel, or a molehill. Questionable: feeding tunnels. The moles may not return to them.
- Place two traps in each tunnel, one in each direction.
- Homeowners can help you by checking lawns daily for new damage.
- Check traps frequently. If the mole is still alive, remove the stake carefully and grasp the wire to pull out the trap. Use a spare NoMol trap to kill the mole. Slide the trap's arms so that the jaws are just behind the mole's front feet, then release the tongs.

Preferred Killing Methods

- A lethal trap

Acceptable Killing Methods

- Pesticides (gel, gummy, and grain-based baits), for those WCOs with commercial pesticide applicator licenses. Baits often don't work well because moles are looking for live prey: earthworms and grubs.

Control Strategies that Don't Work Particularly Well

- Chewing gum, mothballs, thumpers, ultrasonic devices, windmills, and flooding the tunnels—none have proved effective.
- Grub treatments (insecticides) may get rid of the grubs in your lawn, but there will still be plenty of worms for the moles to eat. And the insecticides may not work well in heavy, clay soils, anyway. It's just not a good strategy for discouraging moles.
- Grain-based baits (containing zinc phosphide) don't work that well because moles don't normally eat grain. If they aren't attracted to the bait, they're not likely to ingest the poison.
- Borders of marigolds are thought to repel moles but haven't been tested.
- Castor-oil based repellents haven't been well studied yet, so their effectiveness is unknown.

OPOSSUM *(Didelphis virginiana)*

North America's only marsupial (mammals whose young develop in a pouch). They're more closely related to kangaroos and koalas than to the other animals in the neighborhood!

Size

From 4 to 14 pounds. Body is 15–20 inches long. They often suffer frostbite and lose part of their tails and ears.

Signs of Their Presence

- Sounds: Growl, hiss, screech when threatened.
- Evidence of their feeding: Eggs that have been chewed into many small pieces. (Raccoons usually remove one end of the shell without crushing it. Foxes carry eggs away. Weasels and mink crush the entire egg.) Opossums maul chickens beginning at the rear, while raccoons bite their heads off.
- Tracks: Look like they were made by little human hands, fingers spread wide apart (Figures B-12 and B-13).
- Scats: Are semi-liquid and don't last long. Left everywhere, even in the den. When scared, possums may secrete a smelly, greenish fluid out of their anus (Figure B-14).

Diet

Opportunist. Opossums eat mostly meat (mainly insects or carrion) but they also eat many plants, especially fruits and grains. They may eat garbage, compost, pet food, bird seed, bird eggs, and young birds (turkeys, chickens, geese, and game birds). They also eat voles, shrews, worms, and toads.

Typical Activity Patterns

Social style: Solitary.

Daily activity: Usually nocturnal.

Hibernator? No, but does den up for days at a time when the weather is bad.

Migrates? No.

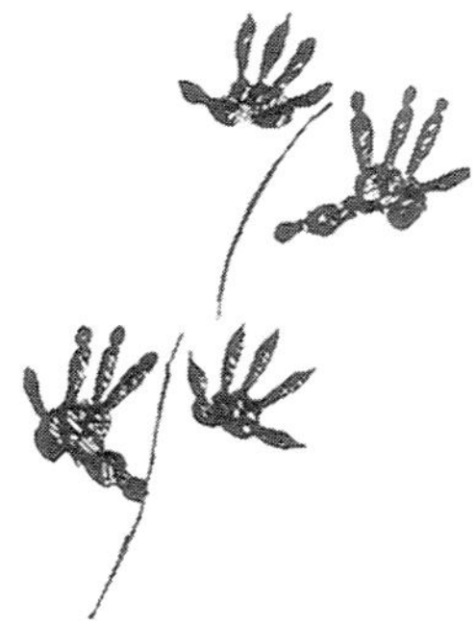

FIGURE B–12 Opossum track pattern. *(Wayne Trimm)*

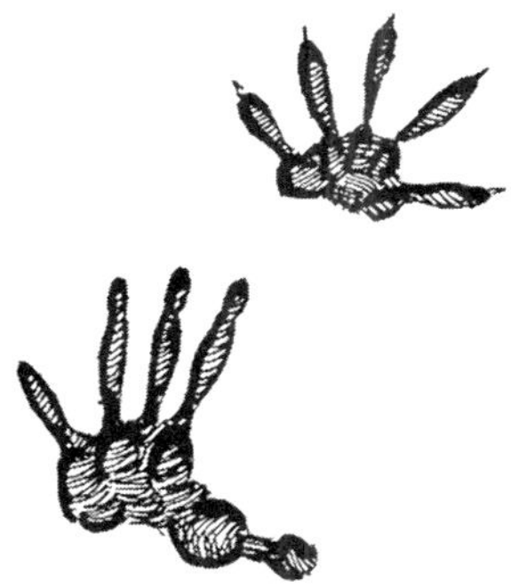

FIGURE B–13 Opossum tracks, close-up. *(Wayne Trimm)*

FIGURE B–14 Opossum scat. *(Wayne Trimm)*

Where Found

Distribution: Throughout the region.

Habitat: Wide ranging—arid to moist, woodsy to open, but more common near streams and swamps. Dens in a different place three out of four nights (except in the cold of winter). Finds shelter under buildings, in brush heaps, hollow logs or trees, old crow or squirrel nests, and rock crevices. Opossums may share quarters with woodchucks, skunks, and rabbits.

Territory and home range: Not territorial. They have constantly shifting home ranges and may be considered nomadic. Home range is usually 10–50 acres.

Breeding Habits

Pair bonding style: Polygamous. Females raise the young alone.

Breeding dates: February through June. Most females, though, have just 1–2 litters per year. The young are born about 13 days after breeding.

Litter size: 6–16, average 8.

Life in a pouch: The tiny (about ½ inch long) young are born blind and helpless. They must crawl into the mother's pouch and attach to a nipple. They'll remain in the pouch for 7–8 weeks, firmly attached to that nipple. Then, for about 2 weeks, they'll begin to explore the world, often riding on the mother's back. They'll return to her pouch to nurse. They're weaned at about 3 months old and are generally fully independent by the time they're 7 inches long.

Amount of time young remain with parents beyond weaning date: 3–4 weeks.

Common Nuisance Situations

Time of year: Any time of year.

What are they doing?

- Raid gardens, chicken coops, bird feeders, pet food, and garbage.
- Sometimes den in garages or attics and make a mess.
- A parasite found in the feces of opossums can contaminate water and food sources for horses (both hay and feed). This parasite can transmit a disease to horses, called equine protozoal myelitis. This disease affects the nervous system and can cause lameness.
- Disease risks to people: mange, rabies (rarely).

Debunking Myths about Opossums

- A hissing or drooling possum is not necessarily rabid. When threatened, a healthy opossum may bare its teeth, make a lot of noise, drool, bite, or leak a nasty fluid out of its anus. Stress may cause them to play dead, which might confuse predators and prevent the possums from being eaten.

Best Practices

Relax:

- Opossums move around a lot, and usually don't stay in one den site. If the problem was caused by an individual possum, it will probably leave on its own. Just realize that the problem could be caused by different animals, which are all attracted by the same source of food, water, or shelter.

Remove food sources and shelter:

- Put trash out in the morning instead of the evening.
- Possum-proof garbage can with a tight-fitting lid, or secure it with straps.
- Don't leave pet food out at night.
- Enclose compost piles in a framed box using hardware cloth; in a sturdy container, such as a 55-gallon drum; or in a commercial compost container.
- Keep the area under bird feeders clean.
- Remove brush piles and debris.
- Close garage doors at night.

Protect vulnerable livestock:

- Close doors to poultry houses, and if birds are caged, keep those doors closed, too.
- To keep opossums from climbing over a wire mesh fence, install a tightly stretched electric wire near the top of the fence, about 3 inches out from the mesh.
- Install an electric fence around the hen house, or use hardware cloth to cover holes and potential entrances.

Trapping Strategies

Live traps:

- Opossums are easily caught with cage traps.
- Foothold traps (#1 or #1½) are also effective.

- Set traps along fence rows or trailways in a dirt hole, cubby, or running pole set.
- They prefer slightly spoiled baits, such as cheese or fruit. If you use a box trap with these baits, you may also capture skunks so be prepared to release them.
- They're slow, so it's possible to capture them by hand, or with the use of a catchpole. Grasp the end of the tail (wear heavy gloves because they have sharp teeth). If you're holding a possum and it tries to climb its tail to reach and bite your hand, lower it to the ground, where it will attempt to crawl away.
- Assume that a female opossum has young in her pouch during the rearing season (March–August). The females are not likely to retrieve young, so make sure that all her babies are either in her pouch or clinging to her before you release her.

Lethal traps:

- Body-gripping trap, #120 or #160, set in a vertical cubby for greater selectivity (see Chapter Five for details).

Preferred Killing Methods

- CO_2 chamber
- Lethal trap
- Lethal injection of barbiturate, if possible
- Shooting, using a shotgun with #6 shot or larger, or a .22 caliber rifle (heart/lungs shot is preferred). Why is just the heart/lungs shot listed as preferred? The head shot is difficult because possums have very small brains located in a relatively large skull—and there's a strong crest on their skull, which can deflect the bullet. See Figures B-15a and B-15b for more information about the head shot.

Acceptable Killing Methods

- Gunshot to the head (this is a difficult target and should be attempted only by WCOs who are more experienced and skilled in the use of firearms)
- Stunning and chest compression
- Stunning and exsanguinations

FIGURE B–15a

FIGURE B–15b

FIGURE B–15a and 15b Properly targeting a head shot for an opossum is challenging because their brains are much smaller than you'd guess, considering the size of their heads. When viewing the opossum from the side, imagine a line between the eye and the ear. Now, aim slightly below that, closer to the base of the ear. Position the gun very close to the head. *(Karen English)*

RABBIT, EASTERN COTTONTAIL *(Sylvilagus floridanus)*

Size

From 2 to 4 pounds. Body is 14–18 inches long.

Signs of Their Presence

- Visual sighting of animal.
- Scat: ⅓ inch in diameter, round to somewhat saucer-shaped pellets. One rabbit leaves 250 to 500 pellets a day. Like hares, voles, and beavers, they eat their feces to extract more nutrients from grasses and tree bark, which are difficult to digest.
- Damaged garden crops, shrubs, and trees. Usually you can tell if this damage was caused by a rabbit, vole, woodchuck, or deer. Rabbits attack smooth bark and gnaw in patches. Their toothmarks are a little less than an inch wide—wider, but less distinct, than the voles'. They often clip twigs, branches, and berry canes with a clean 45°-angle cut. Deer, on the other hand, lack upper incisors, so they leave ragged edges when they browse on branches.
- Tracks: Seen in groups of four. The tracks of the back feet actually imprint ahead of the front feet because rabbits leap, pushing off from their front feet. The front track is almost round, about 1 inch wide, the hind track is about 3–4 inches long and oblong.
- Sounds: Usually quiet, other than a high scream of distress when attacked, the grunt of the mother when her nest is approached, or the high squeal of a female during mating.

Diet

Herbivore. In the winter, they often eat the bark, twigs, and buds of ornamental shrubs and fruit trees because everything else is covered by snow. In the spring and summer, their big menu includes vegetables, legumes, field crops, flowers, and other succulent green plants.

Typical Activity Patterns

Social style: Mostly solitary, although they may have an informal social network.

Daily activity: Nocturnal, and crepuscular. May feed during the day in summer, under or near thick cover.

Hibernator? No.

Migrates? No.

Where Found

Distribution: Throughout the region.

Habitat: Prefer brushy fence rows, field edges, overgrown pastures, sapling stands, and shrub or perennial borders in landscaped backyards. They don't need a water source because they can get what they need from snow or dew. Can reach densities of 3–10/acre; more, if the habitat is favorable (suburban densities are often higher, too). They don't dig holes, but will take refuge in a skunk or woodchuck burrow in bad weather—always staying right near the entry. Normally they rest in small depressions in the grass.

Territory and home range: Not territorial, but they are aggressive and establish a dominance ranking within each gender. Females are generally dominant over males, except during breeding. Rabbits have overlapping home ranges of 1–14 acres (average 5 acres) that may shift as food sources and cover change with the seasons. Males' home ranges are somewhat larger than females.

Breeding Habits

Pair bonding style: Rabbits are polygamous, with dominant males mating the most. Female raises the young alone.

Breeding dates: Late February through September. Gestation is variable but averages 28 days. Females have up to 6 litters per year, giving birth to as many as 35 young. Females may breed again as soon as they've given birth.

Litter size: 4–5. May see as few as 2 or as many as 8. Mothers visit their young only at night, to nurse them.

Weaning dates: Between 4–5 weeks old.

Amount of time young remain with parents beyond weaning date: Not long.

Common Nuisance Situations

Time of year: Any time of year.

What are they doing?

- Eat flowers, vegetables, and agricultural crops.
- Can girdle young trees and shrubs (ornamental and fruit) during winter.
- Disease risks: tularemia.

Best Practices

Rabbits are such prolific breeders, and there are always so many nearby that are ready to move into a vacant territory, that removal won't be effective for long. The best solution combines exclusion and habitat modification.

Reduce their nesting sites:

- Rabbits need dense cover close to their feeding areas to protect them from predators. Remove the cover and you make the area far less attractive to rabbits.
- Remove brush piles.
- Trim shrubs and fencerows.
- Keep paths around gardens and fields closely mowed.
- Clean up overgrown ditches or stream banks that are near crops.

Protect vulnerable plants or areas:

- For a small area, erect a 2-foot-high chicken wire fence with a 1-inch mesh that's either buried a few inches deep or is very tight to the ground. Rabbits won't dig under the fence, but they will try to squeeze through loose spots. Support the fence every 6–8 feet with a strong post.
- Put cylinders of quarter-inch mesh hardware cloth around trees and shrubs until their bark roughens. Keep the mesh an inch or so away from the plant. If you use half-inch mesh, be sure it's far enough away from the plant to prevent the rabbits from nibbling through the mesh.
- Commercial tree wrap can protect young trees. Remember, most tree damage happens during the winter. When there's deep snow, the rabbits can reach 20 inches above snow depth.
- A dome or cage of chicken wire over small garden beds will discourage rabbits.
- A single-strand polytape electric fence will work well. To keep deer from damaging the fence accidentally, hang white cotton flagging on the fence every 6 feet to make it more obvious at night. You can spray the flagging with a deer repellent for extra security (if you have a commercial pesticide applicator license).
- If there's an existing electric fence, add three additional wires at 5, 10, and 15 inches from the ground to keep the rabbits out, too. This also discourages woodchucks.
- A 2-foot-high welded wire fence made of 1-inch mesh, installed in the rat wall L shape with a top wire that's electric also works well, but is more expensive.

WCOs with commercial pesticide applicator licenses:

- Many repellents registered for use against deer are also registered for rabbits. Egg-based repellents have proved effective; other possibilities include capsaicin (hot pepper), and thiram products.

Trapping Strategies

Rabbits are such prolific breeders that trapping—or for that matter, shooting—won't solve the problem for long. More rabbits will gladly move in from other areas.

Live traps:

- Rabbits are protected small game animals. Permits will be needed to trap and transport live rabbits. You can suggest that customers invite beagle club members to live trap and remove rabbits from their properties with the appropriate state permits.
- Rabbits are easy to capture in a box or cage trap (9 × 9 × 18 inches). Traps should be set just

after sunset or just before sunrise, when the rabbits are most active. Winter is the easiest time to trap rabbits because there's less food around, so the bait is often more attractive.
- Place traps close to the hole, feeding area, or trail. A trail of a few pieces of bait leading to the trap will help guide the rabbit into the trap.
- Bait with apples or corn, and add a few rabbit droppings to increase the bait's appeal.
- Don't clean the trap between uses because the scent of a rabbit will attract other rabbits.
- Place traps away from prevailing winds (winter) to keep snow and dry leaves from interfering with the trap door. And cover with dark canvas or other material to make the trap seem like a safe, secure place.
- After a week, if the trap's not working, move to a new site.

Lethal techniques:

- Body-gripping traps, #110, or #120, set in the hole. Cover the trap or take other precautions to prevent the capture of nontargets, as described in Chapter Five.
- In rural areas, you can also suggest that customers invite hunters to hunt on their properties during the legal season. An overall reduction in the local rabbit population may help reduce the chances of conflicts.

Preferred Killing Methods

- CO_2 chamber
- Cervical dislocation
- Lethal dose of barbiturate, if possible
- Stunning and chest compression
- Shooting using an air rifle, shotgun, or .22 caliber rifle (target the head if rabies testing isn't required, or the heart/lungs)

Acceptable Killing Methods

- Stunning and decapitation
- Stunning and cervical dislocation
- Stunning and shooting

RACCOON (*Procyon lotor*)

Size

From 12 to 36 pounds. Body is 26–38 inches long, including 10-inch tail.

Signs of Their Presence

- Visual sightings of the animal.
- Sounds: Cries include a whistle-like tremolo, hisses, soft grunts, barks, growls, and a "churr-churr" noise while feeding. Cry when attacked is a piercing cascade of snarling screams. The young are quite noisy, their chitters are easily heard in the house, and often mistaken for birds. Raccoons can make a lot of noise when they lumber around in your attic.
- Tracks: Flatfooted, like people, so track is big for the animal's size. The length and width of the front paw is about equal, about 2 inches long. The hind paw is much longer than it is wide, about 3¼–4¼ inches long; described as "a miniature human footprint with abnormally long toes" (Figures B-16 and B-17).
- Scat: Likely found at the base of trees, on logs, big rocks, woodpiles, or other prominences (such as roofs). The scat often shows what they've been eating and can give clues about what is attracting the raccoons to the site (Figure B-18).
- Building damage: Black smudges on walls or downspouts; bent gutters; holes in the siding or boards torn off; damaged soffits or louvers; damaged insulation; odors.
- Crop damage: Partially eaten corn ears with the husks pulled back, or broken stalks; hole in the rind of watermelons, through which the contents have been pulled out.

Diet

Opportunist. Eats fruits, berries, and mast (acorns, and nuts and seeds from trees); insects; worms; frogs; fish; turtles; mice; crayfish, clams, and snails; eggs and young of birds and reptiles; garden, orchard, and field crops; bird seed; pet food; garbage; and carrion.

FIGURE B–16 Raccoon track pattern (not to scale). *(Wayne Trimm)*

FIGURE B–17
Raccoon tracks, close-up. *(Wayne Trimm)*

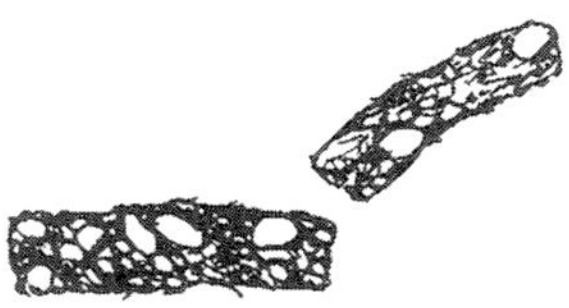

FIGURE B–18 Raccoon scat. *(Wayne Trimm)*

Typical Activity Patterns

Social style: Generally solitary, except female with young.

Daily activity: Nocturnal, but may be active during the day, especially in the spring and summer when the female is nursing her young and needs more food, or when she's looking for a den site.

Hibernator? Sleeps for days at a time during the coldest weather (below 25°F). Adult females (with their young) often den together, especially in a preferred cavity or attic. Raccoons may lose half of their body weight during the winter, as they live off stored fat.

Migrates? No.

Where Found

Distribution: Throughout our region. Can reach densities of 30–40 raccoons/square mile in rural areas, 100+ raccoons/square mile in urban areas.

Habitat: Prefers hardwood forests near streams, rivers, swamps, or ponds. Highly adaptable. Dens in tree cavities and hollow logs, rock crevices, burrows, brush piles, haystacks, beaver lodges, chimneys, attics, crawl spaces, barns, buildings, culverts, storm sewers, and abandoned autos. Usually has a central den (and a few spares) within its range. Females may den together in groups of up to a dozen. Males den by themselves.

Territory and home range: Not territorial, but may fight to establish dominance in common feeding grounds (such as a dumpster). The adult's home range is about a mile in diameter.

Breeding Habits

Pair bonding style: Polygamous. Female raises the young alone. If an adult male comes across the young, he may kill them.

Breeding dates: Peaks in late January to February. Gestation takes about 63 days.

Birthing period: March through May. Late-breeding females may give birth in June, July, or August.

Litter size: 3–5, average 4. May see as few as one kit or as many as eight.

Weaning dates: Between 2–4 months of age.

Amount of time young remain with parents beyond weaning date: Young males leave in the fall, but young females may remain with their mother through their first winter, dispersing the next spring.

Common Nuisance Situations

Time of year: Any time of year. Calls from customers often peak from mid-March through mid-May, when the females are looking for den sites in which to raise their young. From mid-May through July, customers may call about "sick" or "rabid" raccoons that are active during the day (see explanation below). From the late summer through the fall, raccoons may dig through lawns and turf in search of grubs.

What are they doing?

- They den in attics, chimneys, sheds, and barns, annoying people with their noise and odors.
- Their nest materials might block a vent, causing a fire hazard. They also chew on wires.
- Raccoons can damage buildings, either purposefully, to gain entry or create a nesting area, or accidentally, because they're heavy enough to bend gutters as they move through them. Raccoons enter buildings through the roof (using rain gutters, brick chimneys, and overhanging branches to reach the roof); push their way through louvers or soffits; or climb directly up the siding. They may tear shingles, vents, or roofing material to gain entry.
- Raccoons also cause damage as they feed, pillaging gardens and agricultural crops, knocking over and chewing through garbage cans, getting stuck in dumpsters, pulling down and chewing holes in bird feeders, and pulling up turf for worms and grubs.
- Their scat fouls yards and children's play areas and may present a health hazard (parasites found in scat).
- Disease risks: Rabies (they are the primary rabies vector species along the Atlantic seaboard), raccoon roundworm.

Debunking Myths about Raccoons

- A raccoon that's active during the day is not necessarily rabid. It may be a healthy female that's feeding more often than usual, because of the demands of her young. When seeking den sites, coons may also be active during the day. Habitat destruction or development may also prompt increases in daytime activity.
- In raccoons, the symptoms of canine distemper can be easily mistaken for rabies. This leads some people to overestimate the number of rabid raccoons.

Best Practices

Remove artificial food sources (garbage, compost, bird seed, pet food):

- If anyone is feeding the raccoons, persuade them to stop.
- Put trash out in the morning, instead of the evening, if possible, or keep trash in a protected area.
- Raccoon-proof garbage cans or dumpsters with a tight-fitting lid (coons seem to have more trouble opening the type of can that has a 4-inch-high lid that twists on). Secure garbage can with heavy-duty straps or bungee cords, or attach it to a post, or keep it out of reach in the garage (close garage doors at night), or place the can in a covered and secured bin. A top that closes with a latch and snap hook will keep the coons out but will be easy for the garbage collectors to open.
- Feed birds during the winter and gradually stop by April. If the customer really wants to feed birds during the warmer months, install a predator guard on the bird feeder pole. Use sturdy poles. Keep the area underneath the feeder clean.
- Enclose compost piles in a framed box using hardware cloth or welded wire; in a sturdy container, such as a 55-gallon drum; or in a commercial compost container.
- Feed pets indoors. Any food left outdoors should be removed at night. Pet food bowls should also be brought indoors because they retain attractive odors.

Protect children at play:

- Cover children's sandboxes.
- Teach kids to wash their hands thoroughly after outdoor activity.
- Wash toys that were used outdoors with a mild bleach solution (10% chlorine beach, which is one part bleach to nine parts water).
- Keep kids away from typical raccoon latrine areas (base of trees and wood piles).
- As best you can, keep kids from putting things in their mouths. Young children may put raccoon scat, wood chips, soil, or other potentially contaminated objects (including their own dirty hands) into their mouths.
- If there's a known latrine site on the property, you may wish to alter the site conditions to make it less attractive, so the raccoons will stop using it. Remove piles of logs or debris.

Protect vulnerable crops:

- Establish a barrier around gardens and fields with a 2-wire electric fence (if allowed by local ordinances) with the wires placed at 5 and 10 inches above the ground. Fences can be turned off during the day (or left on to also prevent woodchuck damage). Best to install fences at least two weeks before crops reach an alluring stage, so the coons haven't developed the habit of feeding in the garden or field.
- Wrap filament tape around ripening ears of corn (tape should have glass-yarn filaments in it so the coons can't tear through it).
- One scare device, the Critter Gitter, combines a siren and flashing lights. It's triggered by a motion detector. The device switches patterns, so it should be effective longer than a scare device that doesn't vary.

Prevent entry into building: First step: if there are no definitive signs of raccoon activity, determine if coons are still inside by plugging the entry hole with newspaper. If the paper is still there when you return two days later, you can begin exclusion. In the winter, they may be napping, so it may be more difficult to determine whether they're inside or not. Inspect the site as thoroughly as possible.

If this is a preventive action, or there are no young present, you can:

- Replace plastic vents and louvers with metal designs that are securely attached to the building. This is most important for gable louvers, soffit ventilation openings, and roof vents.
- Half-inch hardware cloth (or, even better, welded wire mesh) or galvanized sheet metal may be used to screen holes, decks, or other vulnerable areas. To protect the area underneath a deck or porch, create an L-shaped "rat wall." Attach the hardware cloth to the bottom of the deck. Then bury the bottom 6–12 inches deep, with a 12-inch shelf that sticks out, to prevent animals from digging underneath the barrier.
- Cover chimney flues with commercial caps. Coons can remove some covers, so choose a design that bolts securely to the flue. Raccoons can usually remove the type of chimney cap that simply slips inside the tile liner of the chimney.
- Trim overhanging tree branches 6–8 feet away from the house to make it harder for them to reach the roof (if you also want to foil squirrels, trim to 10 feet away from the building).
- Attach a 2-foot-wide band of metal flashing around trees at chest height, to prevent raccoons from climbing the trees.

If young are present, remove the entire family before blocking the entrance to their den:

- If the raccoons are older and mobile, install a one-way door over the entry hole. The mother and young will leave on their own, but won't be able to re-enter. The mother may bring her young to one of her other dens.
- For trap-and-release strategies to reduce the risk of orphaning wildlife, see Chapter Five. Remove the female at dusk or in the evening.

Trapping Strategies

Live traps:

- Ideally, cage trap should be at least 10 × 12 × 32 inches for a single-door model, longer for double-door models. Bait them with commercial sweet baits and anise oil as an attractant.
- Place a board (or some other sturdy object) underneath the trap to protect the lawn or roof shingles. The board should be 6–8 inches wider than the trap, all the way around. Coons often reach outside traps, grabbing and tearing at anything they can get their paws on as they try to escape. This precaution applies to all live traps, such as cage traps, foothold traps, and foot encapsulating devices.
- New foot encapsulating trap designs specifically for use with raccoons (Lil' Grizz Get'rz, EGG trap, Duffer trap) reduce both the chance of catching the wrong species and the chance of the captured raccoon injuring itself.
- Traditional foothold traps, #1 or 1½, baited with commercial sweet bait.
- Foothold traps are not recommended for use inside a building because the captured raccoon may damage whatever it can reach.

Lethal traps:

- Body-gripping trap, #120, #160, or #220, preferably in a restricted opening set that reduces the risk to dogs and cats (vertical cubby, deep-notch box, or a bucket with a restricted opening). These sets also work well if the entry site is on a building, such as a soffit vent or roof vent. See Chapter Five for details and other tips that reduce the risk of capturing an unintended animal, such as using a one-way trigger.
- Modify the trigger to help ensure a top-to-bottom strike (which is more humane) and to prevent the raccoon from refusing to enter the trap. Raccoons don't like to have anything brush against their eyes or whiskers, so separate the trigger and center it on the top or bottom of the trap. Proper positioning helps to ensure a cleaner, more humane catch.

Preferred Killing Methods

- CO_2 chamber
- Lethal trap
- Shooting, using a shotgun with #6 shot or larger, or a .22 caliber rifle (target the head, if no rabies testing is required, or the heart/lungs)
- Lethal injection of barbiturate

Acceptable Killing Methods

- Stunning and shooting
- Stunning and CO_2 chamber
- Stunning and chest compression, for a smaller raccoon (one that weighs less than 8 pounds)

Control Strategies that Don't Work Particularly Well

- Lights, radios, dogs, scarecrows, streamers, and aluminum pans often don't work.
- Ammonia is dangerous to raccoons and people. Its odor may persuade an adult raccoon to vacate a chimney but there's no guarantee that she'll remove her young—she may simply abandon them. There are better removal methods. WCOs cannot use ammonia even if they have a commercial pesticide applicator license because it's not registered as a repellent.
- Registered repellents that have been tested have proven ineffective.

NORWAY RAT (*Rattus norvegicus*)

Size

This exotic species may weigh up to 1 pound. They're 12–18 inches long, from the nose to the tip of the tail. The tail is somewhat shorter than the body.

Signs of Their Presence

- Sounds: Squeaking, scuttling, scratching, or gnawing inside the walls, ceilings, or between floors of buildings.
- Scat: ½–¾ inches long with blunt ends. Look in kitchen cabinets, drawers, and corners; on counters, under sinks, behind appliances, near food, and in cellars, attics, along walls in barns, warehouses, and feed storage areas. Use an ultraviolet light to look for their urine stains on woodwork (it glows blue-white). Rat urine smells musty and experienced WCOs can distinguish it from mouse urine.
- Runs, smudge marks: Rats use the same route over and over. Eventually, a faint, dark "trail" of body oil and dirt may be noticeable on baseboards and along walls, on beams, rafters, and pipes. Look as well for smooth, worn paths in insulation.
- Nests and burrow holes: They will nest indoors and out. If outdoors, their burrows are usually about 1½–2 feet deep and 3 feet long, with two or more entrances, and usually a well-hidden escape route. Their outdoor burrows are often found in river banks and under sidewalks, platforms, boards, junk piles, foundations, and slabs. They may nest indoors in basements and the lower floors of a building, in crawl spaces, storage rooms, under floors, pallets, junk, and boards, or behind stored items. They may nest in sewers or storm drains. Rat nests are usually 8–12 inches in diameter, made of shredded paper, cardboard, insulation, and bits of fabric or plastic. Their burrow holes are usually 2–4 inches wide.
- Damage to stored goods and buildings: Gnaw marks are often seen on the bottoms and corners of doors, on ledges, in the corners of walls, and on stored materials. Look also for holes and piles of wood shavings. Check cabinets, electric cables, pipes, baseboards, window casings, and foundations. They will gnaw on almost any building material: wood, cinder blocks, aluminum, sheet metal, glass, adobe, asbestos. Their teeth grow constantly, so they gnaw to keep them trimmed.
- Evidence of their feeding: Rats are steady feeders, and will settle down and eat large quantities at a sitting. Their leftovers are usually half-eaten pieces of grain. Rats need water every day. May see and smell scat and urine.

Diet

Opportunist. Norway rats prefer fresh food over garbage, but they'll make do with what's available. They prefer cereal grains and high-protein foods such as meats (sandwich meats, insects, mice, bird eggs, young birds), fish, nuts, insects, and pet food, and garbage. They'll eat some fruits (especially dried fruit), cheese, peanut butter, bird seed, potatoes and vegetables, bacon, butter and lard, compost, and manure. They'll even eat paraffin wax, leather products, and the feces of dogs, cats, or horses.

Typical Activity Patterns

Social style: Generally colonial, with an established hierarchy, although you may find solitary rats.

Daily activity: Nocturnal. If populations are very high, they may be active during the day, too.

Hibernator? No.

Migrates? No.

Where Found

Distribution: Widespread, in urban, suburban, and rural areas throughout the East. Rats are generally found near people. Although many people think of rats as an urban problem, about half of North America's rats live on farms.

Habitat: Any building that provides food and shelter, usually in the basement and on the lower floors of the building. They're found in apartment buildings, homes, kennels, warehouses, stores, slaughterhouses, barns, livestock buildings, silos, granaries, even sewers and dumpsters. Rats will nest underneath buildings and concrete slabs, along stream banks, around ponds, and in dumps. They like to nest near water.

Territory and home range: Territorial, especially among males. Daily, rats will travel through an area that's up to 100–150 feet in diameter, more than ten times the size of the foraging range of a house mouse. Rats generally stay within 300 feet of their burrows.

Breeding Habits

Pair bonding style: Rats are polygamous. Female raises the young alone.

Breeding dates: Peaks during the spring and fall. Females may breed again within a day or two of birthing and may produce 4–6 litters a year.

Litter size: 6–12. Gestation takes about 21–23 days.

Weaning dates: Between 3–4 weeks of age.

Amount of time young remain with parents beyond weaning date: Not long!

Estimating rat populations:

- There are probably about 10 rats in the area for each one spotted at night. Medium population: see one or two rats at night, but none seen during the day; or find old scat, or old gnaw marks. High population: see three or more rats at night, or see rats during the day; find fresh scat; gnaw marks and tracks are abundant.
- Another way to estimate populations: Put out food, then record how much is eaten by the rats to estimate the minimum number of rats in the area. Use finely ground grain, not whole grain or pellets, which the rats can carry off. Remember, rats are cautious, so give them some time to get used to this food source before you begin your data collection. When you're ready to collect data, weigh the food (in ounces) then place it where you believe the rats are active. The next day, weigh whatever's left. That tells you how much food they ate. Multiply that number by two, because one ounce of food/day typically supports two rats, and you'll have an estimate of how many rats are in the area. Of course, rats have other food sources, so this isn't exact.

Example:

a. Day one, place 40 ounces of grain near a rat hole.

b. Day two, measure what's left over (for this example, let's say 12 ounces).

c. Subtract the amount left over from the total bait: $40 - 12 = 28$ ounces eaten.

d. Multiply the amount of grain eaten by 2 (each ounce supports 2 rats): $28 \times 2 = 56$ rats.

Common Nuisance Situations

Time of year: Any time of year.

What are they doing?

- Rats can cause extensive damage to buildings and household goods as they seek food and nest sites. They'll gnaw on or foul siding (even aluminum), woodwork, sheet metal, plasterboard (Sheetrock), insulation, plastic food containers (including garbage cans), papers, packaged goods, clothing, mattresses, furniture, even lead or copper pipes.
- Their nest materials might block a vent, causing a fire hazard.
- They also chew on wires, which in addition to creating a fire hazard could also short-circuit electrical systems, causing alarm systems and refrigerators to fail.
- Their burrowing may cause roads and railroad beds to settle, or damage the banks of irrigation canals and levees. It may also undermine foundations and slabs.

- Rats bite and terrify some people. They transmit several diseases to people.
- Rats may damage crops in the field, in silos, granaries, and warehouses. They contaminate stored foods, especially grains, in commercial settings such as restaurants and warehouses and homes. Rats ruin a large amount of the world's food supply.
- They'll raid bird feeders and pet dishes.
- Their noise and smell may drive you and your pets to distraction.
- They foul items in museums and libraries.
- Disease risks: The diseases that rats are more likely to transmit to people or livestock include murine typhus, leptospirosis, trichinosis, salmonellosis (food poisoning), and rat-bite fever. Bubonic plague is more closely associated with roof rats (*Rattus rattus*) than it is with Norway rats. Rats are often infested with lice, fleas, and mites that transmit other diseases.

Best Practices

It's critical to consider how well rats climb, jump, and swim when planning your control strategy. If you had the physical abilities of a rat, you might be an Olympic athlete. Here's what a Norway rat can do:

- Climb up: Brick buildings (or any building with a rough exterior), wires, conduit, pipes (inside and outside!), vines, shrubs, trees. Rats can climb inside a pipe that's 1½–4 inches in diameter, or along the outside of any pipe that's within 3 inches of a wall or other support. Otherwise, rats can climb up an exterior pipe that's up to 3 inches in diameter (and if the surface is rough, they can climb up an even larger pipe).
- Run along: Telephone wires, power lines, pipes, conduit, and tree branches.
- An adult rat can squeeze through a hole that's about ¾ inch wide. A young rat can squeeze through a ½-inch hole.
- Jump vertically about 3 feet and horizontally 4–8 feet, depending on whether they start on a flat or elevated surface. If people could match that, they'd be jumping about 18 feet up or 24–48 feet out, without poles or a running start.
- Stretch up about a foot on a smooth wall.
- Swim up to a half-mile in open water, underwater for about a half-minute; against strong currents; and up through toilet traps (water seals). Rats can tread water for up to 3 days.
- Burrow 4 feet into the soil.
- Fall from a height of 50 feet without serious injury.
- Gnaw through lead and aluminum sheets, cinder block, plastic, and other materials.

The best way to deal with a rat infestation is to clean up, get rid of the rats, and keep them from finding a way back in. Keep three words in mind: *sanitation, eviction, exile.*

Remove artificial food sources (garbage, compost, bird seed, pet food):

- Garbage is usually the main food source for rats in urban areas. Ideally, garbage would be removed daily, before dusk. This often isn't possible, so make sure that garbage is kept in secure containers.
- Clean garbage cans, dumpsters, and chutes regularly, at least once a week.
- Screen the drainage holes in dumpsters with quarter-inch hardware cloth.
- Steel garbage cans are a better choice than plastic, which the rats may chew through.
- If rats, raccoons, or dogs tip over the garbage cans, either use a spring-loaded fastener or bungee cord to keep the lids on, or put the garbage cans on a platform that's 18 inches above the ground and 3 feet away from buildings.
- At a dump, cover the garbage with soil every day.
- Store food, bird seed, pet food, garbage, compost, and recyclables in metal, glass, or ceramic containers with tight-fitting lids.
- Near buildings, rake up and dispose of fruits and nuts that fall from trees. It's not a bad idea to wrap these trees with sheet metal, so the rats can't climb up and feed in them. Prune low-hanging branches, too.
- Put unfinished pet food in the refrigerator.

- Store large bags of flour, grain, pet food, or livestock feed on open-wire shelves. Bottom shelf should be at least 18 inches off the ground.
- Especially in kitchens and food storage areas, elevate equipment (mixers, stoves, refrigerators) so you can clean underneath it easily. If you can't elevate it, then close it off so the rats can't get underneath the equipment.
- Elevate compost heaps or enclose with half-inch hardware cloth or welded wire mesh.
- Keep livestock feeding areas and feed storage areas as secure as possible.
- Remove dog, cat, and horse feces daily; rats will eat them.
- Keep the area around and underneath bird feeders clean, especially of spilled seed. Use baffles to keep rats (and squirrels) out of feeders.

Remove their nesting sites:

- Keep stored items off the floor and away from walls. In a warehouse, paint a 12-inch white band on the floor all the way around the room to make inspections easier, and to remind people to keep items away from the walls.
- Reduce clutter and remove cardboard boxes.
- Move firewood, junk piles, and garbage cans away from the house.
- Maintain a foot-wide gravel border around the foundation that's free of vegetation (best), or keep the foundation plantings well trimmed. Don't stack anything (such as firewood) against the foundation.
- Rats find shelter in dense ground covers such as ivy. Either keep it well-trimmed or replace it with a more open ground cover.
- Break up large expanses of dense ground cover with exposed pathways. Rats don't like to cross areas where they can easily be seen.

Prevent entry into building:

- Close the door! (Use screen doors.) Install mechanical door-closers in warehouses or other areas where people forget to close doors.
- Add metal kick plates (26-gauge sheet metal) to the bottoms of doors, especially those leading to warehouses and food storage areas.
- Repair every crack and hole that's more than ½ inch wide. Seal openings beneath and behind sinks, stoves, and dishwashers with latex caulk. Fix cracks in foundations and floors with concrete or masonry grout. Remember to check around pipes, sewer outlets, cables, stairs (inside and out), roof joints, and the areas where chimneys and fireplaces come through the floor from the cellar or crawl space. Use strong materials to repair holes, such as half-inch hardware cloth, welded-wire mesh, sheet metal plates, concrete mortar, or coarse steel wool or Stuf-Fit with expanding foam sprayed over it.
- Plug gaps around water, gas, and heating pipes, heat registers, air ducts, electrical chases, and false ceilings with latex caulk.
- For large holes around pipes, use galvanized metal pipe chase covers, sheet metal plates, mortar, plaster of Paris, or cement.
- Wrap pipes that run along exterior walls with sheet metal guards that fit closely to the wall, sticking out 12 inches from the pipe.
- Check vents (sewer, stove, clothes-dryers, roof, ridge-line, soffit, furnace, and air-conditioning ducts, attic fans). If it's damaged, replace the vent with an animal-proof design, or screen it with quarter-inch hardware cloth or welded wire. End caps on ridge vents may loosen, providing access to the attic. Soffit vents are best protected with metal louvers.
- Add guards to power lines to keep rats from traveling along them (consult with your power company first).
- Paint a foot-wide band around the perimeter of block or concrete buildings at a height of 3 feet. Use hard, glossy (slippery!) paint. This technique may also be used to make it harder for rats to climb up vertical pipes.
- Monitor structures routinely for structural cracks and openings. Places that might be overlooked, and

yet are attractive to rats, include elevator shafts, laundry chutes, the compressors of refrigerators or freezers, and the insulated walls of large coolers.

- Trim branches at least 3 feet back from buildings.
- Rats can live in sewers, and may enter buildings through toilets or water pipes. A toilet can be rat-proofed by adding a one-way flap valve called a "rat guard" or by feeding the pipe from the toilet bowl into a wider pipe. Screen drains in basements and shower rooms with half-inch hardware cloth or welded wire.
- Repair broken sewer pipes.
- Add "rat wall" barriers underneath floors, around foundations and footings, or as linings for walls and ceilings. Use quarter-inch hardware cloth or welded wire. Bury it 6–12 inches deep, then bend the bottom edge outward into an L shape that sticks out one foot to prevent the rats from burrowing underneath it. Or install a concrete curtain wall.
- In double-wall construction, add a barrier between the exterior and interior wall. Nail galvanized sheet metal between the studs, joists, floor, and sill.

Rat-proofing tips for new construction:

- Fit pieces together carefully so all joints are tight. These are vulnerable areas that may warrant protection with sheet metal.
- Use concrete when building new grain storage facilities. If you need to build with wood, line the floors, walls, and ceiling with welded wire (quarter-inch mesh) or hardware cloth (19 gauge).
- Ground floors should be 1½ feet above grade, or made of concrete, stone and mortar, or brick and mortar.
- Footings should be buried to a depth of 2 feet, and be protected by "rat walls" and "termite shields," a metal cone that's attached upside down to footings and building piers.
- Build "rat wall" barriers underneath floors, around foundations and footings, or as linings for walls and ceilings (see previous description).
- Install metal kick plates on the outside of doors, and protect door casings with sheet metal. Fasten metal thresholds to floors.
- Steel pipes embedded in concrete create a strong metal door threshold that allows the door to swing freely. This is a good option wherever heavy equipment or livestock travel through the doorway.
- Corrugated metal siding can provide many entry holes. You can butt the siding against a solid material, such as concrete or metal flashing, to seal these holes. (This may cause the metal to rust faster.)
- In double-wall construction, add a barrier between the exterior and interior wall. Fireproof stops of concrete or brick are best.

Trapping Strategies:

- The Norway rat is an exotic species, so please do not release any into the wild (Chapter Two explains why).
- Rats are often hard to trap because they avoid new objects. Place baited, unset traps for up to a week, until the rats are used to the sight of the traps. Be persistent.
- To increase your success, trap intensively for several days. More is better. A good rule-of-thumb is to place one trap for every rat. In homes with moderate to heavy infestations, use 12–24 traps. In barns or large buildings, 50–100 traps may be in order.
- Place the traps in their runways, in dark corners, along rafters, near food sources, nests, or holes—wherever the rats are most active (look for droppings and gnaw marks). You may be able to lift some ceiling tiles to place traps in a dropped ceiling.
- Set traps at night, when rats are most active, and check them in the morning.

Live traps:

- Norway rats are generally too smart to enter a live trap. Because they're an exotic species, this technique is not recommended. Lethal trapping will likely prove challenging enough! If a customer insists, use a squirrel-sized cage trap (6 × 6 × 24 inches) for adults. You may be able to capture younger rats in a chipmunk-sized cage trap (16 × 6 × 6 inches). Set live traps parallel to the wall.

Lethal traps:

- The familiar mouse trap is called a "snap trap." There are larger models for rats. Don't try to use the mouse-sized traps because they're too small. Look for some of the newer designs, such as traps that have bait covers, which are triggered when that lid is lifted; or traps with expanded "triggers" or a clothespin design (Figure B-19). The design with the bait cover is more selective, while all of these newer models are easier to set than the traditional rat trap.
- Place traps right against the wall, every 5–10 feet.
- Set snap traps in pairs. This is much more effective. Two sets that work well:
  - Side-by-side, perpendicular to the wall, with the trigger snapping towards the wall
  - Parallel to the wall, with the triggers snapping to the outside (not into the center)
- Traps may be attached to rafters with nails, or to pipes with wire or Velcro strips.
- Bait with bacon, hot dog, liver, peanut butter, or nut meats. You can sprinkle oatmeal around the trap to make it even more attractive. You may want to put the bait on the bottom of the trigger, which increases the chance that the rats will spring the trap.

FIGURE B–19 A snap-back trap showing the clothespin design. *(Susan MacKay)*

- To protect young children and pets, place traps in bait stations, a cage trap with 1 by 2 inch mesh, a coffee can with both ends cut out, or in PVC pipe (remember to test that the trap will spring within its container). Or use a commercial tunnel trap. Alternatively, keep your pets locked in a safe room at night and don't release them until you've sprung all of the traps.
- Body-gripping trap, #55. Use a one-way trigger to increase the selectivity of the trap.
- The use of glue boards, which the authors do not consider a best practice, may be warranted in the case of a *severe* infestation. A glue board is a layer of long-lasting adhesive spread over a surface, usually cardboard or plastic. The rats get stuck in the adhesive (glue boards are used to capture mice, rats, and snakes). Although some call the glue board a live trap, it's not often used that way. In fact, some biologists believe that you cannot remove the animal from the trap unharmed because the oil that is used to loosen the glue may harm the animal. In practice, rats are frequently left to die on glue boards. Snap traps are often as effective as glue boards and are more humane, although setting them does take more effort. If using glue boards, set them in protected areas, such as within a dropped ceiling. Check them frequently (at least every 12 hours) and kill any captured rats by stunning them. Do not leave dead rats to rot on glue boards because the carcasses will stink and likely attract other pests.

Other lethal techniques, for WCOs with a commercial pesticide applicator license:

- Poisons (in various forms, such as baits, fumigants, and tracking powder) can be effective and may be warranted in some situations. Rodenticides can be hazardous to children, pets, and animals that eat poisoned rats. The rats may die in walls and stink, while providing a breeding place for flies. Trapping is often a better solution.
- Rats are generally cautious when approaching a new food, including poison baits. They sample

just a bit at first, and it takes several days for them to overcome their fear. Nor is any bait a universal favorite. Young rats may imitate their mother's food preferences, so if their mother avoids poison baits, the young may, too. This can lead to bait shyness in large populations of rats. This is why testing various baits and pre-baiting with non-toxic baits is so helpful in controlling rats.

Preferred Killing Methods

- Lethal traps
- CO_2 chamber
- Shooting, using an air rifle or .22 caliber rifle

Acceptable Killing Methods

- Stunning and chest compression
- Pesticides (with the proper license)

Control Strategies that Don't Work Well

- Ultrasonic and electromagnetic devices don't work against rats. Loud or unusual noises will frighten them and may drive them off for a short time.
- Mothballs and ammonia don't do much, and may not be registered.
- Cats may kill some rats, but the rats may also kill the cat, especially if it's vastly outnumbered. Other predators that kill rats include snakes, owls, dogs, coyotes, and birds of prey. They'll help to reduce rat populations but shouldn't be relied on as the sole source of control.
- Glue boards can capture rats but they may cause other pest infestations if used improperly. There are more humane options that are effective.

SKUNKS

Species' Names

- Striped skunk (*Mephitis mephitis*)
- Spotted skunk (*Spilogale putorius*)

The striped skunk is more common, especially in urban and suburban settings. The far less common spotted skunk may be identified by a break in the white stripes running along its back. Skunks have a wide variation in coat color, ranging from almost completely black to almost entirely white.

Size

Striped skunk is 20–30 inches long, including 10- to 15-inch tail. They weigh 6–12 pounds. Spotted skunks are somewhat smaller.

Signs of Their Presence

- Visual sighting of animal.
- Sounds: Adults are generally quiet, although you'll hear them stamp their feet. Young skunks are more vocal, especially when playing. You may hear teeth clicking, hissing, grunts, growls, purrs, squeals, and shrill screeches.
- Odor is nauseating, penetrating, acrid musk.
- Tracks: Small relative to body size, 5 toes on all feet, smooth continuous palm pads, long front nails (Figures B-20 and B-21).
- Scat: Includes mostly insect body parts, some fur, and seeds. (May be slightly curved, not shown in Figure B-22.)
- Evidence of their feeding: Funnel-shaped holes in lawns, 3–4 inches in diameter, where skunks dig

FIGURE B–20 Striped skunk, track pattern. *(Wayne Trimm)*

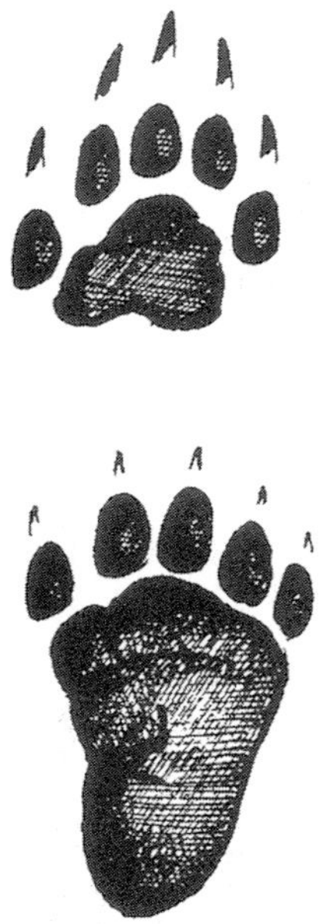

FIGURE B–21 Striped skunk tracks, close-up. *(Wayne Trimm)*

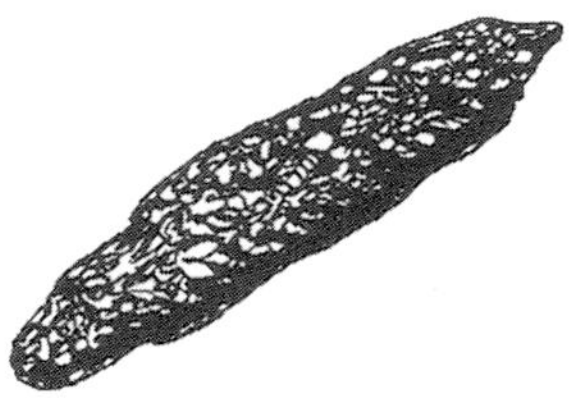

FIGURE B–22 Striped skunk scat. *(Wayne Trimm)*

for grubs. Eggs that have been crushed at one end, with shell fragments pushed inside. (Raccoons usually remove one end of the shell without crushing it. Foxes carry eggs away. Weasels and mink crush the entire egg.) Another sign that skunks have been raiding the chicken house is if only one or two birds have been killed, because most other predators will kill several birds.

Diet

Opportunist. Their diet changes seasonally. Skunks eat primarily insects (including ground bees and wasps), as well as earthworms, snakes, mice, moles, fruit, nuts, fish, amphibians, crustaceans, birds, the eggs of birds and turtles, poultry, garbage, pet food, and carrion. They're particularly fond of grubs and occasionally raid vegetable gardens.

Typical Activity Patterns

Social style: Generally solitary, except for females with dependent young, and when in winter dens.

Daily activity: Nocturnal. During the summer, may see daytime activity, as females forage with their young. May bed down during the summer in open sites away from the den.

Hibernator? Skunks sleep deeply for up to $3\frac{1}{2}$ months at a time, but are not true hibernators. They'll emerge periodically during warm spells and during the mating season. Skunks den alone, or in a group of 2–7 females and 1 male. They may lose up to 38% of their body weight during the winter.

Migrates? No.

Where Found

Distribution: Common throughout the East. Can reach densities of 50 skunks/square mile in suburban areas.

Habitat: Widespread, from coastal habitats to mature woodlands and small woodlots. Prefers open fields, lawns, and agricultural areas with areas of mixed shrubs and forest edges, near buildings, barns, or porches.

Territory and home range: Skunks rarely travel more than one mile from their den except during the breeding season.

Breeding Habits

Pair bonding style: Polygamous. Females raise young alone. (Male skunks will kill their young.)

Breeding dates: Late February through March. Gestation takes about 62–75 days.

Birthing period: May through early June.

Litter size: 2–10, often 4–7.

Weaning dates: At 2 months old.

Amount of time young remain with parents beyond weaning date: Kits forage with their mother when they're 7 weeks old. They're independent at 3 months, and disperse in the fall.

Common Nuisance Situations

Time of year: Calls peak in February and March, when they're mating. In May and June, calls are usually related to their grubbing in lawns. This picks up again in late July, continuing through mid-October. Also during that period, you may get calls about "rabid" skunks that are active during the daytime (see explanation in "Debunking myths about skunks").

What are they doing?

- Seeking a sheltered place in which to raise their young. They may den under porches, decks, foundations, garages, barns, or sheds.
- Stinking up the place. Skunks can be very smelly, especially from the mating season through the whelping season, if the female fights off a male. If the smell seems to come and go and is more noticeable at dawn or dusk or with a shift in wind direction, or if the smell seems to be coming from an area with evergreen trees, it might be the odor of a great horned owl. These owls commonly eat skunks.
- Defending themselves. Skunks are a mild-mannered, slow-moving, mind-your-own-business kind of animal. If provoked, they may spray people or pets. Their spray can reach up to 16 feet. Skunks can spray once they're 2–4 weeks old. They can spray up to six times in a row, but would then need a day to "recharge."
- They'll fall into window wells while searching for insects and toads, and then become trapped.
- Skunks dig in lawns for grubs. They'll sometimes scratch beehives in search of honey and insects, or raid poultry houses for eggs and chickens (but that's rare, and such damage is more likely the work of a raccoon).
- Disease risks: Rabies (they are a rabies vector species in our region), distemper.

Debunking Myths about Skunks

- A skunk that's active during the day isn't necessarily rabid. It may be a healthy female that's feeding more often than usual, because of the demands of her young.
- Adult skunks are not trigger-happy but "teenaged" skunks may be. Very young skunks squirt small amounts of fluid as they walk because they're not yet mature enough to have control of the "spray muscles." You can usually tell if an adult skunk intends to spray (described later).
- Grubbing by skunks is sometimes blamed on other animals because there's no skunk smell. Skunks spray in defense.

Best Practices

When dealing with skunks, a new factor enters into the choice of capture, transport, and dispatch—how to keep the skunk from spraying. WCOs who have handled skunks successfully for decades advise those less experienced with handling skunks to relax. Move slowly and quietly, and don't wave your arms around. Be patient and gentle. Learn their habits and use those to your advantage.

For example, skunks like to see their targets. If they can't see, they're not likely to spray. So, if you use a plastic box trap or cover the sides of a cage trap, you'll reduce the risk of being sprayed. For a more securely covered cage trap, attach quarter-inch plywood to the sides and top. Leave room to reach the trigger release and the carrying handle.

Some WCOs have found that it's better to work with a covered trap than to cover the trap after you've caught the skunk. You may want to create a partially covered wire trap to use during hot weather, when the plastic traps could cause the skunk to overheat and die.

Here are some tips for capturing a skunk that's indoors. Set up a covered trap, and then slowly and quietly approach the skunk from behind. Guide the skunk toward the trap by gently pushing it with a broom or occasionally squirting it with water from a spray bottle.

Another option for nervous skunk handlers: don't do it! You are not required to handle all nuisance species. Some WCOs specialize in one, or perhaps a few species, and have successful businesses. If you've spent some time with experienced skunk handlers and still don't feel your skills are adequate, consider referring customers to someone else.

Remove food sources and shelter:

- Put trash out in the morning instead of the evening, if possible, or keep it in a protected area.
- Skunk-proof the garbage can with a tight-fitting lid, or secure the lid with straps.
- Don't leave pet food or their food bowls out at night.
- Enclose compost piles in a framed box using hardware cloth or welded wire; in a sturdy container, such as a 55-gallon drum; or in a commercial compost container.
- Treat lawns to reduce grub populations. (Biological controls are preferred. In the southern part of the region, where it's been proven to work, try Milky Spore.)
- Keep mice out of buildings. Skunks eat them, and will go inside buildings looking for rodents.
- Remove brush piles and debris.

Protect vulnerable areas and crops:

- Close garage doors at night.
- Cover window wells. There are readily available commercial window well covers that are inexpensive.
- Close basement windows at night, and keep them in good repair.
- Fence beehives or poultry areas with 1-inch chicken wire, 1 by 1 inch or 1 by 2 inch vinyl coated or galvanized welded-wire mesh, or hardware cloth (quarter-inch or half-inch mesh). If there's already an electric fence, add a wire at 5 inches off the ground, or place the hives 3 feet above ground.

Keep them from denning under buildings: If this is a preventive action, or there are no young present, you can:

- First, ensure that the skunks have left the den. Close all of the entrances to the den except the main hole. You can place a one-way door over that hole for 3–4 days to give the skunks time to leave, or use the soft plug method. Sprinkle flour, talc, or nontoxic tracking powder on the ground inside the den area near the hole, and then cover the hole with hardware cloth. Return the next day to check for tracks. Once you're sure the skunks are gone, you can permanently seal the hole.
- Screen areas under decks, porches, and houses (foundation skirt) with a "rat wall." Use 1 by 1 inch or 1 by 2 inch vinyl-coated, welded-wire mesh or hardware cloth (quarter-inch or half-inch mesh). The fence must be buried 3–6 inches deep, with the bottom edge bent outward at 90° into an L shape that sticks out 6–12 inches to prevent the skunks from burrowing underneath it. If you can't bury the fence, the 90° bend extending along the ground can be effective. This design also works for a freestanding fence. If the top isn't attached to the deck or porch, the fence should be 3 feet high.
- Skunks can squeeze through small openings in buildings. Seal any hole or crack that's 3–4 inches across with sheet metal, concrete, or hardware cloth.

If young are present, remove or evict the entire family before blocking the entrance to their den:

- For trap-and-release strategies to reduce the risk of orphaning wildlife, see Chapter Five. Capture and release the mother and young at dusk or in the evening.
- If the young are older and mobile, install a one-way door over the entry hole. They'll leave but won't be able to re-enter. Wait 3–4 days before sealing the entry permanently.

How to avoid being sprayed:

- Skunks give a warning before they spray. They turn to face the aggressor, arch their backs, raise their tails, stamp the ground, and shuffle backwards. Then, just before spraying, they bend into a "U" shape, so both their head and tail faces the target. Should you see any of these signals, back away slowly and quietly, and don't wave your arms around.
- Take precautions before letting dogs out at night, or keep them on leashes and maintain control.

And how to get rid of that lovely "eau de skunk":

- First, ventilate the area.
- A mix of equal parts of tomato juice and vinegar will clean a dog, but most people don't realize that you'd have to soak your pet for an hour—and then wash with soap.

  Here's an easier recipe, developed by chemist Dr. Paul Krebaum:

  1 quart of 3% hydrogen peroxide
  ¼ cup baking soda
  1 teaspoon liquid soap

  Mix the ingredients together and immediately wash your pet (or soak your clothing) while the solution is bubbling. Rinse. Don't try to bottle this mix because it generates a lot of oxygen and could explode.
- To clean clothing or objects, two household products will work: ammonia or bleach (oxygen or chlorine bleach). *You must choose one or the other. Don't mix them because together they form toxic fumes.* So, either pour a little ammonia into the water, or a little bleach into the water. Soak the clothes for several hours, then wash as usual. You may have to wash the clothes a few times, and there may be some discoloration. You could also try Dr. Krebaum's recipe, listed above.
- There are many commercial deodorizers that can neutralize or mask the odor. For information, see the publication "Removing Skunk Odor" (http://www.ianrpubs.unl.edu/epublic/live/nf646/build/nf646.pdf).

Trapping Strategies

Live traps:

- Cage trap should be 9 × 9 × 24 inches for a single-door model, longer for a double-door model.
- Bait with commercial fruit-based bait or peanut butter.
- The skunk will be calmer if it enters a dark space. Use a plastic box trap, except during hot weather, or cover the sides of a cage trap with boards.
- Foothold trap, #1 or #1½ (double-jawed traps preferred, but can also use standard jaw or laminated traps), or the Lil' Grizz Get'rz, Duffer, or EGG foot encapsulating traps designed for raccoons.
- How to get a skunk out of a window well: Skunks are poor climbers. Many people will suggest placing a board in the well to allow the skunk to climb out on its own, but this will work only if the window well is large enough so the board can be placed at a shallow angle of less than 45° (if you try this, give the skunk some traction by nailing a few boards or some cleats across the board). Unfortunately, most window wells are too small for this technique. Instead, try restraining the skunk with a Cat Grasper (akin to a catchpole) and then lifting it out, or place a small cage trap or small cardboard box in the well and guide the skunk into it using a long stick. Work calmly and you should be able to remove the skunk without being sprayed.

Lethal traps:

- Body-gripping trap, #120, #160, or #220, preferably in a restricted opening set that reduces the risk to dogs and cats (vertical cubby, bucket with restricted opening, or a deep-notch box set). The skunk will likely spray. See Chapter Five for details about these sets, and other tips, such as the use of a one-way trigger, that reduce the risk of capturing an unintended animal.
- Modify the trigger to help ensure a top-to-bottom strike (which is more humane) and to prevent the skunk from refusing to enter the trap.

Skunks don't like to have anything brush against their eyes or whiskers, so separate the trigger and center it on the top or bottom of the trap. Proper positioning helps to ensure a cleaner, more humane catch.

Preferred Killing Methods

- The skunk will probably spray, so be prepared
- CO_2 chamber (let the animal settle down before turning on the gas, and use a lower flow rate to avoid frightening the animal)
- Lethal trap
- Lethal injection of barbiturate, if possible
- Shooting, using a shotgun with #6 shot or a .22 caliber rifle (Target the heart/lungs. The skunk will almost certainly spray if you use a head shot.)

Acceptable Killing Methods

- Stunning followed by chest compression (with care, to avoid rabies exposure). The skunk will probably spray, so be prepared.

A Killing Method that's Under Debate

Some people inject acetone into skunks to kill them. Is it a best practice? This method looks good from the outside. The skunk seems to "just go to sleep," falling over quietly before it dies. There aren't any signs of struggle, discomfort, or pain and the skunk rarely sprays.

Unfortunately, there's no scientific data to adequately explain how acetone injections kill skunks. You'll hear opinions both for and against this method. If acetone is as bad as some people believe, then why don't the skunks spray? And even if it is traumatic, if it reliably kills skunks as quickly as some people report, could it still qualify as a best practice? We prefer facts—and although we tried hard, we didn't get satisfactory answers. So for now, we can't condone this method.

Why not trust your eyes? Like wildlife biologists, successful WCOs are careful observers of wildlife behavior. In fact, to use the best practices method, you must make decisions based on what you see, smell, and hear. Consider the drug succinylcholine hydrochloride, which was once used to immobilize animals. This drug paralyzes muscles, so the animal would be quiet and unable to move—but it would also be fully alert and able to feel pain. How do we know that? Some people volunteered to take the drug so they could describe the experience, which they said was painful and frightening. This drug is no longer recommended for immobilizing wildlife.

Kind? Cruel? You may not be able to tell just by watching. We recommend you choose a method that is well understood and has been properly studied.

Control Methods that Don't Work Well

- Mothballs aren't registered for this use, and could be dangerous to people if used in the quantity that would be needed.
- Other repellents haven't worked against skunks.
- There are no toxicants registered by EPA for skunk control.

TREE SQUIRRELS

Species' Names

- Gray squirrel, *Sciurus carolinensis*
- Red squirrel, *Tamiasciurus hudsonicus*
- Fox squirrel, *Sciurus niger*
- Northern flying squirrel, *Glaucomys sabrinus*
- Southern flying squirrel, *Glaucomys volans*

Size

Gray: 18–20 inches long; tail half that length; 1–1½ pounds

Red: 12 inches long, same with the tail; about 5½ ounces

Fox squirrel: 21 inches long, includes 9½ inch tail; nearly 2 pounds

Northern flying: 10–11 inches, including 4½ inch tail; 3–4 ounces

Southern flying: 9–10 inches, including 3½ inch tail; 1½–2½ ounces

Signs of Their Presence

- The animals themselves. Don't be surprised if people report seeing black or white squirrels. They're really gray squirrels—just a color variation.
- Sounds: Red squirrels are loudest, with their sometimes birdlike, sometimes scolding, but seemingly endless chatter. Gray and fox squirrels also chatter, and during the mating season, they'll make a chucking bark as they chase each other. You may hear chewing, pattering, scampering, scratching sounds in attic, eaves, and walls from early morning throughout the day—unless the residents are flying squirrels, which are nocturnal.
- Scat: Oval, smooth, roughly ¼ inch long. The scat of flying squirrels is often found in distinctive piles.
- Nests: Gray, fox, and flying squirrels make leaf nests, usually placed in a tree crotch, that are used summer and fall. The flying squirrel's nest is about 8 inches in diameter; those of the gray and fox squirrel are larger.
- Evidence of their feeding: Nipped twigs of spruce, hemlock, and pine trees; piles of gnawed hickory nuts and walnuts, or strips of acorn shell, between attic joists or in wall cavities (gray, fox, red, or flying); piles of pine cones, acorns, hickory nuts (red).
- Garden and crop damage: They eat flower bulbs and seeds, raid bird feeders, damage the equipment used for maple syrup collection, eat cherry blossoms and ripe pears or apples, and chew on the bark of fruit trees. They may also strip bark, which they use in their nests.
- Building damage: Holes in vents, eaves, soffits, and fascia boards. Claw marks on siding. Tunnels in insulation. Chewed wires. Damage to stored household goods from their chewing, urine, or feces.

Diet

Opportunists, primarily herbivores. The flying squirrels are the most carnivorous of the group, although all tree squirrels will eat bird eggs and nestlings. All of these tree squirrels store food for the winter. Red squirrels create one large cache, while gray and fox squirrels bury nuts singly, all over the place.

Gray and fox squirrels prefer the same foods: fall through winter, they eat fruits and nuts (especially acorns, hickory nuts, and walnuts) and bird seed, if available. In early spring they switch to tree buds, then in summer to fruits, berries, and succulent plants. They'll also eat insects; bird eggs; mushrooms; corn; garden, orchard, and field crops; and when very hungry will chew tree bark and lick the sap.

Flying squirrels tend to eat the same foods as the gray and fox squirrel, but they're more likely to eat bird eggs, nestlings, insects, and carrion.

Red squirrels prefer pine seeds and buds, but will eat many of the foods listed above. They're more carnivorous than gray and fox squirrels, but not as likely to eat meat as flying squirrels.

Typical Activity Patterns

Social style: Gray and fox squirrels are somewhat sociable. Red squirrels are solitary, except for females with dependent young. Flying squirrels are social, with dozens (perhaps as many as 50) nesting together.

Daily activity: All are diurnal except for the flying squirrels, which are nocturnal.

Hibernator? No.

Migrates? Not typical, but when food supplies crash, they may migrate in large numbers.

Where Found

Distribution: Throughout the region. The fox squirrel has the most limited distribution of the group. The gray squirrel is the most common and adaptable but they're all comfortable in cities and suburbs.

Habitat: Wooded areas. Gray and fox squirrels prefer hardwood forests (fox squirrels like the forest edge); red squirrels prefer softwood forests or mixed hardwoods and conifers; flying squirrels also prefer softwood or mixed forests, but aren't as picky as red squirrels. Squirrels den in tree cavities, rock crevices, burrows, brush piles, deserted buildings, chimney flues, attics, and barns. Gray, fox, and flying squirrels also make leaf nests for use in the summer and fall. Red and flying squirrels prefer old woodpecker nest holes and hollow tree limbs.

Territory and home range: The red squirrel is strongly territorial, defending both food sources and den trees. Gray and fox squirrels are not, but may fight to establish dominance in common feeding grounds, such as around a bird feeder. Their home ranges are broadly overlapping and variable, generally about an acre. Flying squirrels are often found in large groups and are most likely not territorial.

Breeding Habits

Pair bonding style: Polygamous. Females raise young alone.

Breeding dates: Gray and fox squirrels: mid-December through January (fox squirrels mate in January). Red and northern flying squirrels: late winter. Southern flying squirrels: early spring. Gestation takes 40–45 days. Five to ten percent of older female gray squirrels may breed again in June. In more southern locales, some squirrels may breed twice during the season.

Birthing period: Gray and fox squirrels: February through March. Gray squirrels may have a "second" litter in late summer. Red and northern flying squirrels: April through May (red squirrels may continue into June). Southern flying squirrels: May through June.

Litter size: Gray squirrel: 2–4 young; fox squirrel: 2–4; red squirrel: 3–6; flying squirrels: 2–7 young.

Weaning dates: Gray squirrels begin leaving the nest at 10–12 weeks.

Amount of time young remain with parents beyond weaning date: Not long. Young female gray squirrels may stay with their mother for several months, although they won't necessarily remain near the den site.

Common Nuisance Situations

Time of year: Any time of year.

Customers calling from fall through winter (September through February) often complain about dens in attics or walls. Typically, an attic den could be home to 8–10 squirrels (red or gray squirrels) or dozens of flying squirrels (perhaps up to 50).

From March through May, most calls relate to their breeding, as females seek places to raise their young. That's when you typically find one female and her young in the attic or wall.

What are they doing?

- They den in attics, walls, sheds, barns, and chimneys, annoying people with their noise and odors.

Squirrels usually gain access via overhanging branches, power lines, or by climbing up the siding. They may fall into chimney and furnace flues, thus gaining entrance to the basement or interior of the house.

- Their nest materials might block a vent, causing a fire hazard.
- They chew on and scratch wires (another fire hazard) and also damage attic vents, eaves, screens, bird feeders, siding, insulation, household goods, and the tubing used for maple syrup production.
- They run along power lines and sometimes short out transformers.
- Squirrels also eat garden, field, and orchard crops; bird seed; and newly planted vegetable seeds.
- They'll strip the bark from trees, especially fruit trees, maples, and cedar.
- Disease risks: mange, cat scratch disease, typhus, rabies (rarely).

Best Practices

Remove artificial food sources (bird seed, pet food):

- If anyone is feeding the squirrels, persuade them to stop.
- There are metal bird feeders that close once the squirrel jumps onto them, which are effective. Other feeder designs can be modified to make them more squirrel-proof. Place a stovepipe baffle (min. length 2½ feet) or an inverted cone on the pole, at least four feet off the ground.
- Keep the area underneath the feeder clean.
- Enclose compost piles in a framed box using hardware cloth or welded wire; in a sturdy container, such as a 55-gallon drum; or in a commercial compost container.
- Feed pets indoors.
- Just so you know (and for those WCOs with commercial pesticide applicator licenses)—there are brands of sunflower seed and suet that are treated with a repellent. The active ingredient is capsaicin, the chemical that makes hot peppers taste hot.

Protect vulnerable crops:

- Plant bulbs within a cylinder of 1-inch poultry wire. Lay the wire in a trench, then plant the bulbs in it. Add some dirt, finish wrapping the wire around the bulbs, then cover with soil.
- Another option for bulbs is to plant them and then lay a piece of half-inch hardware cloth over the soil surface to reduce the squirrels' ability to dig up the bulbs. The hardware cloth should extend at least a foot around the plantings, and be covered with soil. Its mesh must be large enough for the stems to grow through, so you may need to experiment with different sizes for different plants.
- Establish a barrier around gardens and fields with fences (wire mesh, electric, or combination wire/electric fence). Use half-inch hardware cloth or welded wire. The fence must be 30 inches high, buried 6–12 inches deep, with a foot-wide L-shaped shelf that sticks out to prevent the squirrels from burrowing underneath it. Or use a 2-wire electric fence (if allowed by local ordinances) with one wire placed at 2 inches above ground, and the other at a height of 6 inches. A combination fence should have a wire at 2 inches off the ground, and along the top of the fence.
- If there aren't any mammals nesting in the tree, wrap two-foot-wide bands of sheet metal around fruit trees at 6–8 feet, to prevent squirrels from climbing the tree. This will work only if the squirrels can't leap from another tree or other object onto this tree. Attach the band loosely so the tree has room to grow. Don't staple the band onto the tree because that can prove dangerous if someone needs to cut down the tree.
- For WCOs with commercial pesticide applicator licenses: There are several repellents for use on maple sap collection equipment, lawns, gardens, outdoor furniture, and buildings. Their effectiveness is quite variable.

Prevent entry into building:

First step: Remove any current residents. Exclude them with a one-way door when the young are old enough to be mobile. (For squirrels in a chimney, see the rope trick, described in the second bullet below under "If young are present . . .")

If this is a preventive action, or there are no young present, you can:

- Replace plastic attic vents with metal designs that are securely attached to the building, or screen them with half-inch hardware cloth. Attic vents are a common entry point for squirrels.
- Seal openings at the joints of siding, overhanging eaves, and where pipes and utility lines enter buildings. Plug gaps around water, gas, and heating pipes with latex caulk. For large holes around pipes, use galvanized metal pipe chase covers, sheet metal plates, mortar, plaster of Paris, or cement.
- Cover chimney flues with commercial caps, and seal any gaps in the chimney's flashing.
- Wrap 2-foot-wide bands of sheet metal around trees that are within jumping distance (10 feet) of the building (see notes above).
- Trim overhanging tree branches so they're 10 feet away from the house.
- Screen gutter pipes, downspouts, and foundation drain pipes with quarter-inch hardware cloth.

If young are present, remove the entire family before blocking the entrance to their den:

- If the young are older and mobile, install a one-way door over the entry hole. They'll leave but won't be able to re-enter. Make a squirrel excluder of 4-inch diameter plastic pipe, 18 inches long, mounted over the opening, pointing down at a 45° angle.
- If the squirrels are caught in a chimney, give them a way to climb out. Place a small weight on a rope that's 1 inch in diameter. Drop the rope down the chimney. The weight helps you drop the rope all the way down, and then keeps the rope taut so the squirrels can climb it. Once the squirrels have left, cap the chimney so they won't enter it again.
- For trap-and-release strategies to reduce the risk of orphaning wildlife, see Chapter Five. Release red, gray, and fox squirrels on-site, during the day. Flying squirrels should be released at dusk or in the evening because they are nocturnal. Locate the young by following trails made in attic insulation. Flying squirrels don't show signs of nursing, so assume young are present during the breeding season.
- A caution about relocating squirrels. All of these squirrels rely on a cache of food to survive the winter, so if you move them too far away during that time, they'll probably starve to death. Not possibly—probably. Limit relocation to times when food is readily available.

Trapping Strategies

- Set traps over the entry hole, or as close as you can.

Live traps:

- Install one-way doors, especially with the smaller squirrels, such as flying squirrels.
- Cage traps should be 6 × 6 × 24 inches. Set and bait the trap, then prop it open for 2–3 days so the squirrels will grow accustomed to feeding in it.
- Bait with apples, nuts, peanut butter, sunflower seeds.
- Multiple-capture cage traps are available for use with flying squirrels. Place food in the trap; this may reduce their level of stress and the risk that they'll fight.

Lethal traps:

- Body-gripping traps: many new products have been released recently, so scan the markets. Options include: #110 (and a newer, slightly smaller version based on the #110, the #50-2); #120 (and its smaller cousin, the #60-2); #55; the 5 × 5 Buckeye; 3 × 3 Eradicator; and smaller Koros traps.
- Tunnel traps (cylinder with body-gripping trap inside). Less obvious to viewers. You can place a standard body-gripping trap in a wooden box or other container for an equally discreet effect.
- Modify the trigger to help ensure a top-to-bottom strike (which is more humane) and to prevent the

squirrel from refusing to enter the trap. Squirrels don't like to have anything brush against their eyes or whiskers, so separate the trigger and center it on the top or bottom of the trap. Another option is to bend the trigger into a circle; you can add a piece of thin wire or monofilament line to complete the circle, if necessary. Proper positioning helps to ensure a cleaner, more humane catch.
- Rat-sized snap traps for the smaller squirrels (flying and red squirrels). Look for models that have "covers" over the bait (see Chapter Five for details).

Other lethal techniques: In most states in our region, gray squirrels and fox squirrels are game species. Hunting may reduce the population and help alleviate some concerns.

Preferred Killing Methods

- CO_2 chamber
- Lethal trap
- Shooting, using an air rifle, a shotgun, or a .22 caliber rifle

Acceptable Killing Methods

- Stunning and cervical dislocation
- Stunning and chest compression

VOLES

Species' Names

- Meadow vole, *Microtus pennsylvanicus*
- Pine vole, *Microtus pinetorum*
- Prairie vole, *Microtus ochrogaster*

Size

4–7½ inches long; ½–2½ ounces. Meadow vole is larger than the pine vole. The meadow vole's tail is longer than its hind foot; the pine vole's tail is shorter than its hind foot.

Signs of Their Presence

- High vegetation, when mowed, reveals a network of small, crisscrossing tunnels, 1–2 inches wide, "roofed over" by vegetation, at the soil surface. Similar tunnels occur in mulched garden beds and tree and shrub borders. If the lawn's mowed, you will see runways, not tunnels. They also tunnel under plastic and paper mulch. Runways are particularly well displayed during winter thaws.
- Scat piles at tunnel crossroads and scattered along runways, ¼ inch long, cylindrical (mouse scat fits same description).
- Plant cuttings, ¼–½ inch long, scattered through tunnels.
- For the pine vole, a subterranean burrower, small holes mark the entryway to their burrows. Burrows are 3–4 inches below ground, or occasionally just below the soil surface; in this case, they resemble small mole tunnels. Pine voles may take over the abandoned burrows of moles or short-tailed shrews—and even make surface tunnels at times. Where pine vole infestation is heavy, the ground will have an almost "spongy" feel.
- Girdled trees and shrubs, especially seedlings and saplings up to about 15 years old. Ornamental and orchard plantings are equally at risk. Tooth marks (⅛ inch wide, ⅜ inch long) make a cross-hatch pattern near the ground or snow line.

(Rabbit gnaw marks are larger and not as distinct; they clip right through branches with a clean, oblique cut.)

Diet

Green plants, roots, tubers, bark, mushrooms, and occasionally snails, insects, carrion, and each other's young. They store food for the winter (grains, tubers, bulbs, and rootstock). Pine voles generally eat roots and tubers. Like rabbits, hares, and beavers, they eat their feces to extract more nutrients from grasses and tree bark, which are difficult to digest.

Typical Activity Patterns

Social style: Generally social, especially female with young.

Daily activity: All day and night, with alternating periods of rest and feeding.

Hibernator? No. In fact, voles may even breed and bear young through the winter if snow cover is deep enough to provide sufficient insulation for their nests.

Migrates? No.

Where Found

Distribution: Abundant and widespread in rural and suburban areas throughout the region. Population densities vary wildly, often in four-year cycles. Prairie voles are found primarily in the Midwest and central United States.

Habitat: Fields and moist, grassy bottomlands, but adapt well to suburban woodlots, gardens, and ornamental plantings as well as orchards. Pine voles prefer deciduous forests, brushy areas, and orchards with dense vegetation (especially apple orchards). They are excellent swimmers and decent climbers (although the pine vole is a bit clumsy).

Territory and home range: Females are scrappy fighters and territorial toward other females; males are not territorial. Females' home ranges cover roughly 75 square yards, males' about 200 square yards. The home ranges of the males may overlap those of several females and other males as well.

Breeding Habits

Pair bonding style: Polygamous.

Breeding dates: Year-round as the weather and food permits. Gestation takes about 20–23 days.

Birthing period: Year-round as the weather and food permits.

Litter size: 3–5, average 4. May see as few as 1 or as many as 9 pups.

Weaning dates: Between 2–3 weeks of age. Females may breed within days of being weaned. Males and females are sexually mature at 45 days old.

Amount of time young remain with parents beyond weaning date: Not long!

Common Nuisance Situations

Time of year: Any time of year. The type of damage changes seasonally:

- Early spring (mid-April through end of May): can ruin lawns, golf courses, some perennial bulbs (especially tulips and irises), newly planted vegetables (peas, beans), and some ornamental shrubs.
- Spring and summer: they damage hay, leafy vegetables, and legumes (beans and peas).
- Summer and fall: voles eat root crops (carrots, beets, potatoes, as well as kohlrabi).
- Fall (September through November): they damage lawns, golf courses, fruit trees, and some perennial bulbs.
- Fall and winter: they will girdle trees and shrubs (especially fruit trees and some ornamental shrubs). Pine voles will girdle the roots of many trees and shrubs during this period.

Look for damage up to the level of the deepest snow cover.

What are they doing?

- Burrow through and damage lawns and golf course turf.
- Girdle some fruit trees and ornamental shrubs.
- Eat flower bulbs, especially tulips and irises.
- Eat some vegetables in gardens and farms, especially legumes (peas, beans) and root crops (carrots, beets, potatoes).
- Eat hay crops. A population of 100 voles per acre may reduce the crop by a half-ton over the course of a season.
- Disease risks: minimal because of their infrequent contact with people, but voles can carry tularemia and hantavirus.

Debunking Myths about Voles

- Voles are often confused for moles. Here's how to tell them apart:

Voles have:	Moles have:
Small eyes	Very small eyes
Small, but definitely noticeable ears	No external ears
Furry noses	A naked, pointy snout
Small, mouse-like feet	Large front feet that are turned sideways, and big claws (excellent shovels)

Best Practices

If your strategy includes lethal control, plan to reduce vole populations before the first winter snow.

Protect ornamental plantings and lawns:

- Mow closely under and around ornamental trees and shrubs; remove vegetation and pruned branches.
- Pull mulch away from the bases of trees.
- Make vole guards for trees. The guards must be large enough to allow 5 years of growth. Circle the tree with quarter-inch hardware cloth that's buried 3–6 inches deep. The tree guards should be taller than the anticipated snow depth by about 3–4 inches.
- Mow lawns regularly.

Protect garden crops:

- Remove vegetation, ground covers, and brush piles or other plant litter near crops.
- Tilling before planting annual crops destroys tunnels and removes cover.
- Small areas may be fenced with quarter-inch mesh hardware cloth that's buried 3–6 inches deep.

Protect orchard crops:

- Follow recommendations for ornamental plantings.
- Consider the relative economic and environmental value of tilling or close mowing between rows and applying herbicide in rows to reduce cover. Rotary mowers cut closer than sickle bar mowers.
- Mow adjacent strips and drainage ditches; work to reduce vole populations in older orchard blocks (where trees are too big to be vulnerable) that border younger blocks. Clean up windfall apples.
- Trap intensively over a 5-day period. Trapping can reduce vole populations by 90%.
- Encourage predators. Voles provide 85% of a hawk's or owl's diet. All the other carnivores—foxes, skunks, weasels, coyotes—rely on them, too. However, voles are so prolific that predators alone won't usually provide effective control.

For WCOs with a commercial pesticide applicator license:

- Repellents may give short-term protection against meadow voles (they don't do much against pine voles). There are thiram and capsaicin-based repellents.
- Toxicants (zinc phosphide) will work, and may make economic sense in some situations. Some states also have chlorophacinone registered for field use in orchards, nurseries, and vineyards.

Trapping Strategies

To increase your success, trap intensively for several days. More is better.

Live traps:

- Set cage traps in their runways. Bait with apple chunks.
- Place larger multiple-capture traps (Ketch-All) in the runways with the door facing the runway. For meadow voles, use a larger multiple capture trap such as the Ketch-All or a 3 × 3 × 8 inch Sherman trap. Traps set above ground can be effective for meadow voles, but successful trapping of pine voles almost always requires that traps be placed underground, perpendicular to the runway systems. Cover the traps with a board or roofing shingle.

Lethal traps:

- The familiar mouse trap is called a "snap trap." Look for ones with an expanded "trigger" (properly, it would be called the "pan" but you're more likely to hear it referred to as a "trigger") or a clothespin design, because they're easier to set.
- For pine voles, use the mouse-sized snap trap: preferably, a design such as the Victor Quick-Kill trap, which has a lid over the bait cover. Only animals that are motivated to seek the bait will lift the lid. This means that an animal can accidentally step on the lid without triggering the trap. The trap will not fire if it's picked up. In addition to being more selective than the traditional mouse trap, this design is also more effective, because the location of the bait cup positions the mole in the perfect strike position.
- For the larger meadow vole, you may want to switch to a larger trap.
- Place traps in the runways, spaced every 15–20 feet. Cover the trap with a bent shingle "roof."
- Set snap traps in pairs. This is much more effective. Place them side-by-side, perpendicular to the runway, with the trigger snapping into the runway.
- Bait with apple chunks.
- To protect young children, place traps in a cage with 1-inch mesh, a bait station, a coffee can with both ends cut out, or in PVC pipe (remember to test that the trap will spring within its container).
- Wildlife rehabilitators may appreciate donations of voles, which are used to feed some snakes, birds of prey, and other animals. Be sure that no poisons have been used during previous control efforts. You can double-bag the voles and freeze them.

Preferred Killing Methods

- Lethal trap
- CO_2 chamber
- Cervical dislocation

Acceptable Killing Methods

- Pesticides, in appropriate settings (often not recommended for urban or suburban areas)

Control Strategies that Don't Work Particularly Well

- Ultrasonic devices don't work.
- Repellents don't work well against pine voles.

WOODCHUCK or GROUNDHOG *(Marmota monax)*

Size

20–27 inches long, excluding tail; 5–12 pounds

Signs of Their Presence

- Adults often seen basking in the sun, in a grassy area, on a fence post, stone wall, large rock, or fallen log—always near its burrow.
- Sounds: Occasional sharp whistles and low churrs, given at times of danger.
- Odor is distinctive. Will often see flies around an active burrow.
- Scat: Rarely seen (woodchucks excavate a privy off their main burrow).
- Evidence of their feeding: Chewed wood. Chewing on fresh plants similar to that of rabbits; difficult to identify as woodchuck's without supporting evidence.
- Dens: Will see a large mound of dirt and stones by the main entrance to their burrow; the secondary entrances, which were dug from the inside, generally don't have a dirt mound by their opening. Well-worn trail from entrance to entrance, or to the garden.

Diet

Herbivore. Woodchucks eat succulent grasses, weeds, clover, fruits (apples, cherries, pears), berries, field and garden crops (cabbage, lettuce, beans, peas, carrots, alfalfa, soybeans), and ornamental plants (they love phlox). They'll climb trees to take fruits such as cherries, apples, and pears.

Typical Activity Patterns

Social style: Generally solitary.

Daily activity: Diurnal, most active in the early morning and evening. They rely on dew as their water source. Woodchucks have good eyesight and are good swimmers. They'll climb trees up to a height of about 20 feet, although more usually they keep to 8–12 feet.

Hibernator? Yes. Hibernates deeply from the time of the first heavy frost through early spring. Occasionally hibernates in small groups.

Migrates? No.

Where Found

Distribution: Throughout the region.

Habitat: Meadows, woodlots, hay fields, pastures, hedgerows, idle fields, parks, suburbs. Dens usually found in open fields; near fence rows or woodland edges; under barns, sheds, porches, decks, stone walls, and wood piles.

Territory and home range: Territorial. Woodchucks may skirmish to establish dominance. Subordinate woodchucks avoid dominant ones. Home ranges overlap and are usually small. Woodchucks rarely travel more than 50 yards from their den, even to feed. Their burrows can be 2–5 feet deep and as much as 60 feet long. There are usually 2 or 3 (but perhaps as many as 5) entrances, possibly including a well-hidden, straight-down "plunge hole."

Breeding Habits

Pair bonding style: Polygamous. Females raise young alone.

Breeding dates: Late February through March.

Birthing period: Late March to early May. Gestation takes about 31 days.

Litter size: 3–4.

Weaning dates: at 5–6 weeks.

Amount of time young remain with parents beyond weaning date: Young stray from burrow alone at 6–7 weeks, mid-June to early July. Mother drives young from her burrow by July.

Common Nuisance Situations

Time of year: Calls peak in July and August, although their damage may begin in spring and last into the fall.

What are they doing?

- Feeding, or just filing down their front teeth, which never stop growing. Woodchucks raid gardens, fields, lawns, orchards, and nurseries, and may gnaw or claw on shrubs and fruit trees. Occasionally chew on outdoor furniture, decks, and siding while scent-marking or filing their teeth.
- Marking their territories: They may strip off the bark at the base of a tree that's near their burrow entrance.
- Burrowing. Look for burrow entrances among shrubs near vegetable and ornamental gardens; under woodpiles, brush piles, and stone walls; under sheds, porches, decks, and crawl spaces. Burrows in fields may damage agricultural equipment, while those in pastures may trip livestock, resulting in injuries.
- Disease risks: Low. Mange, rabies (rarely), raccoon roundworm.

Best Practices

Year-old woodchucks will search for and occupy abandoned burrows. You can try filling in the entrances, but woodchucks may re-open the holes. Burrows systems are difficult to excavate, so there are often younger woodchucks looking for empty burrows. After lethal control, burrows may be reoccupied in 2–3 weeks, on average.

Remove artificial food sources and shelter:

- Remove brush piles and debris, and keep areas well-trimmed.

Protect vulnerable crops:

- Woodchucks are excellent climbers and diggers. They can climb trees and may be found 20 feet up in a fruit tree. For this reason, use a combination woven-wire and electric fencing. Erect a "rat wall" fence around gardens and fields. Use 2 by 4 inch welded wire that's 2 feet high, buried 1 foot deep; if you prefer, you can bury it only 1–2 inches down, if you bend the edge outward into an L shape that sticks out at a 90° angle to prevent the woodchucks from burrowing underneath it. Also bend the top 15 inches of the fence out at a 45° angle to keep them from climbing over it, or add an electric wire strung 4–5 inches above ground level, and 4–5 inches from the outside of the fence.

Keep them from creating dens under buildings: First step: Remove any current residents. Exclude them with a one-way door when young are old enough to be mobile.

If this is a preventive action, or there are no young present, you can:

- Screen areas under decks, porches, and houses with the rat wall fence, as described above. Attach the top of the fence to the structure.

If young are present, remove the entire family before blocking the entrance to their den:

- If the young are older and mobile, install a one-way door over the entry hole. They'll leave but won't be able to re-enter.
- For trap-and-release strategies to reduce the risk of orphaning wildlife, see Chapter Five. Cover cages during transport to minimize stress. Release them on-site, preferably in the morning. Use a larger box with a 7-inch hole. (One WCO recommends a 2 × 2 × 1 foot box.)

Trapping Strategies

Live traps:

- Cage trap should be at least 10 × 10 × 24 inches. Double-door traps should be at least 10 × 10 × 30 inches.
- Conceal the trap, using grass or canvas.
- Choose the size of trap based on the size of the burrow's hole, but realize that woodchucks can destroy a small or lightweight trap.
- Foothold traps, #1 or #1½.
- Bait with apples, cantaloupe, cabbage, carrots with their green tops, fresh peas, or lettuce.

Woodchucks may ignore the bait if food is plentiful. Use a cage trap that's already held a woodchuck, because the scent will attract other woodchucks, especially males.

- Check traps twice daily, and provide shade and protection from weather. Woodchucks overheat easily.
- Clean brush away from the opening of the trap so it won't interfere with the door.
- You can also set trap without bait, placing it directly in front of the hole. Dig down a bit and use fencing to guide the woodchuck into the trap.

Lethal traps:

- Spring is the best time for control, when the adults are active but before the young are born. It's also easier to see the burrows then, and other animals are less likely to be inside. Woodchuck burrows provide shelter to several species.
- Body-gripping traps, #160, #220, #120, or a 5 × 5 Buckeye, placed at the entrance to the burrow. To reduce the risk of catching pets or unintended wildlife, cover the hole and the trap with a weighted box or hardware cloth. Another option is to add a one-way trigger to the trap, so it fires only when the woodchuck is leaving its burrow. See Chapter Five for details.
- Modify the trigger to help ensure a top-to-bottom strike (which is more humane) and to prevent the woodchuck from refusing to enter the trap. Woodchucks don't like to have anything brush against their eyes or whiskers, so separate the trigger and center it on the bottom of the trap. Proper positioning helps to ensure a cleaner, more humane catch.

Other lethal techniques:

- For WCOs with commercial pesticide applicator licenses: Aluminum phosphide gas, a restricted-use pesticide, may be used to kill woodchucks in their burrows. The gas is extremely toxic and must be used carefully, but it doesn't pose a fire hazard.
- Carbon monoxide gas cartridges, a registered product, may be used to kill woodchucks in their burrows. These gas cartridges do pose a fire hazard so don't use them near buildings, under sheds, or near stumps. They could ignite grass, buildings, gasoline, and other flammable objects.

Preferred Killing Methods

- CO_2 chamber
- Lethal trap
- Shooting, using a shotgun, a .22 caliber rifle, or a centerfire rifle where safe (target the head, if no rabies testing is needed, or the heart/lungs)
- Lethal injection of barbiturate, if possible

Acceptable Killing Methods

- Stunning and chest compression
- Pesticides (carbon monoxide fumigants) for WCOs with a commercial pesticide applicator license

Methods that Don't Work Well

- There aren't any registered repellents for woodchucks.
- Commercial deer and rabbit repellents, as well as some pesticides thought to repel woodchucks, weren't that effective at keeping them away from crops.

SNAKES

Species' Names

This account will focus on the four nonvenomous species that are often found near homes:

- Common garter snake, *Thamnophis sirtalis*
- Eastern milk snake, *Lampropeltis t. triangulum*
- Black rat snake, *Elaphe o. obsoleta*
- Northern water snake, *Nerodia s. sipedon* (common around homes with nearby ponds)

How to tell a nonvenomous snake from a venomous one:

Nonvenomous snake:	Venomous snake:
Pupil: round	Pupil: like a cat's eye, vertical
No pit between the eye and nostril	Pit between the eye and nostril
Shape of head variable, often slender	Broad, triangular-shaped head
Scales underneath tail, toward tip, are divided	Scales underneath tail, toward tip, undivided

Size

Garter snake: 2–4 feet, usually smaller.
Milk snake: Up to 4½ feet long. Sometimes confused for the copperhead or water moccasin.
Black rat snake: Up to 4½ feet long.
Water snake: Up to 4½ feet long. Sometimes confused for the water moccasin or copperhead.

Signs of Their Presence

- Water snakes often sun themselves on boat docks.
- Milk snakes and black rat snakes are frequently found in barns.
- On cool days, you may find snakes (especially the black rat snake) resting on top of the heating ducts in buildings heated with forced hot air.
- Sounds: Silent, except some snakes may rattle their tails when annoyed, including the rattlesnakes, milk snake, and black rat snake.
- Scat: Elongated, whitish. The scat of a black rat snake may be large.
- Large shed skin (over 2 feet long): Probably from a black rat snake.
- Evidence of their feeding: Hard to identify, because they swallow their prey whole.
- Garden and crop damage: None, because they are strictly carnivores.
- Building damage: None, because they use only existing holes and entryways, and don't create others.

Diet

Carnivores. Garter snakes eat mostly earthworms but also slugs, amphibians, fish, crayfish, insects, small birds, other snakes, and carrion. Milk snakes eat rodents and other snakes. Black rat snakes eat mostly small rodents and birds. Water snakes eat mostly fish, also amphibians, insects, and crayfish. Other snakes add spiders, bird eggs, and rabbits to the menu.

Typical Activity Patterns

Social style: Solitary, but may hibernate with other snakes, even those of different species.

Daily activity: Mostly diurnal. Milk snakes are usually nocturnal.

Hibernator? Depends on the locale. In the northern parts of our region, snakes will often hibernate, usually beginning in October or November and lasting through March or April, in a large group that may include snakes of different species. They don't hibernate in the southern portions of the region.

Migrates? No, but they do move to their hibernating site.

Where Found

Distribution: The common garter snake, milk snake, black rat snake, and water snake are common throughout the region.

Habitat:

- Garter snake—wide variety of moist areas, from woodlands to marshes to fields.
- Milk snake—usually seeks brushy or woody cover in many of same habitats favored by the black rat snake. Also often found in barns or other areas that are home to rodents and other snakes.
- Black rat snake—woods, fields, rocky hillsides, and river bottoms. Often found in barns, homes, and garages, or other areas that are home to rodents.
- Water snake—rivers, brooks, wet meadows, ponds, and swamps, preferring still or slow-moving water, in areas with overhanging branches and rocks (for cover and basking). Common near dams and bridges. Often suns on boat docks.
- Black rat, garter, and milk snakes will follow their prey into barns and houses, usually in basements but sometimes in attics. That's especially true for the black rat snake, because it's an excellent climber. Most snakes prefer sunny areas where rock or wood piles and other debris provide cool, shaded hiding places. They move from sunny to shady areas to regulate their body temperatures.

Territory and home range: Not generally territorial, but snakes are faithful to den sites ("hibernacula") in their home ranges. They'll reuse these sites from year to year, and are sometimes found in large numbers. This makes them vulnerable to habitat destruction and persecution.

Breeding Habits

Pair bonding style: Polygamous.

Mating dates: Garter: first few warm days after emerging from hibernation, usually mid-March through May, then mates again in the fall before entering hibernation. Milk snake: June. Black rat: May through June. Water snake mates in April through May and again in the early fall.

Egg-layers: Black rat and milk snakes lay their eggs in loose soil, decaying wood, or sawdust or manure piles. Black rat snakes lay their eggs from May through early July. Milk snakes lay theirs in mid-June through July.

Live young: Garter and northern water snakes.

Birthing/eggs hatch dates: Garter snakes give birth July through early September. Northern water snakes give birth in August through early October. The eggs of the black rat snake hatch between July and September, those of the milk snake from late August through October.

Clutch size: Garter: avg. 14–40 (3–85). Milk snake: avg. 13 (6–24). Black rat snake: avg. 14 (6–24). Northern water snake: avg. 20–40 (10–76).

Weaning dates: Young are able to fend for themselves at birth or upon hatching.

Common Nuisance Situations

Time of year: Spring through fall.

Snakes don't damage buildings or eat crops. They enter buildings only through existing holes, cracks, or "doors" (such as an open window). Some people are afraid of snakes. Others welcome them, because some snakes eat mice and rats and help to control those pest populations. Remember, however, that if a snake can get into a home, so can other creatures.

What are they doing?

- These snakes sometimes hibernate in buildings, especially the basements of old houses with stone foundations. They usually enter houses through torn screens, open basement windows, cracks in the foundation, or through gaps next to pipe and cable entrances.
- They follow prey (mice, insects) into cellars, crawl spaces, attics, barns, sheds, and garages. They may also be found in wood piles and debris, in heavily mulched gardens, and under shrubs, tarps, or planks. They seek cool, damp, dark places.
- Their presence may frighten or annoy people. Several species, including the garter snake, may emit a foul and musky smell when handled.

- Disease risks: salmonellosis (food poisoning).
- Injury risks: nonvenomous snakes have tiny teeth. They leave a faint, U-shaped bite mark. Their bites rarely hurt much or cause problems, with the exception of the northern water snake, which is known for its nasty bite. If bitten, remain calm and get medical help. Do not use a commercial snake bite kit; they tend to do more harm than good.

Debunking Myths about Snakes

- If bitten by a venomous snake, *do not try to suck out the poison and do not slice the wound.* Get medical help.
- Snakes don't dig. They can't make holes.
- The milk snake (nonvenomous) is often confused with the copperhead (venomous).
- The northern water snake (nonvenomous) is often confused with the water moccasin (venomous).

Best Practices

Remove their food sources:

- Grains, pet food, and bird seed will attract mice, insects, and other species, which then attract snakes. Keep these foods in mouse- or insect-proof containers. Exclude insects and mice from your buildings.
- Reduce the amount of mulch in your garden, around trees and shrubs (again, this will discourage voles and other potential snake food).

Reduce their shelter:

- Mow closely around the building.
- Remove wood piles, junk, and piles of rocks.
- Don't plant right next to the foundation, because that provides cover for snakes and many pests.

Prevent them from entering building:

- Seal all openings that are larger than ¼ inch with mortar, expanding foam, copper mesh (Stuf-Fit), quarter-inch hardware cloth, or sheet metal.
- Fences may keep them out. Use quarter-inch hardware cloth. The fence should be 3 feet high, buried 1 foot deep, with the bottom edge bent outward into an L-shaped shelf that sticks out at a 90° angle to prevent the snakes from slipping under the fence. Fences are more likely to work well around a small area. Otherwise, high maintenance needs may make this impractical, because some snakes would be able to travel through chipmunk tunnels that pass under a fence.
- Although snakes cannot create holes, they will use holes that were made by rodents and other animals. To prevent the problem from happening again, you may need to exclude the species that created the hole.

Trapping Strategies

Direct capture methods and live traps for nonvenomous snakes:

- Pick them up, wearing heavy leather gloves for protection. Support the snake's entire body to keep it calm. Hold snakes behind the head to keep them from biting you.
- With care, snakes can also be captured with a "snake stick," which is a catchpole modified for snakes. A forked stick can also be used (carefully!) to pin down a snake.
- They can be scooped into a garbage can using a scoop or shovel.
- To live trap water snakes, add a brick-sized piece of plastic foam (Styrofoam) to a minnow trap (so the trap will float, allowing the snake to surface for air). Bait with about a half-dozen minnows. Attach a rope to the trap for easy retrieval, then float it in near the shoreline.
- If the snakes can't be found, you can lure them to a spot where they can be easily captured. Place piles of damp towels or burlap sacks on the basement floor, near the walls. Cover the pile with a dry burlap bag to keep it moist. In a few days, return to the pile during the middle of the day, when the snakes are most likely to be there. Scoop up the pile with a large shovel, put it into a large garbage can, and carry it outside.

- Several variations on the above technique: Use a board or a piece of plywood instead of the pile of towels or burlap sacks. Place some decomposing grass clippings on top of the board, which should be 1 inch off the ground. Dead mice or mouse droppings placed underneath the board will help to attract snakes. Check every few days. This technique works well with garter snakes and black rat snakes, but is not as effective with water snakes.
- Create a reptile tube trap, based on a technique suggested by Humane Society animal capture consultant Dave Pauli. Inside this trap, the temperature should be just right for the snake—more appealing than the surrounding area. The trap is a piece of thin-walled PVC tubing that's 2–3 feet long. Drill a few ⅛-inch air holes along the length of the tube. Cap one end. In cold weather, place a disposable hand warmer, battery-operated electric sock, or heating pad in the far end of the tube, along with some soft cotton rags. If you have fresh rodent droppings, you may want to toss a few in there, too. (In hot weather, substitute an ice pack or cold wet rag for the heat source.) Then drill a 1-inch hole into a cap and use it to cover the other end of the tube. Although the snakes can leave the trap, they usually stay inside because it's more comfortable for them. You can install a one-way valve by affixing a 1½-inch stiff plastic circle over the inside of the cap. The snake can push its way in, but can't leave easily. This trap also works with other reptiles, such as lizards.

Venomous snakes: Only experienced snake-handlers should capture venomous snakes. They're often handled using a catchpole and then transferred into a sturdy container.

Using a one-way door to exclude snakes from a building: Technique developed by wildlife consultant William Bridgeland: Roll aluminum insect screening into a tube, then attach it over the entry hole, which is usually found in the foundation. Angle the tube up slightly and flatten its outer end a bit. Leave the tube in place for at least two weeks while snakes are active (summer). Don't substitute another material for the insect screening. Snakes may use scent to find their entry holes. Insect screening scatters odors, which would make it harder for the snakes to locate the entry hole, but other materials might retain their scent and direct the snakes back to the opening. If they find the opening, they may be able to get back inside.

For WCOs with a commercial pesticide applicator license: There's a registered repellent, Snake-A-Way. Its effectiveness varies greatly depending on the species, and it has been ineffective for nonvenomous snakes.

Preferred Killing Methods

First, discuss the situation with your customer. Is it really necessary to kill the snake? And remember, is it legal? Would removal and exclusion be sufficient? If not:

- Shooting, using bird shot (target the head)
- Stunning and decapitation

Strategies that Don't Work Well

- Large glue boards are sometimes intended for use with snakes. A glue board is a layer of long-lasting adhesive spread over a surface, usually cardboard or plastic. The snakes get stuck in the adhesive (glue boards are used to capture mice, rats, and snakes). Although some call the glue board a live trap, it's not often used that way. In fact, some biologists believe that you cannot remove snakes from the trap unharmed because the oil that is used to loosen the glue may injure their scales.
- The authors do not consider the use of glue boards a best practice for vertebrates, especially in regards to nonvenomous snakes. In most cases, snake problems are the result of a customer's intolerance. Snakes don't damage buildings. Unlike pest mice and rats, which can pose a significant health and safety hazard to people, the presence

of snakes in a building is unlikely to result in disease transmission. There are effective and humane removal methods.

For More Information

- Species identification: www.kentuckysnakes.org
- Massasauga fact sheet: www.dec.state.ny.us/website/dfwmr/wildlife/endspec/massfs.html
- Timber rattlesnake fact sheet: www.dec.state.ny.us/website/dfwmr/wildlife/endspec/tirafs.html

PIGEON (*Columba livia*)

Species' Name

Previously also known as the "rock dove," the pigeon's common name has recently been changed to "rock pigeon."

Size

This exotic species is about 11 inches long, and 13 ounces.

Signs of Their Presence

- The bird itself is the most obvious sign.
- Sounds: Distinctive cooing, clicking as wing tips touch during take-off.
- Droppings: May accumulate on rafters, building ledges, public areas.
- Nests: Crude-looking platform nests of sticks, twigs, and grasses. Find them near dormer tie-ins (the joint where the dormer meets the roof), on building ledges, in steeples, attics, and lofts, on top plates of pole barns, on the girders of bridges, and on ornamental architectural features of classic buildings.

Diet

Mostly herbivorous. Pigeons eat mostly seeds and grains, but will eat garbage, livestock manure, insects, and the bread and crackers that people feed them.

Typical Activity Patterns

Social style: Sociable, nesting and foraging in large flocks.

Daily activity: Diurnal.

Hibernator? No.

Migrates? No.

Where Found

Distribution: Common in cities and around farms and certain agricultural businesses throughout the region.

Habitat: Prefer domestic environments such as town and city parks, buildings, and bridges; grain elevators, feed mills, and farm yards.

Territory and home range: Pigeons will defend a small area immediately around their nest site.

Breeding Habits

Pair bonding style: Monogamous, but both sexes may occasionally mate with others, too. The male guards the female and nest. Both sexes feed the young.

Breeding dates: Year-round, with peaks in the spring and fall. Females lay more eggs even before the young leave the nest.

Clutch size: 1–2 eggs. Young hatch in about 8–12 days.

Fledging dates: Young leave the nest about 10 days after hatching.

Amount of time young remain with parents beyond fledging date: Very little, if any. If the young return to the nest site when the parents are raising a second brood, they will be driven off.

Common Nuisance Situations

Time of year: Any time of year.

What are they doing?

- Droppings deface and corrode building facades and may kill plants. They're unpleasant on park benches, statues, and cars. Under certain conditions, the droppings can promote the growth of the fungus that causes histoplasmosis, an airborne disease that affects people.
- Eat or contaminate stored grain.
- One of their parasites, the northern fowl mite, is also a major pest of poultry. Other pigeon parasites (mites, fleas, lice) will bite people. Some of their parasites destroy fabric or stored foods.
- Like other birds, flocks of pigeons may cause plane crashes.
- May transfer disease from one livestock facility to another.
- Disease risks: histoplasmosis, salmonellosis (food poisoning), cryptococcosis, pigeon ornithosis, encephalitis, and Newcastle disease, among others.

Best Practices

Remove artificial food and water sources (bird seed, pools):

- If anyone's feeding the pigeons, persuade them to stop. This can be challenging because many people love to feed pigeons.
- Clean up spilled grain around feed mills, grain elevators, railcar clean-out areas, and barns.
- Clean up spilled bird seed around feeders.
- Eliminate pools of standing water.

Make outdoor roosts less appealing:

- An overhead grid-wire or netting system will keep pigeons from landing in a courtyard. Use 80-pound monofilament wire or ¼- to 1-inch polypropylene netting spaced in a square grid, with the wires one to two feet apart.
- To keep pigeons off support cables, narrow ledges, conduit, and other narrow areas, use a commercial "post-and-wire" system (Bird Barrier, Birdwire). With care, you can create a homemade version, but it will probably be much harder to install, so it's probably not a practical approach. But if you'd like to try, here's how. Stretch steel wire (16- to 18-gauge) or monofilament line (80-pound) in parallel lines across the area. The lines must be very tight, so fasten the wires to L-brackets with turnbuckles to remove slack. Attach the brackets to the wall using cable clamps or aircraft hose clamps, which can handle the high torque load on the wires.
- Some WCOs report great success repelling pigeons with the Avian Dissuader. This laser is powerful and can damage your eyes; seek proper safety training before using this product.

- Dousing the birds with water from hoses or sprinklers that are mounted near their roost may work. Be persistent.

Keep them out of, and off of, buildings:

- Seal all openings to eaves, lofts, steeples, and vents. Many materials work: metal, wood, glass, masonry, galvanized ¼- to 1-inch hardware cloth, and plastic or nylon netting.
- To keep them out of sheds, barns, garages, hangars, and warehouses, staple ¼- to 1-inch polypropylene netting to the underside of the roof beams.
- To keep them off ledges:
  - Install a "post-and-wire" system, as described above.
  - Fasten wood, stone, sheet metal, plastic foam (Styrofoam), or acrylic plastic plates (Plexiglas) to the ledge at a 45° angle so the birds can't comfortably perch there.
  - Attach a sharply pointed steel device to the ledge. There are a few variations, including porcupine wire (prongs point out in many angles), Ecopic (vertical rods), and a steel coil that looks like a Slinky. Pigeons don't like to land on these objects because they hurt, but some will foil these devices by layering nesting materials over them. If that happens, remove the nest. If the prongs are too widely spaced apart, the pigeons will find it easier to perch on them.

    Install electric shock devices on the ledge (Avi-Away, Flyaway, and Vertebrate Repellent System). When the bird lands, it receives a painful shock but is not killed.

WCOs with a commercial pesticide applicator license: There are registered pesticides for use against pigeons:

- Nontoxic repellent: Polybutenes are sticky and pigeons don't like to land on ledges, signs, or other surfaces that have been treated with them. Polybutenes can affect other species, and they can be messy and hard to remove. For these reasons, consider restricting your use of this tool to indoor applications.

  Toxic repellent: Avitrol, a restricted-use pesticide, is available in a whole-corn bait mixture. Here's how it works: Pigeons that eat the treated grain will behave erratically or give warning cries, frightening the others in the flock. The birds that eat the Avitrol usually die. Like any lethal technique, this pesticide must be used carefully. Make sure there are no threatened or endangered raptors feeding in the area because if they ate a poisoned pigeon, they might also die.

Trapping Strategies

The pigeon is an exotic species, so please do not release any into the wild (Chapter Two explains why). If a few pigeons are causing some trouble and your customer prefers nonlethal techniques, rest assured that releasing a few birds is not going to make a significant difference in your state's pigeon population. Unfortunately, they are both abundant and well established.

Live traps:

- Pigeon traps vary in size. If you need to capture more birds, use a bigger trap or several smaller traps.
- Trap near roosting, loafing, and feeding sites. If they're roosting in a barn, you can trap them at night using nets.
- In hot weather, trap near their water sources, such as a rooftop air-conditioning system.
- Pre-bait the traps for 3–4 days by placing corn or millet around the outside of the trap.
- Commercial pigeon trap. Various sizes. This is basically a box trap with a one-way door.
- Cage trap designs include the funnel trap, lily-pad trap, and clover-leaf trap. The name describes the shape of the trap. These portable traps are made from screen and direct the birds inward. They work best if you leave 4–5 pigeons inside as a decoy to attract other pigeons (of course, leave

food and water for the birds daily, and partially cover the trap to protect them from weather extremes).

If the birds become trap-shy, leave the traps open for 2–3 days, then reset for 4–5 days. If it's still not working, choose another site.

For construction details on these traps, see the "Pigeons" chapter in *Prevention and Control of Wildlife Damage*. (See the full citation in resource list.)

Lethal techniques:

- Shooting, with a .22 caliber rifle, shotgun, or air rifle, can be used to remove a small flock.
- Destroy nests and eggs. If you're persistent, this may be effective. Make a sweep of the area every two weeks, and use this technique in conjunction with other management methods.
- Avitrol, the toxic repellent mentioned previously, does kill the birds that eat the treated bait. There is a risk of poisoning birds of prey, too, if they eat poisoned pigeons. Use the product carefully. Use of Avitrol is limited in some cities by local laws or regulations.

Preferred Killing Methods

- CO_2 chamber
- Cervical dislocation
- Shooting, using an air rifle, a shotgun, or a .22 caliber rifle with bird shot

Acceptable Killing Methods

- Stunning and chest compression

Control Strategies that Don't Work Well

- Ultrasonics don't work—the birds can't hear them.
- Pigeons quickly become accustomed to noise-making devices of many types. After all, they live happily under eight-lane bridges. At best, expect only a short-term effect.
- Lights, flags, balloons, rubber snakes, and owl decoys may amuse your neighbors but they won't scare the pigeons for long.

EUROPEAN STARLING *(Sturnus vulgaris)*

Size

This exotic species is about 3 ounces. Body is 11 inches long.

Signs of Their Presence

- The bird itself is the most obvious sign. It's a hunch-backed, short-tailed, and robin-sized bird with swift, direct flight, not rising and falling like many blackbirds. Starlings are often seen in large, noisy flocks.
- Sounds: Cries and songs range from a raucous to nearly melodic chatter, made up of dozens of variable (and loud) wavery, whiney, wheezy sounds mixed with high whistles and imitations of other bird calls.
- Droppings: Buildup of droppings on rafters, building ledges, public areas.
- Nests: Coarse-looking nests of sticks and stems in any hole or cavity. They'll nest in such places as trees, birdhouses, cracks in buildings, within eaves, on cliff faces or building ledges, on roof beams inside barns and warehouses, and on shopping center signs.

Diet

Omnivore. Seeds and fruits (native and cultivated), and insects—especially grubs, which are essential during breeding season. Insects and other invertebrates make up about half of their diet. Starlings will gladly feast on every orchard and berry crop. They eat livestock rations, picking out the high-protein supplements mixed into the feed. Starlings often contaminate more than they actually eat. They'll eat garbage, too.

Typical Activity Patterns

Social style: Sociable outside the breeding season. Fall flocks are smaller (up to several thousand birds) and spread over a large area. In winter, starlings gather in much larger flocks (sometimes over a million birds) that are concentrated in smaller areas (few acres). They may use the same winter roost year after year.

Daily activity: Diurnal.

Hibernator? No.

Migrates? Some do, some don't. Yearlings are more likely to migrate. The starlings that do migrate may travel up to several hundred miles.

Where Found

Distribution: Common in cities and around farms throughout the region.

Habitat: Urban, suburban, and rural areas that offer nest sites (holes in trees, buildings) and foraging areas (parks, lawns, fields, pastures, livestock facilities, dumps).

Territory and home range: Starlings are territorial during the nesting season.

Breeding Habits

Pair bonding style: Monogamous. Both parents build the nest, incubate the eggs, and feed the young.

Breeding dates: Early to mid-spring.

Clutch size: 4–7 eggs. Young hatch in 11–13 days. Females may lay a second clutch, but it's apt to be less productive.

Fledging dates: Young leave the nest after about 3 weeks.

Amount of time young remain with parents beyond fledging date: They don't.

Common Nuisance Situations

Time of year: Any time of year.

What are they doing?

- Nesting in attics, under the eaves, and in soffits and other openings in buildings.

- Where enormous flocks (up to a million starlings) gather, they can be intensely noisy. Their droppings smell bad, and are corrosive and slippery to walk on. Under certain conditions, the droppings can promote the growth of the fungus that causes histoplasmosis, an airborne disease that affects people.
- These "feathered bullets" can cause plane crashes. Starlings travel in large flocks that can collide with a plane or get sucked into the engine.
- Eat (and contaminate) livestock feed, grains, fruits (grapes, peaches, blueberries, strawberries, figs, apples, cherries), and garbage.
- Take over nesting sites of native songbirds (purple martins, flickers and other woodpeckers, bluebirds) and wood ducks. If nest sites are limited, starlings may severely hurt the populations of these native birds.
- May transfer disease (transmissible gastroenteritis) from one livestock facility to another.
- Disease risks: histoplasmosis to people, transmissible gastroenteritis to livestock, especially pigs.

Best Practices

Remove artificial food and water sources (bird seed, grains, pools):

- If anyone's feeding the starlings, persuade them to stop.
- Clean up spilled grain.
- Store grain and bird seed in bird-proof containers or structures.
- Use bird-proof livestock feeders: flip-top pig feeders (constant banging keeps starlings uneasy); lick wheels for liquid supplements; auto-release feeders for high-protein rations.
- Livestock feed that's compressed into cubes or blocks larger than ½ inch across are too big for starlings to swallow. Avoid $\frac{3}{16}$-inch pellets because starlings eat them six times more quickly than granular meal. And don't feed your livestock on the ground—that's like setting a place for the starlings.
- Starlings really like high-protein supplements, so mix the supplements into the feed thoroughly to make it harder for the birds to pick it out.
- Delay feeding livestock until late afternoon or nighttime, if possible.
- Feed livestock in a covered area, such as a shed, which is less attractive to the birds.
- Starlings are attracted to water. You have two choices with pools, troughs, and other containers that catch water: either drain them, or keep the water level out of the starlings' reach. Do that by keeping it low enough so they can't dip in easily while perching on the edge, and deep enough so they can't stand in the bottom.

Make outdoor roosts less appealing:

- In a dense grove, thin trees. If a tree is a preferred roost site, trim out about a third of its branches, concentrating on the inside center of the crown. This will reduce the number of available perches and increase the birds' exposure to weather. With less and poorer shelter, fewer starlings will congregate.
- A combination of frightening techniques (noises and visual deterrents) may convince the starlings to leave a roost. Some WCOs report excellent results after 3–5 days, if the birds are roosting in 1 or 2 trees. As always, your chance of success increases if the techniques are used together and in an unpredictable fashion. Try noisemakers such as tape-recorded distress and alarm calls, shell crackers, propane cannons, shotguns, and beating on tin sheets or barrels. Eye-spot balloons, hawk kites, lights, and Mylar reflectors, and dousing the birds with water from hoses or sprinklers that are mounted nearby, may also work. Starlings that are used to people and city noises may not respond.
- Use bird boxes with openings that are too small for starlings. Modify wood duck boxes to make them less attractive by placing them horizontally instead of vertically. Build them out

of a 2-foot-long piece of stove pipe that's 1 foot in diameter. For details, see the "European starling" chapter in *Prevention and Control of Wildlife Damage*. (See resource list for full citation.)

Keep them out of and off of buildings:

- Remove the nest by hand, if possible. Then seal the entry hole.
- Seal all openings that are bigger than 1 inch in diameter. Many materials work: metal, wood, glass, masonry, galvanized quarter-inch hardware cloth, and plastic or nylon netting.
- Cap chimneys. (A cover that slips inside the tile liner is adequate.)
- To keep them off ledges:
    - Fasten wood, stone, sheet metal, plastic foam (Styrofoam), or acrylic plastic plates (Plexiglas) to the ledge at a 45° angle so they can't comfortably perch there.
    - Attach a sharply pointed steel device to the ledge. There are a few variations, including porcupine wire (prongs point out in many angles), Ecopic (vertical rods), and a steel coil that looks like a Slinky. Birds don't like to land on these objects because they hurt, but some will foil these devices by layering nesting materials over them. If that happens, remove the nest. If the prongs are too widely spaced apart, the starlings will find it easier to perch on them.
    - Install electric shock devices on the ledge (Avi-Away, Flyaway, and Vertebrate Repellent System). When the bird lands, it receives a nasty shock but is not killed.
- To keep them out of farm buildings and warehouses, hang 10-inch-wide vinyl or rubber strips over open doorways (with no more than a 2-inch gap between strips). You and your equipment will pass through, but the birds won't.
- To keep them from nesting and roosting in sheds, barns, garages, hangars, and warehouses, staple ¼- to 1-inch polypropylene netting to the underside of the roof beams.

For WCOs with a commercial pesticide applicator license:

- Nontoxic repellent: Certain grape-like flavorings (methyl anthranilate), when sprayed on fruit, repel starlings but don't harm the fruit or people (Rejex-it, BirdShield). This may be impractical because of cost.
- Nontoxic repellent: Polybutenes are sticky and starlings don't like to land on ledges, signs, or other surfaces that have been treated with them. Polybutenes can affect other species, and they can be messy and hard to remove. For these reasons, consider restricting your use of this tool to indoor applications.
- Toxic repellent: Avitrol, a restricted-use pesticide, is available in a whole-corn bait mixture. Here's how it works. Starlings that eat the treated grain will behave erratically or give warning cries, frightening the others in the flock. The birds that eat the Avitrol usually die. Like any lethal technique, this pesticide must be used carefully. Make sure there are no threatened or endangered raptors feeding in the area because if they ate a poisoned starling, they might also die.

Protect valuable crops:

- Cover berries, cherries, and grapes with netting.
- Methyl anthranilate (the nontoxic repellents described above) may help protect fruit.

Trapping Strategies

The European starling is an exotic species, so please do not release any into the wild (Chapter Two explains why). If a starling has fallen into someone's chimney and your customer prefers nonlethal techniques, rest assured that releasing one bird is not going to make a significant difference in your state's starling population. Unfortunately, they are both abundant and well established.

Direct capture techniques and live traps:

- If the starlings are roosting on a low perch, you may be able to capture them at night, using spot lights and dip nets.

- Nest-box trap is useful only during nesting season. A bird box that's modified to close the opening once the starling hits the trigger panel on the bottom of the box. A mouse snap-back trap can be used to create the trigger mechanism for this starling trap.
- Decoy trap, for use during the fall and winter when starlings are flocking. This trap may capture as many as 100 starlings a day. This trap is big: 6 × 8 × 6 feet, or even bigger. It can be mounted onto a farm wagon for easy movement to the starlings' preferred roosts. Leave a few starlings (with lots of water and food) in the trap as decoys. The trap can be used with bait instead of decoys, but be more patient, because that method is less effective.

For construction details on both traps, see the "European starling" chapter in *Prevention and Control of Wildlife Damage*. (See the full citation in resource list.)

Lethal techniques that require a commercial pesticide applicator license:

- Avitrol, the repellent described earlier, is usually fatal to the birds who eat it.

Preferred Killing Methods

- CO_2 chamber
- Cervical dislocation
- Shooting, using an air rifle, a .22 caliber rifle with bird shot, or a shotgun
- Stunning and chest compression

Acceptable Killing Methods

- Pesticides

Control Strategies that Don't Work Well

- Netting over a doorway isn't as wise a choice as plastic strips, because the netting will probably tear.
- Ultrasonics don't work—the birds can't hear them.
- Some WCOs have modified their techniques and report some success using lasers to repel starlings (Avian Dissuader). This product works great with other birds, but isn't as effective with starlings.
- Avitrol has been used to repel starlings from feedlots but the results have been mixed.

Appendix C
Management Accounts for Species That Are Beyond the Routine

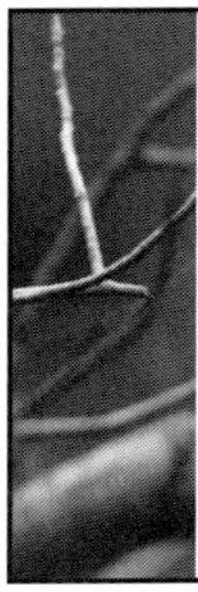

APPENDIX C: Management Accounts for Species That Are Beyond the Routine

Situations involving protected wildlife that may require additional permits

About a dozen additional protected wildlife species may come into conflict with people in our region, including crows, Canada geese, gulls, woodpeckers, black bears, beaver, white-tailed deer, muskrat, and Eastern wild turkey. Dogs and cats, which aren't even classified as wildlife, often cause problems.

As a WCO, chances are you can't just pack your truck and hit the road when a customer calls with a complaint about one of these species because they are either managed by a state agency or are federally protected. You may also receive calls about waterfowl, cormorants, and songbirds such as robins. They are also federally protected migratory birds.

You may need to secure permits—perhaps at both the federal and state levels—before using certain control techniques. The focus of this appendix is on those protected species that more routinely cause conflicts with people in our region. The species are grouped by the level of protection they receive, to help you understand which agencies to contact if you choose control activities that require permits.

Federally protected wildlife

Endangered species (on the national list), threatened species (national list), and migratory birds are all federally protected wildlife.

You must take special care to make sure that activities intended to control other species do not accidentally harm an endangered or threatened species. So here's what to do. First, review the lists of state endangered and threatened species to see if any are found in your work area. If so, learn how to identify those species. Then take special precautions, especially if applying pesticides or setting traps.

Migratory birds that most commonly cause conflicts with people include the American crow, Canada geese, gulls, double-crested cormorants, and woodpeckers. The Migratory Bird Treaty Act protects the birds, their feathers, nests, and eggs. You may not take, possess, or transport a migratory bird without permits from the U.S. Fish and Wildlife Service (50 CFR Depredation Permit) and your state wildlife agency.

Bird control methods that require state and federal permits include:

- any attempt to capture, relocate, injure, or kill migratory birds (except for those waterfowl species which may be taken during the hunting season by people with a state hunting license, a federal waterfowl hunting stamp, and HIP registration).
- any attempt to destroy bird eggs.
- any attempt to destroy bird nests that currently have eggs or young within them.

On April 15, 2003, the U.S. Fish and Wildlife Service changed its policy regarding the nests of migratory birds to allow for the destruction of nests that lack eggs or young—as long as those nests are not protected by other laws, which is true for bald eagles, golden eagles, and other endangered and threatened migratory birds. You can read the memo that describes this policy change, MBPM-2, at http://www.fws.gov/policy/library/m0208.pdf.

Although this policy change now makes it possible (in some cases) to destroy an unoccupied nest, make sure you do not violate the Migratory Bird Treaty Act by accidentally taking eggs or birds. For example, with a ground-nesting or cavity-nesting species, such as a bank swallow, it can be difficult to tell if there are eggs in the nest.

Legal controls that do not require state or federal depredation permits include harassment, exclusion, habitat modification, and the use of repellents, unless you're dealing with a bird that's currently nesting or has dependent young. Then you'd need a permit, even for harassment or exclusion activities. If you have any questions, contact your state wildlife agency. They can offer advice about management strategies and information about permits.

A combination of exclusion and habitat modification techniques will likely result in the best and most long-lasting control. Expect only short-term benefits from harassment. Use harassment techniques in varying combinations, preferably integrating both visual scare devices and frightening noises to increase your success. This approach works best when the population is small to mid-sized. It's a labor-intensive approach that requires skillful use.

In this appendix, you'll find accounts for these federally protected migratory birds:

- American crow
- Canada geese
- Gulls
- Woodpeckers

Game species

The following five wildlife species are categorized as game species in many states in our region. In those states, these species can be hunted or trapped during open seasons, as spelled out by the hunting and trapping laws and regulations. Each year, some aspects of these regulations may be modified to help the state wildlife agency achieve its management goals for that species. For example, the number of permits issued for the taking of antlerless deer will vary from year to year, based on population levels and management goals.

In this appendix, you'll find accounts for these game species, which are managed by the state wildlife agency:

- Black bear
- Beaver
- White-tailed deer
- Muskrat
- Eastern wild turkey

Domestic dogs and cats are not wildlife, so they probably do not fall under the jurisdiction of your state wildlife agency. In some states the laws related to the control of stray and feral dogs and cats are complex and it's risky for WCOs to become involved with the control of these species. For this reason we'll address some of the issues surrounding feral cats and dogs in this appendix.

Additional permits would also be needed for these species, which occasionally cause problems

Federal and state permits required:
Any migratory bird species, such as songbirds, hawks, owls, cormorants, waterfowl.

State permits would likely be required:

- Bobcat
- Fisher
- Marten
- Mink
- Moose
- Otter

AMERICAN CROW (*Corvus brachyrhynchos*)

Size

About a pound. Body is 17–21 inches long.

Signs of Their Presence

- The bird itself is the most obvious sign.
- Sounds: Loud, raucous, throaty, "caws."
- Nest: Sticks, twigs, and coarse stems lined with shredded bark, feathers, grass, cloth, and string. Nests are 18–60 feet above the ground in tall trees, especially evergreens. Lacking trees, crows may nest on the crossbars of telephone poles or on the ground.
- Evidence of their feeding: Poor sprouting of corn may be due to crow damage.
- Crows will sometimes poke holes in garbage bags, insulation, or through roofs (especially on flat roofs covered with a rubbery membrane).

Diet

Opportunist. Crows are excellent hunters, pirates, and scavengers. They eat over 600 foods. A third of their diet consists of insects, frogs, salamanders, fish, snakes, bird eggs and young, small mammals, carrion (including roadkill), and garbage (in landfills); the rest is plants, mostly corn or waste grains in fields that have been harvested. They'll also eat peanuts, sunflowers, pecans, various fruits, and sorghum. During the winter they survive mostly on waste grain. From spring through summer crows eat mostly worms, insects, and other invertebrates. One crow often acts as a sentinel while others feed.

Typical Activity Patterns

Social style: Sociable, living in family groups with the young of previous years, who may help bring up the current brood. Crows gather and roost in flocks of ever-increasing size from summer through winter.

Daily activity: Diurnal.

Hibernator? No.

Migrates? Some do, some don't.

Where Found

Distribution: Throughout the region.

Habitat: Prefer open agricultural lands, old fields, meadows, with nearby orchards, woodlots, or hedgerows for nesting and cover. Crows are adapting to using parks and tree-lined streets in suburban and urban areas as large roosting areas and nest sites. They will fly up to 12 miles during the day to feeding sites, then return to the roost at night.

Territory and home range: Territorial during the nesting season.

Breeding Habits

Pair bonding style: Monogamous. Crows have strong pair bonds that endure over the years, but they probably occasionally mate with others. Both sexes help build the nest and rear the young. Female incubates the eggs by herself, but the male will feed her and guard the nest when she leaves. (Other females may feed the incubating bird, too.)

Breeding dates: From March–June. They lay their eggs in the end of March. One brood/year.

Clutch size: 4–6. Young hatch in about 18 days.

Fledging dates: About 35 days from hatching. The young leave the nest at 5 weeks to forage with parents.

Amount of time young remain with parents beyond fledging date: Often one or more years.

Common Nuisance Situations

Time of year: Any time of year.

What are they doing?

- Huge flocks—up to 25,000 crows in some areas—may roost together in the winter, especially in urban areas. With so many birds present, they can make an astounding racket and their weight may even break tree limbs. Noise, droppings, property damage, and possible public health hazards are common complaints.

- Eat farm crops, both vegetables and grains, especially corn. Crows will also damage home gardens, again, often for sweet corn. Ironically, crows may be attracted to the garden by the insect pests, but then stay to feast on the crop they protected.
- Eat bird eggs and young chickens, pheasants, and turkeys.
- Large flocks may spread diseases to other species: histoplasmosis to people, avian cholera to other birds (when the crows roost near wetlands), and gastroenteritis to swine (when they roost near farm buildings).
- They'll sometimes poke holes in roofs, insulation, and garbage bags left outdoors.
- Rarely, groups of crows will attack newborn lambs, goats, calves, and pigs, but that's more often the work of magpies or ravens.
- During the nesting season crows may store food in trees, in the grass, or even in birdbaths or rain gutters—whatever's handy to the nest. This collection of carcasses may not be appealing.
- Crows are very protective of their young before they fledge, and will attack predators that threaten the nest. Usually, this is a cat or dog, but if people are too close to the nest, the crows might mob them. This behavior stops as soon as the young can fly.
- Rarely, young crows will play with, and damage, windshield wiper blades. (Crows are very curious, and young crows may play with objects that older crows ignore.)

Debunking Myths about Crows

- Forget Alfred Hitchcock's classic horror movie "The Birds." Crows don't attack people (unless you seem to be threatening their young).
- Many crows have died from West Nile virus but they may play only a small role in spreading the disease. Mosquitoes are the main vector.
- A crow roost nearby does not increase your risk of catching West Nile virus.
- Crows are often mistaken for ravens (also a federally protected bird). Ravens are larger and have a wedge-shaped tail.

Legal Status

Although crows are a federally protected migratory bird, an exception to the Migratory Bird Treaty Act allows the taking of crows without a federal permit when the birds are "found committing or about to commit depredations upon ornamental or shade trees, agricultural crops, livestock, or wildlife, or when concentrated in such numbers and manner to constitute a health hazard or other nuisance." In these cases, you don't need permits beyond your WCO license. In some states in the region, crows are also a game species.

Best Practices

Sometimes, damage blamed on crows is caused by other species such as blackbirds, starlings, pheasants, grackles, or mice. Often, community cooperation is critical for effective solutions to nuisance problems caused by a large urban roost.

Make roosting sites less appealing:

- In a dense grove, thin trees. If a tree is a preferred roost site, trim out about a third of its branches, concentrating on the inside center of the crown. This will reduce the number of available perches and increase the birds' exposure to weather. With less and poorer shelter fewer crows will congregate.
- A combination of frightening techniques (noises and visual deterrents) may convince the crows to leave a roost. As always, your chance of success increases if the techniques are used together and in an unpredictable fashion. Try noise-makers such as tape-recorded distress calls (crow squalling while under attack), clappers, shell crackers, propane cannons, shotguns, beating on tin sheets or barrels, and alarms. The Avian Dissuader, a laser, has proved effective against crows

(it's a powerful tool so use it cautiously). Mylar tape, lights, bright objects, animated owl models (complete with a dummy crow attached to its talons), and dousing the birds with water from hoses or sprinklers that are mounted nearby may also work. Crows that are used to people and city noises may not respond. Check with local authorities before using noise devices in urban areas. Some communities categorize bird bangers as firearms.

- If young crows are damaging windshield wiper blades, try to chase them away, using the harassment techniques mentioned above. Or try changing the car's appearance a bit to frighten them, by hanging a tassel on the antenna, for example. A car cover will protect the wipers, too.

Protect vulnerable crops:

- Are you sure the crows are causing more harm than good? They eat insect pests; mice; waste grain that could result in unwanted "volunteer" corn the next season; and carrion, which could attract other pests. Assess whether the damage justifies control. And remember to determine the real culprit.
- Exclusion is often impractical but for a valuable crop, especially one in a small area, it may be cost-effective. Create a barrier with nylon or plastic netting.
- In a home garden, place paper cups or sacks over ripening sweet corn, once the silk has turned brown.
- Try frightening them, as described earlier.

Remove artificial food sources (garbage, compost, bird seed, pet food):

- If anyone is feeding the crows, persuade them to stop.
- Enclose compost piles in a framed box using hardware cloth; in a sturdy container, such as a 55-gallon drum; or in a commercial compost container.
- Feed pets indoors.

Trapping Strategies

Trapping is probably impractical for large flocks. There may be some situations in which it makes sense, especially if there aren't many crows in the area. Crows are very wary and often difficult to capture.

Live traps:

- Australian crow trap. A large trap (8–10 feet square, 5–6 feet high) that uses both bait and a decoy to lure in the birds. This trap can be used with bait only, but be more patient, because that method is less effective. Bird eggs or carcasses are used as bait if there's little risk of attracting carnivores such as dogs; otherwise, try whole corn, watermelon, or poultry feed. The decoy is a few live crows, which of course need food, water, and shelter. See *Prevention and Control of Wildlife Damage* for details.

Preferred Killing Methods

- CO_2 chamber
- Cervical dislocation
- Shooting (air rifle, shotgun, or .22 caliber rifle)
- Stunning and chest compression

Control Strategies that Don't Work Particularly Well

- Ultrasonics don't work. Birds can't hear them.
- There have been mixed results with the grid-wire system.

For More Information

- See the Cornell Laboratory of Ornithology website, http://www.birds.cornell.edu/crows.
- "Controlling urban crow roosts with taped distress calls," Univ. of California publication 21561, includes a tape of the calls. Contact:

  University of California
  ANR-Communication Services
  6701 San Pablo Ave., 2nd Fl.
  Oakland, CA 94608
  (510) 642-2431
  Email: jerry.hendrex@ucop.edu
  http://anrcatalog.ucdavis.edu

CANADA GEESE (*Branta canadensis*, several races)

There's an added twist to solving nuisance problems caused by Canada geese. Some geese migrate, while others remain as year-round residents in many Eastern states. Resident, or local-nesting, Canada geese are actually a totally different subpopulation.

You need to know whether you're dealing with migratory Canada geese (*Branta canadensis*, several races) or resident Canada geese (*Branta canadensis maxima*) before you plan your strategy. For example, summer roundups will work only if you're dealing with a resident population because the migratory geese will be in Canada all summer long. Migratory and resident Canada geese look the same, so the best way to tell them apart is to determine if they migrate or if they breed locally. In urban areas, the problem is probably caused by a resident flock, but migratory Canada geese may cause problems in rural areas, and they may temporarily join resident flocks in spring and fall.

Size

From 8–15 pounds. Males tend to be larger than females.

Signs of Their Presence

- The bird itself is the most obvious sign. They are often seen flying in V-shaped flocks.
- Sounds: The most familiar call is the honk, but they also make hinking, hissing, and snoring sounds.
- Nest: Their bowl-shaped nest is about 1½ feet in diameter, made from plants, lined with goose down. Most nests are very close to water, almost always within 150 feet of water. When choosing a nest site, Canada geese seek a good view of the surrounding area. They prefer islands and peninsulas, but may nest in seemingly unusual places such as on ledges or rooftops, in the woods, flower gardens, or parking lots.
- Droppings: A single goose produces a pound of droppings each day. The volume of droppings may be more of a concern than their content, which is mostly digested grasses.
- Remember to ask your customers if the birds are seen all year long. Are they around during the summer? Have your customers seen nests and young?

Diet

Primarily herbivores. They prefer tender young plants. In urban areas they usually graze on lawn grass but they also eat grains, cattails, pondweed, and clover, and some small aquatic insects, snails, clams, and mussels.

Typical Activity Patterns

Social style: Canada geese form strong social bonds. They're usually seen with a mate, family group, or in a flock—rarely alone. If nesting close together, it's common to see "gang broods": 2–20 adults followed by 20–100 goslings. They'll eventually separate into family groups.

Daily activity: Diurnal.

Hibernator? No.

Migrates? Yes, but it's complicated. Migratory Canada geese migrate in the spring (from the U.S. to their summer breeding grounds in Canada) and the fall (from Canada to their wintering sites in the U.S.). Resident geese usually don't go far. They may migrate within the state in which they were born, or fly to a neighboring state. But some resident geese, usually young birds and those adults that didn't breed successfully, migrate to Canada right before the molting period (early June). They remain there during the summer, returning in September. This "molt migration" (first reported in the late 1990s) can dramatically alter the approach to goose management.

Molting: Adult Canada geese molt (shed their old feathers and regrow new ones) each summer, during which they are flightless for about a month. In the northern parts of our region, this usually begins in mid-June, while further to the south the molt generally begins during the first week of July. The geese congregate in large numbers near ponds and lakes, which provide a safe place to rest and feed, and the open water they need to escape danger. Most adult Canada geese molt in the same place they've chosen for raising their young (although subadults, and adults that didn't breed successfully, will often "molt migrate" as mentioned above). During the molt, geese cannot fly. This is critical—it's the easiest time to capture Canada geese. Some techniques, such as barrier fencing, may be highly effective during the molt (see "Protect vulnerable crops and sites," below).

Where Found

Distribution: Throughout the East.

Habitat: In urban areas, they prefer lawns next to a body of water. These areas function as "runways"; Canada geese need a fairly large area that's free of obstacles for their takeoffs. The clear sight lines provided by open water become especially important during their flightless period, helping the vulnerable geese spot and escape from predators. Canada geese prefer fertilized, mowed lawns found at parks, golf courses, industrial parks, and playing fields, especially near ponds.

Territory and home range: Both parents are aggressively territorial around their nest until the young hatch. Both parents will defend their young until the goslings can fly, usually at about 10 weeks old.

Breeding Habits

Pair bonding style: Monogamous. They pair for life. Canada geese usually begin breeding at 2–3 years old.

Breeding dates: Canada geese return to their nesting areas in late February or March (as soon as the water opens up).

Egg laying dates: Peaks in early April–early May.

Clutch size: 1 brood of 1–10 eggs, average 5. They don't begin incubating until all of the eggs are laid. Incubation takes 26–28 days. If the nest fails before the young hatch, they may renest.

Hatching: The eggs generally hatch within a period of 8–36 hours. Within one day of hatching, the adults may move their brood up to 2 miles away, seeking a grassy feeding area next to water, for safety. In urban areas, a variety of open areas, such as parks or golf courses, may substitute for the natural wetlands sites sought by Canada geese.

Fledging dates: at about 10 weeks old.

Amount of time young remain with parents beyond fledging date: They'll migrate with their parents in the fall.

Common Nuisance Situations

Time of year: Any time of year. Severe problems, such as the accumulation of large quantities of feces or feathers or excessive turf damage, are often associated with their molt, when large numbers of geese may gather on lawns that are next to water. Some birds will travel hundreds of miles to a favored molting ground.

What are they doing?

- Large flocks may gather or nest in parks, playing fields, yards, or near ponds or other water sources. Some people find the resulting piles of droppings and the feathers disgusting, and avoid these public areas. The birds can be quite noisy, too.
- Their droppings contain nitrogen, which in high concentrations can promote algal blooms in ponds and lakes. Bacteria and viruses are also found in their droppings. Canada geese are sometimes blamed for beach closures (health departments often close swimming areas when there's a lot of bacteria in the water), yet it's unclear whether they transmit any diseases to people.

- Canada geese aggressively protect their nest sites and young, and may attack pets, children, and adults.
- They may overgraze grass, causing large dead spots on lawns.
- They trample turf. In medium-heavy soils, this can cause the surface to harden, which may stop other plants from growing and lead to erosion.
- They eat farm crops, mostly grains. Heavy grazing can damage newly planted alfalfa and grain fields. They may pull up or eat seedling corn.
- They may collide with airplanes, or be sucked into their engines, causing crashes.
- Disease risks: they can infect other birds with coccidiosis, avian influenza, schistosomes, chlamydiosis, salmonella, and avian cholera. When Canada geese concentrate in very large numbers, this risk increases dramatically. Canada geese may also transfer salmonella to cattle. It's not known if they transfer diseases or parasites to people.

Control techniques marked by the Ø symbol require a federal and state permit.

Obtain the federal permit first. The U.S. Fish and Wildlife Service automatically sends copies of the permits they issue to the state wildlife agency, so you don't have to do this. In most cases, once the state agency has received the federal permit, it can approve the activities through the existing general Canada goose depredation permit.

Control measures that affect wetlands may require other permits. Those techniques are marked by the Ω symbol. Contact your state agency for information.

To apply for the federal permit, contact the regional office of the U.S. Fish and Wildlife Service. Include your telephone number in the text of any email message.

Region 5 (New England and Mid-Atlantic): Connecticut, District of Columbia, Delaware, Maine, Maryland, Massachusetts, New Hampshire, New Jersey, New York, Pennsylvania, Rhode Island, Virginia, Vermont, and West Virginia
U.S. Fish and Wildlife Service
Migratory Bird Permit Office
PO Box 779
Hadley MA 01035
(413) 253-8643
Fax (413) 253-8424
Email: permitsR5MB@fws.gov

Region 4 (Southern and territories): Kentucky and North Carolina; this US FWS region also includes Alabama, Arkansas, Florida, Georgia, Louisiana, Mississippi, South Carolina, Tennessee, Virgin Islands, and Puerto Rico
U.S. Fish and Wildlife Service
Migratory Bird Permit Office
PO Box 49208
Atlanta GA 30359
(404) 679-7070
Fax (404) 679-4180
Email: permitsR4MB@fws.gov

Region 3 (Midwest and Plains States): Illinois, Indiana, Michigan, Ohio, and Wisconsin; this US FWS region also includes Iowa, Minnesota, and Missouri
U.S. Fish and Wildlife Service
Migratory Bird Permit Office
One Federal Drive
Fort Snelling MN 55111
(612) 713-5436
Fax (612) 713-5393
Email: permitsR3MB@fws.gov

Allow 2 months for processing of the request. There is an application fee. To read the regulations and apply online, see http://www.fws.gov/permits/mbpermits/birdbasics.html. Questions about controlling nuisance Canada geese should be directed to your state wildlife agency or to the state director of USDA APHIS-Wildlife Services.

Best Practices

Canada geese strongly prefer short grass, the same kind of buzz-cut lawn that many people consider ideal. To increase the success of your management strategy, discourage intensive lawn management. If you want to reduce local goose conflicts, let the grass grow, especially grass that borders open water. Opt for smaller lawns away from open water. Such approaches may reduce other environmental problems as well.

Another aspect of goose life history that must be considered when planning your management strategy is their strong homing instinct. Canada geese will usually return to the area in which they fledged, and use the same nesting and feeding grounds year after year. That means that without control, the local population will almost certainly increase and problems will likely grow worse over time.

It's best to address problems when goose populations are low. Some techniques that work for smaller populations fail with larger flocks; and most techniques work better when populations are of low to moderate size. The best time to begin your control activities is in the late winter—soon after the geese have shown up, but before they start nesting. Plan ahead! You'll need about two months to secure any needed federal and state permits.

Nuisance problems associated with Canada geese usually affect an entire community, not just the residents of one property. The community may be willing to work with your state wildlife agency and USDA APHIS-Wildlife Services staff to craft a solution. WCOs will sometimes be hired to assist with aspects of the management plan.

Most long-term solutions to goose conflicts involve the use of several techniques in combination, such as hazing, habitat modification, summer roundups, shooting, egg oiling, and persuading people to change their habits (most often, you must convince them to stop feeding the geese). Don't rule out any technique that may be feasible because every little bit may help. Persistence is critical in goose management. Most of these techniques work well only with repeated applications. The use of one technique by itself is less likely to work. Also, it may convince the geese to move to another site where they may cause similar problems.

With a community-wide problem, long-term solutions usually involve lethal techniques to stabilize or reduce the goose numbers, at least initially. Nonlethal techniques used alone may reduce the problem at an individual site, but this approach rarely solves large-scale problems. The most effective way to reduce the size of a flock is to remove adult geese. Canada geese can live for more than 20 years. A female goose may have more than 50 young during her lifetime.

Lethal measures can be controversial. As with any community-wide nuisance wildlife problem, a public information campaign to explain management goals and methods is usually critical to success.

How molt migrations may influence your strategy
One other major point to consider is the possible effect of molt migrations on your goose management strategy. This behavior isn't consistent geographically, or from year to year, and likely depends on the birds' local nesting success.

If some birds in the flock you're trying to manage "molt migrate" in early June, then they may be offsite when certain control techniques, such as harassment and roundups, have traditionally been most successful. In these situations, it may take more than one year of roundups to solve the problem. At some sites, however, a molt migration might help solve the problem! How's that? Perhaps geese cause problems only during the summer. If some of the birds leave, that helps reduce the nuisance.

Here are options for goose management. Remember, start when the populations are low, be persistent, and combine methods for a better chance of long-term success.

Reduce their food sources (especially young grass shoots):

- Let grass grow to a height of 6 inches. This makes it harder for the geese to find the young grass shoots they like to eat. Right around the water's edge, it's better to let plants grow as tall as they will. Consider planting cattails, rushes, or short shrubs near the water's edge, which may deter geese in some cases.
- Reduce fertilizer use and stop watering lawns, to make the grass grow more slowly.

- Make lawns smaller while retaining the appeal of open space by leaving sections of the lawn unmowed.
- If short turf is not essential (as it is for a playing field), either let the grass grow higher or switch to another ground cover. This works well for picnic areas and parks and even for rough areas in golf courses.
- Switch to grass and plant species the geese don't like to eat as much. They tend to avoid tall fescue, periwinkle, myrtle, pachysandra, English ivy, hosta or plantain lily, euonymus, and ground junipers.
- If anyone is feeding the geese, persuade them to stop. Consider installing signs at popular feeding sites such as public parks, which explain the problems associated with these handouts.

Lure them to another site:
If migrant geese are feeding on crops, or resident geese are damaging a certain site, such as a park, you may be able to divert them to a different spot during the time of year when they'd cause trouble at the vulnerable site. Frighten them away from the vulnerable site, then lure them to an alternative area. Obviously, you need a nearby site where the geese will be tolerated, and that site must be attractive. This method won't work if people are feeding the geese at the vulnerable site. It may even attract more geese. At the site where geese will be tolerated, keep the grass well-mowed near a pond to make the field particularly attractive to the geese. If there are many geese, they may trample the grass, which might cause them to leave. If that's a problem, plant some clover, which is more resistant to trampling.

Protect vulnerable crops and sites:

- Offer an alternative gathering site, as described above.
- Install a grid of wires over a smaller crop or garden area (described in the section, "Block access to open water ..." below).
- Install fences. For small, high-use areas, such as a picnic site or play area, this may be a terrific solution. Fences work best before the geese nest, and during the early summer, when they're molting and can't fly. The fence must be long enough so the geese can't walk around it. It can be made from many materials including welded wire, chicken wire, snow fencing, corn cribbing, chain link, netting, Mylar tape, monofilament lines, or an electric fence. These materials differ in their costs and durability.
    - The mesh should be 3 inches or smaller, and the fence must be at least 30 inches high. If dealing with really aggressive birds, switch to a fence that's 48–60 inches high.
    - Fences may be even more effective when combined with a barrier of plants or rocks.
    - Monofilament lines (at least 20-pound test): set on poles that are 6 feet apart. Can use two lines, spaced at 7 and 12 inches above the ground, or five lines, set at 4, 8, 12, 18, and 24 inches above the ground (the lower wires deter goslings, too). Consider adding markers to the monofilament line so it's visible to other birds, which might collide with it.
    - Mylar tape: strands need support every 20 feet, secured with duct tape or electrician's tape. Twist the tape at least once within that span. The tape is prone to tearing in high wind.
    - Electric fences: Usually, two strands of polytape or 10-gauge high-tensile wire are attached to fiberglass or plastic posts at 8 and 16 inches above the ground. Use only UL-approved fence chargers and set the fence back at least 5 feet from a pond edge.

Frighten the birds away:

- If the geese aren't nesting, you may harass them without a federal or state permit as long as the geese are not touched.
- Visual scare devices: the most effective ones include the Avian Dissuader (a laser effective at night for roosting geese); laser pointers; Mylar tape or streamers strung along the water's edge (leave some slack in the line and twist the tape, to make it noisier and more reflective); and flagging or balloons on

poles that are 6 feet high or taller, installed around the vulnerable area (make sure the materials don't become tangled in tree branches or power lines).

- Frightening noises: gunshots, sirens, air horns, and various pyrotechnic devices (bangers, screamers, whistle bombs, shell-crackers, propane cannons).
- Hazing with dogs (especially border collies) or radio-controlled boats, aircraft, or cars. Geese will not overcome their fear of being chased by dogs, so this technique will not lose its effectiveness over time. However, it is labor-intensive and can be expensive. At first, the dogs must chase the geese several times a day. Eventually, you can reduce the number of patrols. In general, don't use dogs when the geese are nesting, molting, or have goslings. Dogs may not be able to chase geese away from large areas of water or properties subdivided with fences. Roads can cause a problem, too. Don't let the dogs herd geese into traffic.
- As always, use an unpredictable combination of frightening techniques.

Ω WCOs with a commercial pesticide applicator license: There are chemical repellents registered for use against Canada geese, but results have been mixed. Methyl anthranilate (a nontoxic, biodegradable food ingredient) may make grass taste bad to geese. The product is costly and needs frequent application, so its use is best limited to small lawns. Remember that this repellent is intended to stop geese from eating the grass, not from using a site. If they're just loafing, it would be useless. A second repellent, anthraquinone (Flight Control), includes a chemical that makes the geese sick. It also coats the grass with an ultraviolet stain, which the birds can see, but people can't. The idea is that the geese learn to associate UV-colored grass with an upset stomach, and avoid the area.

Block access to open water and the paths between the water and grassy areas (both flight paths and walking paths):

- Geese like to rest on open water, or on the nearby shore, where they feel safe from predators. If they're feeding on a lawn next to water, they tend to take off and land on the water. Deny them access to that water, and they may move away.
- Install a grid of wires above the water's surface to keep the geese from landing on the pond (or on land). You can use #14 wire, 80–100-pound monofilament line, Kevlar twine, or cotton rope. Secure the wires 1–1½ feet above the water, in parallel lines that are 10–15 feet apart. Attach brightly colored rope, flagging, or some other marker to the wires to keep the geese from flying into them. Install the grid before the geese arrive. For more certain success, add a perimeter fence to keep the geese from walking down to the water underneath the grid. Wire grid systems work best on golf course ponds, reflecting pools, wastewater ponds, and newly seeded lawns with limited public access. This isn't a practical method for ponds that are used for swimming, fishing, or other recreation, especially if the visual appeal of the water is important—for example, a reflecting pool.
- "Bird ball" systems include brightly colored floating balls that suspend ropes above a pond surface. The ropes work as a grid system to keep birds from landing. These systems often work best for small ponds near airports, retention ponds, and wastewater facilities.
- Ropes can be strung between trees to block flight paths and prevent geese from landing. Ropes should be loose enough to move in the wind, and highly visible. Polypropylene rope and Mylar tape have been used.
- With small ponds, tall trees in the flight path between the water and grazing area may prevent geese from landing.
- Create a visual barrier along the shoreline, with plants or rocks that will keep the geese from walking onto the land. This technique is less effective as geese become more adapted to suburban living.
- Locate playing fields at least 450 feet away from water to reduce their use by geese during the molting period. When the birds are unable to fly, they're reluctant to leave preferred sites that offer

open water with great views (for safety) located near grassy areas (for feeding, resting). When the geese can fly, they'll use fields that are more than a mile away from water, so this won't give year-round protection to the fields.

- Before the molting period, install a fence around the pond, to keep geese from walking down to the water. (See details under "Protect vulnerable crops and sites," above.)

Reduce the number of favored nesting sites (islands and peninsulas with quiet, straight shorelines):

- It's very hard to eliminate goose nesting sites, because geese adapt to nesting in some unusual places. Certainly, you can make sure that you don't accidentally create prime nesting sites for them while building or landscaping around ponds. You may be able to prevent them from nesting at a site by harassing them in the early spring. Try to block their path to the water or their line of sight, which makes geese nervous.
- Ω You may also need permits from the other departments with your state's environmental agency and the Army Corps of Engineers to modify some protected waterways.
- Remove any artificial nest sites, such as platforms, tubs, and up-ended hay bales. Even debris, such as tires, found near shorelines may provide good nesting sites.
- Plant (or protect) wetlands plants that grow along the shoreline, such as cattails and bulrushes, to create a visual barrier that may deter geese from walking onto the land.
- Stop mowing grass along the shoreline.
- Ω Eliminate islands and peninsulas that provide nesting sites by flooding the pond or reducing the water level. This is usually the most effective way to reduce the number of potential nesting sites, but it's expensive.
- Ω Add boulders or shrubs every 10–20 yards along the shoreline to block the pathway and line of sight. The boulders should be at least 2 feet wide. Plants must be dense and tall enough (30 inches) to prevent the geese from seeing through or over them. Shrubs, wildflowers, or prairie grasses may provide an effective barrier. Wide plantings tend to work better than narrow ones. A combination of a rock barrier with a hedge of thick plants may be more effective. This method works best with smaller flocks.
- Nest materials can also be removed to discourage nesting (works best in small areas where the nests are easily found, and when labor is cheap or free). This is very labor-intensive, requiring daily trips to the nest sites, which can be hard to find and hard to reach. Another problem is that individuals within a flock may begin nesting during a period that lasts for several weeks. It takes the birds about a week to build their nest, but once it's done, they lay their eggs quickly—and once there are eggs, you'd need a federal permit.

Reduce the amount of good winter habitat for the geese:

Turn off fountains to encourage earlier freeze-up of ponds.

Control their reproduction by disturbing their eggs so they don't hatch:

- These techniques are most effective at keeping a small population small. For example, if there are one or two pairs of geese on a pond and that's tolerable, you may be able to maintain the status quo using these techniques. With larger populations, these techniques are probably impractical because they're labor-intensive and time-consuming. Also, you'd need to tamper with nearly every egg to ensure success, and that grows more challenging with larger flocks. If nests are primarily located on private lands, access may be difficult or impossible.
- One disadvantage of these techniques is that they take a long time to work (10–15 years)—if they work at all. New birds might join the flock, increasing the numbers you're trying to reduce. Birds that fail to hatch eggs successfully might

move to a new breeding area and cause a nuisance there, so this approach might not be neighborly. Other lethal techniques, such as hunting or summer roundups, will almost certainly work faster and more effectively.

- The techniques of oiling, puncturing, and addling eggs should be done as early in the incubation as possible. They may require two trips to the nest to treat any eggs laid after the first visit (so you'll have to mark the nests). The second visit should be 7–10 days after the first trip. The male goose will probably defend the nest more aggressively on the second visit, so you may need a partner to fend off the gander. With all three methods, after the eggs are treated, they are put back into the nest so the female will continue incubating them. If the eggs are removed or destroyed, especially early in the incubation, the geese may lay another clutch.
- Ø Oiling eggs: Coating eggs with corn oil prevents gases from passing through the shell, so the embryo suffocates. The eggs are either sprayed with oil or dipped into a container of oil. In some states, this technique must be performed by certified pesticide applicators.
- Ø Puncturing eggs: Use a barbeque skewer, turkey lacer, or sharpened nail to pierce the shell. The best place to make the hole is slightly off center, on the bottom of the egg (opposite the point). The tool must be strong enough to puncture the egg. Don't crack the egg, because that might cause the birds to lay more. One problem is that the smell of the punctured eggs might attract predators, such as raccoons, skunks, coyotes, foxes, and crows. Even if your actions are unnoticed, if the geese realize that predators raided the nest, they may renest and lay more eggs. Some rotten eggs may explode when punctured, which can be quite unpleasant!
- Ø Addling eggs: Addling, or vigorously shaking, the eggs kills the embryo. Many people choose oiling or puncturing eggs over addling because it's easier to tell immediately whether the technique worked.
- Ø If you intend to addle or puncture eggs, wait until the female's been incubating for 1–2 weeks. By then, she's well established. She's less likely to renest and may be less biologically able to lay another clutch of eggs.
- Ø Removal of eggs or replacement with dummy eggs: If it's at least 18 days into the incubation, the eggs can probably be removed without causing the birds to lay more eggs. Or the eggs can be replaced with dummy eggs (wood or plastic eggs, or real eggs that are unfertilized or hard-boiled). This is simpler than oiling, addling, or puncturing eggs. It requires only two visits to the nest: once after all the eggs have been laid (sometime after the first week of incubation) to place the dummy eggs, and then later to retrieve them. You may need a partner to defend from the gander. The geese tend to continue incubating the dummy eggs.
- Once the geese are off the nest, try to move them. If there are no goslings, you can harass them with such techniques as hazing. If there are immobile young or goslings, you cannot harass them without federal and state permits.

Allow hunting (during the legal Canada goose season):

- Hunting is one of the most cost-effective ways to reduce goose populations. In addition to directly removing some adult birds, it may scare off others. Hunting also increases the effectiveness of noisemakers, because the birds learn that some loud noises really mean trouble.
- Hunters must have a state hunting license and the federal Migratory Bird Hunting Stamp. They must have the landowner's permission, be in an area that permits hunting, and hunt only during the legal season for Canada geese.
- Even in areas where hunting is prohibited, a controlled hunt may still be possible. This has been done successfully on golf courses, for example. Such controlled hunts may require variances to local laws.

- Hunting is also an effective way to target the specific birds that are causing the problems. How's that? Most goose problems are associated with resident geese, not migratory ones. Although you probably can't tell the two apart by looking at them, you don't need to because the migrants leave. Geese removed in September hunts are usually resident birds, because very few migratory geese will be in the U.S. before October 1st. (Check state regulations for early goose hunting seasons and bag limits.)

Remove the birds using direct capture methods (the birds are usually killed afterwards):

- Ø If you're dealing with only a few birds, you may be able to catch them with a dip net or use a super-sized funnel trap for ducks baited with grain, corn, or oat bait. Contact your regional wildlife staff for information about using the duck trap.
- Ø In the early summer when the geese are molting and can't fly, the adults can be easily herded into a holding pen. This is called a roundup. The pen (a moveable fence made of netting) is set on dry, flat land, usually about 20 yards from the water. It should be 4 feet tall, made of snow fence or plastic or cotton net, supported every 5–10 feet with poles. The side facing the water should have V-shaped "wings" that funnel the birds into the pen. People in canoes or boats herd swimming geese toward the capture area. Then others walk slowly behind the geese with outstretched arms, herding the birds into the pen. Once the geese are in the pen, the open side is closed. The birds can then be handled. Young should be removed first, because if there are many birds in the pen, they might be trampled. Remember, if some of the birds migrated just before the molt period, you may need to return the next year to conduct another roundup. (Check your state regulations about this technique, which in some areas is available only to state or federal wildlife agency staff.)
- Ø Cannon nets can capture many geese at once. Generally, a large, open area is baited with grain. Once enough geese are feeding, the net is fired over the flock. The birds are then hand-captured. Special training is required to use this technique. (The nets are propelled by explosives, so you may need to check local noise ordinances.)
- Ø USDA APHIS-Wildlife Services staff can capture geese using an immobilizing agent, alphachloralose (a highly restricted drug). The drug is usually put in a bread bait, which the birds eat. It may take 1½ hours to sedate the birds. This product can't be used during, or one month before, goose hunting season. Local, state, and federal permits are required.

Ø Preferred Killing Methods

- Follow the conditions of your federal permit.
- Gunshot (using a shotgun with nontoxic shot)
- Carbon dioxide chamber
- Stunning and decapitation
- Commercial poultry processor

Ø Acceptable Killing Methods

- Cervical dislocation

Control Strategies that Don't Work Particularly Well

- Ultrasonics don't work. Birds can't hear them.
- Installing a fountain or aerator in the pond won't work well, and it may actually attract geese to the area. (Ouch!)
- Dead goose decoys or swan decoys don't work, especially with resident geese.
- Scarecrows don't scare them, either.
- Ω Joggers on paths along the shoreline are supposed to scare geese away, so some sources suggest that you add a jogging path to an area. Think about this for a minute. Are these geese afraid of people? Probably not. This is an expensive technique and it hasn't worked well in many areas.

- Birds of prey are used to chase other birds, but we don't know if this works to scare away Canada geese. Right now, it's also impractical because there just aren't enough trained falconers around, and the service can be expensive.
- Another technique that works well for other birds is the use of distress calls, but resident geese often pay little attention to the calls.
- Ø Releasing mute swans to keep Canada geese out of ponds doesn't work. Swans are aggressive and territorial, and have even driven geese off their nests, but the swans may cause worse problems than the geese. Swans and geese like the same habitats, so the presence of swans might even attract geese. Mute swans are an invasive exotic species, and many states have restrictions on releasing swans into the wild.
- Ø In the past, geese were captured and moved to new locations, sometimes even in different states. No more. Most places are full up and would not welcome problem geese. Moving adult geese usually doesn't work because they tend to return to their nesting areas. However, relocating young geese with their parents during the summer molt has shown promise in some cases.
- Ø Some communities might want or be willing to tolerate a certain number of geese, but not a breeding population. One suggestion was to create a single-sex flock for that area. Unfortunately, creating and maintaining a single-sex flock is difficult and costly. The birds must be captured and examined carefully to tell their gender. Then, all members of one gender must be killed or moved. But adult geese tend to return to their birth sites, and other geese would join the single-sex flock. This is a high-maintenance approach that has not worked well.
- Ø Vasectomy (surgical sterilization) of male Canada geese is expensive. It's possible that the treatment changes the behavior of the goose. If other, fertile male geese mate more successfully than the neutered males, the technique would work well only if many geese received vasectomies.
- Ø The USDA National Wildlife Research Center is studying the use of a drug (nicarbazin) that reduces the number of eggs that hatch successfully. One product currently has federal registration as a chemical for goose fertility control.
- In a variation of the diversionary feeding technique, bait (loose grain) is provided to lure the geese to an alternative location. Some sources suggest that a lure crop, such as Kentucky bluegrass or grain, actually be planted at that site. These preferred foods may draw the geese away from the vulnerable site, but you could end up attracting other geese, too. Also, how would you persuade people to stop feeding the geese at other sites, if they know you're feeding them? This is not recommended.

GULLS (family *Laridae*)

Species

Most common in our region:

- Ring-billed gull, *Larus delawarensis.* Most common of the four species described in this account. Likely to nest on flat roofs or be found in agricultural fields.
- Herring gull, *L. argentatus.* Common, especially around coastal areas.

Less common, but may cause some complaints:

- Laughing gull, *L. atricilla.* Of these four gulls, they're the most likely to be found near airports. There's an infamous colony at Kennedy Airport in New York.
- Great black-backed gull, *L. marinus.* Although any of these gulls may eat the eggs and young of other seabirds, including endangered species, the great black-backed is so much larger than the others that it poses more of a threat.

Size

Laughing gull is the smallest, at 16–17 inches high. Next is the ring-billed, then the herring, with the great black-backed gull standing 28–31 inches tall.

Signs of Their Presence

- The bird itself is the most obvious sign.
- Feathers and droppings.
- Sounds: Varied, depending on species. May hear a "kleew, kleew," "hiyak," or "kyow" call, a plaintive mewing, and a stacatto "gah gah" alarm call. The laughing gull is known for its "ha ha ha" call.
- Nest: Size is proportional to the size of the bird. They'll nest in low trees and on flat roofs, especially those covered with gravel or rocks. The nests are often a mere scrape in sand or gravel, but they will add natural materials and bits of trash. Herring gulls make the most elaborate nests of the four, using sticks, other plants, and debris. Ring-billed gulls also use debris, but they favor lighter plant materials, such as dried grasses and weeds.

Diet

Opportunist. Gulls eat fish; shellfish; bird eggs and nestlings (they prey mostly on seabirds); insects; worms; grubs; mice; carrion; and garbage. They will steal food out of a person's hand.

Typical Activity Patterns

Social style: Gregarious. Most are colonial nesters.

Daily activity: Diurnal.

Hibernator? No.

Migrates? In the spring, they'll migrate north as the ice breaks open on lakes and rivers. In late spring, they'll seek a more secluded area, such as an island, for breeding. In late summer, they'll gather along the coast and then migrate south with the onset of cold weather. Some gulls remain all year, spending the winter near the open water of oceans or estuaries, the Great Lakes, and the Niagara River. Most gulls no longer migrate far because people provide abundant, year-round food sources.

Where Found

Distribution: Widespread in our region, from coastal to inland areas.

Habitat: Lakes, rivers, beaches, estuaries, mudflats, islands, harbors, ponds. Gulls adapt well to rural, suburban, and urban environments and will use agricultural fields, fish hatcheries, airports, landfills, reservoirs, parking lots, flat roofs, parks, malls, and athletic fields. In the winter, gulls seek open water, moving to the ocean, estuaries, and the Great Lakes. The Niagara River is a major wintering area for gulls.

Territory and home range: Highly territorial on their breeding ground, defending their nest sites, which they'll likely return to the next year. Prime territories are in the center of the colony.

Breeding Habits

Pair bonding style: Monogamous. Both parents care for the young.

Breeding dates: April–May.

Egg-laying dates: May–June. Most have one brood/year.

Clutch size: 3–5 eggs.

Eggs hatch: 21–28 days after they're laid.

Fledging dates: 4–5 weeks.

Amount of time young remain with parents beyond fledging date: Remain with colony.

Common Nuisance Situations

Time of year: Any time of year.

What are they doing?

- Steal fish from boats and hatcheries.
- They are involved with more aircraft collisions than any other group of birds, because they're plentiful, widespread, gather in large flocks, and are large birds.
- Eat livestock feed and fruits (such as cherries).
- Gather in large numbers in parking lots, near restaurants, marinas, food-processing plants, and parks. Their droppings foul objects and buildings. They can be raucous.
- If they gather in large numbers, their droppings can contaminate public water supplies.
- Mob people, trying to steal food from them.
- They may eat the eggs and nestlings of endangered seabirds, such as the piping plover.
- They sometimes cause a nuisance when they nest on rooftops.
- Disease risks: cryptosporidiosis, *E. coli* bacteria

Legal Status

Federally protected migratory birds (under the Migratory Bird Treaty Act). Federal and state permits are required to capture, handle, or kill gulls, or disturb their eggs or nests (if there are eggs or young in the nest). Most gull management is handled by USDA APHIS-Wildlife Services or state agencies, under the direction of the USDA APHIS-Wildlife Services. A landowner may chase or disperse gulls at any time without a permit, as long as the gulls are not physically harmed. Questions about controlling nuisance gulls should be directed to your state wildlife agency or to the state director of USDA APHIS-Wildlife Services.

Ø symbol: indicates a control technique that requires a federal and state permit.

Ω symbol: indicates a control technique that may require other permits, because it affects wetland. Contact your state agency for information.

Best Practices

Often, community cooperation is critical for effective solutions to nuisance problems caused by gulls. If you're confronted with a large colony nesting on a rooftop or at a landfill or airport, work with government wildlife biologists because they have the option of using additional techniques that require federal permits. In some cases, these techniques are far more effective or are an important part of the strategy. If this is a new problem, you may be able to deal with it successfully using only the techniques that don't require permits.

Remove artificial food sources (garbage, livestock feed, fish from hatcheries and boats):

- It's not easy to control their food sources because gulls are highly adaptable, but you don't have to put out the fine china, either. Focus on the areas that provide their most favored foods and restrict the gulls' access as much as you can.
- If anyone is feeding the gulls, persuade them to stop.
- Clean up any garbage piles. Keep garbage cans and dumpsters closed securely, and the areas around the containers clean.

- Gulls also feed at fish docks, sewer outflows, food processing plants, trawlers, and feedlots. Keep those areas clean and try to frighten the birds away.
- Use a grid-wire network of highly visible stainless steel wire (28-gauge) or 80-pound nylon monofilament line to protect large outdoor areas, such as fish hatcheries, garbage dumps, landfills, reservoirs, livestock feedlots, and fields. String the lines parallel to each other, about 15 feet apart and about 8 feet off the ground (a 15 by 15 foot grid also works well). This technique is highly successful with gulls.

Make roosting and loafing sites less appealing:

- Turn off fountains to encourage earlier freeze-up of ponds.
- Let the grass grow to a height of 8 inches or more to discourage gulls from resting in parks, playing fields, airports, and around ponds. This may work for ring-billed and laughing gulls but not herring gulls.
- Ω Filling or draining ponds, such as those near malls and office parks, may discourage gulls. With natural wetlands, this would require additional permits.
- To keep them off ledges:
    - Fasten wood, stone, sheet metal, Styrofoam, or acrylic plates (Plexiglas) to the ledge at a 45° angle so they can't comfortably perch there.
    - Install one of the sharply pointed, steel exclusion devices, such as porcupine wire (prongs point out in many angles), Ecopic (vertical rods), Bird Coil (a steel coil that looks like a Slinky), or nets.
    - Stretch steel wire (28-gauge) or monofilament line (80-pound test) in parallel lines across the area. The lines must be very tight, so fasten the wires to L-brackets with turnbuckles to remove slack. Attach the brackets to the wall using cable clamps or aircraft hose clamps, which can handle the high torque load on the wires. (Commercial versions are available, too, and may be easier to use.) Steel wire is more permanent and requires less maintenance than monofilament line. However, steel wire needs to be adjusted seasonally as it expands and contracts with changes in temperature (use spring tension devices).
- To keep gulls off rooftops or away from parking lots and other flat areas, install a 15 by 15 foot grid-wire network (described earlier) or nets.

Frighten them away:

- Visual scare devices, such as helikites (a kite with an attached balloon) or a laser (at night use the Avian Dissuader) may frighten the birds away from the site.
- Try noisemakers such as tape-recorded gull distress and alarm calls, shell crackers, and propane cannons. They're most effective when the birds are airborne.
- Hazing, with trained birds of prey (usually falcons) or radio-controlled aircraft that look like falcons, may also work. This technique is often used at airports.

WCOs with a commercial pesticide applicator license:

- Nontoxic repellents: There are chemical repellents registered for deterring gulls. Rejex-it Fog Force (containing methyl anthranilate, which is a nontoxic, biodegradable food ingredient) can be used to disperse nuisance gulls at airports, docks, and landfills. The product is costly and needs frequent application, so its use is best limited to small areas. This repellent is intended to stop gulls from using a site. Apply Rejex-it Fog Force undiluted using a thermal or mechanical fog generator that will deliver a dry fog with droplets within the aerosol size range (below 30 microns) that stay suspended and do not settle to the ground. For fast and lasting results, expose as many birds as possible to the fog. Withhold treatment until favorable temperature, wind direction, and wind speed conditions arise.

Gulls don't like to land on surfaces that have been treated with sticky polybutene, which is another nontoxic repellent. But polybutenes can affect other species and are messy and hard to remove. For these reasons, consider restricting your use of this tool to indoor applications. And how often do gulls cause problems indoors? You have better options.

- Ø Toxic repellent: Avitrol, a restricted-use pesticide, is available in a whole-corn bait mixture. Here's how it works. Gulls that eat the treated grain will behave erratically or give warning cries, frightening others in the flock. The birds that eat the Avitrol usually die. The erratic behavior may distress onlookers, so this isn't recommended for use in public areas such as beaches or around restaurants. Its use is best limited to places with restricted access, such as landfills. Like any lethal technique, this pesticide must be used carefully. If another animal eats a poisoned gull, it might sicken or die.

Control their reproduction by removing their nests or disturbing their eggs so they don't hatch:

- WCOs are unlikely to be involved with these efforts, but here's an overview. Many factors influence the control strategy, including the size of the colony, how long the birds have nested at the site, and whether the goal is to chase them away or to stop them from breeding. Let's say the wildlife biologists will be removing eggs as part of their gull management. If the gulls have just chosen a new site, the wildlife biologists may remove eggs as soon as they're laid because the gulls may just fly off and seek a better breeding site. But if it's a large colony that's well established, the gulls will not easily abandon the site. In this case, the biologists may focus on trying to break the gulls' breeding cycle, instead. They may wait until the birds have been incubating for a week or two before they remove the eggs, because then the gulls will be less likely to lay more eggs. The biologists may repeat the egg removal after another two weeks.
- Ø Egg disturbance techniques (oiling, addling, puncturing, or removing eggs, or substituting dummy eggs) are most effective when the colony is small. With larger populations some of these techniques, such as addling, puncturing, and substituting dummy eggs, are probably impractical because they're labor-intensive and time-consuming. Also, you'd need to tamper with nearly every egg to ensure success, and that grows more challenging with larger flocks.
- One disadvantage of these techniques is that they may take several years to work—if they work at all. New birds might join the flock, increasing the numbers you're trying to reduce. Birds that fail to hatch eggs successfully might move to a new breeding area and cause a nuisance there, so this approach might not be neighborly. Some biologists believe that gulls that have taken to nesting on rooftops will continue to seek rooftops, for example. In such cases, they recommend removing the adult birds.
- Ø Of all these egg disturbance techniques, the only ones that are really practical in most situations involving gulls are removing the eggs outright or oiling them. Generally, the colony is just too large for the other techniques.
- Ø Oiling eggs: Coating eggs with corn oil prevents gases from passing through the shell so the embryo suffocates. The eggs are either sprayed with oil or dipped into a container of oil, then put back into the nest so the parents will continue incubating them. If the eggs are removed, the gulls usually seek a more secure area in which to lay another clutch. In an established colony, if used by itself, this technique may not eliminate the problem. In some states, this technique must be performed by certified pesticide applicators.
- Ø Removal of eggs: If it's at least 1–2 weeks into the incubation, the eggs can probably be removed without prompting the female to renest. She may be less biologically able to lay eggs, but don't count on it. Return in two weeks to remove any new eggs. Once the gulls are off the nest, try to disperse the flock. If there are no chicks, you can harass the gulls with such techniques as hazing. If there are chicks, you cannot harass them without

federal and state permits. Then install a barrier, such as a net or grid wires, to keep the gulls from landing in the area. If you can't install an exclusion device you may need to repeatedly remove the eggs; in time this treatment may convince the gulls to abandon the site.
- Ø Nest removal: If there are no eggs or young in the nest you would not need a federal permit as long as you do not accidentally injure or kill birds. The gulls will often attempt to find a more secure nesting area and start again, so expect to repeat this treatment every two weeks. Eventually, this may convince the colony to abandon the site.

Trapping Strategies

It's unlikely that a WCO will trap gulls to solve a nuisance problem, because of several practical issues. Permits would be required, from both the U.S. Fish and Wildlife Service and your state wildlife agency. You need specialized equipment, and it tends to take a lot of time and effort. Gulls are likely to return to the site, too. The nonlethal methods described in this account are a much more practical approach to dealing with the problem, especially in urban areas. (**Ø** USDA APHIS-WS staff may use a highly restricted drug, alpha-chloralose, to capture gulls or to disperse flocks.)

Ø Preferred Killing Methods

- Requires a federal depredation permit from the U.S. Fish & Wildlife Service and a state permit
- CO_2 chamber
- Shooting, using a shotgun or rifle

Ø Acceptable Killing Methods

- There are toxic pesticides registered for the control of nesting gulls (herring, ring-billed, and great black-backed gulls) in some areas.
- Stunning and cervical dislocation
- Stunning and decapitation

Control Strategies that Don't Work Particularly Well

- Ultrasonics don't work. Birds can't really hear them.

WOODPECKERS (family *Picidae*)

Species that are More Likely to Cause Nuisances in Our Region

- Downy woodpecker, *Picoides pubescens*
- Hairy woodpecker, *Picoides villosus*
- Yellow-bellied sapsucker, *Sphyrapicus varius*
- Red-bellied woodpecker, *Melanerpes carolinus*
- Northern (a.k.a. "common") flicker, *Colaptes auratus*
- Pileated woodpecker, *Dryocopus pileatus*

Size

The downy woodpecker is the smallest, at 6½ inches. The sapsucker and the hairy woodpecker are both 8–9½ inches (hairy looks like a larger version of the downy). The red-bellied woodpecker is 9–10½ inches, and the flicker is 12–14 inches. The pileated is the largest, about the size of a crow, at 16–19½ inches.

Signs of Their Presence

- The bird itself is the most obvious sign.
- Sounds: drumming, drilling, and calls. The sound of the cartoon character Woody Woodpecker was based on the call of the pileated woodpecker.
- Holes in trees, utility poles, and buildings.

Diet

Mostly wood-boring insects, but they'll also eat berries, beechnuts, acorns, seeds, fruits, and suet. Sapsuckers also feed on tree sap (surprise) and the inner bark of trees. Poison ivy berries are a winter staple for them. Woodpeckers eat insect pests, such as carpenter ants (a winter staple for the pileated), woodboring ants and beetles, bark lice, wasps, and carpenter bees. Flickers feed on ants on the ground. The other woodpeckers feed on trees. Red-bellied woodpeckers cache food.

Typical Activity Patterns

Social style: Pairs may interact for about half the year.

Daily activity: Diurnal, with peaks at dawn and dusk.

Hibernator? No.

Migrates? The sapsucker and flicker may migrate. The other woodpeckers do not.

Where Found

Distribution: Widespread in our region, but abundance varies by species.

Habitat: Open mixed woods with dead trees (some prefer bottomlands), woodland edges, orchards, rural, suburban, and urban areas with trees (especially if there's a good supply of suet), parks. Some woodpeckers use wooded swamps, fields and meadows, and some need large trees. The pileated likes a larger area that combines second-growth and mature trees, often near a river or wooded swamp. Red-bellied woodpeckers prefer to nest in dead limbs in living trees, competing with starlings for these cavities. Sapsuckers prefer to nest in aspens, and in trees with rotten heartwood. Some woodpeckers create new cavities each year (downy and hairy) while others reuse their holes (flicker).

Territory and home range: Most are territorial just in a small area around their nests.

Breeding Habits

Pair bonding style: Monogamous.

Breeding dates: April–June.

Egg-laying dates: April–June. May have 2–3 broods/year.

Clutch size: 3–7, usually 4–5.

Incubation lasts: 11–18 days.

Fledging dates: 20–30 days after the eggs are laid.

Amount of time young remain with parents beyond fledging date: Downy woodpeckers stay with their parents for 3 weeks after fledging. Hairy woodpeckers feed their young for a few days after they fledge. The young leave as soon as they can feed themselves.

Common Nuisance Situations

Time of year: Most damage occurs from March through June, and then again during September–October in the fall. In the early spring, woodpeckers drum to proclaim their territories and attract mates. Then they'll create nest cavities. From summer through fall, they seek insects and roosting sites. They may forage, nest, or roost in buildings.

- Buildings that are near wooded areas with high woodpecker populations are more vulnerable. Woodpeckers tend to avoid open lawns with few or no trees.
- They prefer cedar and redwood siding but will damage pine, fir, and other woods.
- They don't often damage composite wood or Masonite.
- Woodpeckers will frequently damage wood shakes, plywood siding such as T-111 (a mimic of board-and-batten siding), tongue-and-groove and board-and-batten siding, and wood clapboards.
- Untreated or stained wood (especially in earth tone colors) is preferred over painted wood.
- Wood painted in earth tones (dark brown, red, or green) is more attractive to them than white, pastel, or brightly colored wood.
- Try to distinguish between the signs of woodpecker drumming and feeding. Why? Drumming behavior generally causes little damage and is seasonal. Foraging can cause more damage, and it means that the building is infested with insects, usually carpenter bees, leaf-cutter bees, or grass bagworms. Ask customers about the noise. A rapid-fire, loud, and continuous drill—or any banging on metal—is the sign of drumming, which happens most frequently in the spring. They don't excavate wood while drumming. When woodpeckers search for insects, the sound is often a bit more sporadic and may be quieter.

What are they doing?

- Drumming. In the spring, woodpeckers will peck on a variety of resonant objects, such as a hollow or dry tree; aluminum siding; metal roofs, gutters, drainpipes, chimneys, chimney caps, vents, stove and pipes; cars; canoes; mailboxes; garbage cans; the trim and fascia boards of a wood, brick, or stucco building; and even metal road signs. Drumming may leave small, shallow dents in wood, usually no more than 1 inch across. Look for the dents clustered along the corners of buildings, or on the trim or fascia boards. Some people may find the noise annoying.
- Chiseling foraging holes in wooden buildings (siding, eaves, trim boards, even window frames) or utility poles. Certain building materials are more prone to insect invasion, such as T-111 grooved plywood siding, so they're more vulnerable to woodpecker damage, too. Woodpeckers also prefer anything made of cedar.
    - In T-111, look for this sign of woodpecker feeding damage: horizontal rows of small holes across the siding or the fascia.
    - Cedar clapboards, resawn shakes, tongue-and-groove, or board-and-batten siding: very small holes clustered on the fascia boards (could also be caused by drumming).
    - Cedar shakes or shingles: in addition to the very small holes clustered on the fascia boards, you may see vertical rows of 3–6 small holes that follow the crack between two shingles up into the overlapping shingle; or large and small holes along the corners of the building, at the bottom of the shingle; or holes clustered near wires, which the woodpeckers used as perches.
- Drilling larger holes into wooden buildings to create roosts and nests. These holes are usually only slightly wider than the bird, and are either round, rectangular, or gourd-shaped. Woodpeckers seem to prefer a hard outer shell and soft inner cavity, which they usually find in a dead tree—or cedar-sided building. They will drill into the insulation, in which they'll hollow out their nest or roost. Woodpeckers often create several holes before settling down to business. Nesting holes are excavated late April–May. Roosts are usually created in the late summer through fall, as the birds prepare for winter. Woodpeckers are more likely to excavate roosts and nests in buildings with board-and-batten, clapboard, or tongue-and-groove siding than those with shingles or shakes.
    - In cedar clapboards: nest and roost holes are usually drilled on the seam between two clapboards, and may be found throughout the siding.
    - In board-and-batten: these holes are usually seen on the batten between two boards, throughout the siding (some preference for corners).
    - In tongue-and-groove: corner holes are more typical, but you may see them at the seam of two boards.
    - In resawn shakes and shingles: generally, corner holes, usually between two shingles, where the top and bottom meet.
- Sapsuckers may also feed on ornamental or fruit trees, boring a series of small holes (¼–⅜ inch) in parallel rows in the trunks or branches of healthy trees. (Nearby trees of the same species may be left untouched.) These holes sometimes girdle the tree, and create wounds that may be invaded by insects or diseases.
- Woodpeckers will sometimes feed on fruits and nuts from orchards and backyard trees, but this isn't much of a problem in the northeast.

Legal Status

Federally protected migratory birds (under the Migratory Bird Treaty Act).

Best Practices

- If your customer plans on building in a forested suburb or rural area, or in an area that's suffered previous woodpecker damage, tell them to use brick, aluminum, vinyl, stone, or steel siding when possible.

- For better results, begin your control program as soon as damage starts.

To stop drumming:

- First, talk to your customers. The simplest solution is to live with the noise. Padding placed behind the area where the birds drum will soften the noise. The drumming will stop.
- Hang strips of aluminum foil or Mylar tape (3–4 inches wide, 4 feet long) or aluminum pie tins in the area that's been damaged. The strips need to hang freely so they can blow in the wind. Brightly colored wind socks can be hung at the corners of the building.
- Playing a radio, banging on pots and pans, shouting, using noisemakers, or spraying the birds with water from a garden hose may work temporarily.
- Anecdotal reports from some WCOs indicate motion-activated spider models may deter woodpeckers in some cases. This technique needs further evaluation.
- Attach lightweight nylon or plastic bird netting or ¼-inch hardware cloth to the outer edge of the eaves, and then angle it down and attach it to the wood siding. The netting needs to hang out at least 3 inches from the building, or the woodpeckers will be able to reach through it.
- To seal a hole in siding and stop more damage: cover with flashing (aluminum sheets) soon after damage has begun.

To stop foraging:

- You must get rid of the insects and prevent them from re-infesting the building. First, remove any insects. Then seal the gaps that let them in. Paint the siding for even more protection using white, pastels, or bright colors—avoid those dark brown, red, and green earth tones. Use an oil-based paint or polyurethane to seal gaps and create a hard finish that deters carpenter bees.
- Board-and-batten: run a wire through the gaps to remove any insects. Then seal the gaps with wood putty or caulk.
- Cedar siding: remove loose knots, then fill holes with wood putty.
- T-111: scrape along the vertical grooves to remove any bugs, then caulk along the grooves. Painting the siding helps a lot.
- Shakes/shingles: a tough situation! It's hard to prevent insect infestations in a shingled building and to remove the bugs. The house can be sprayed with pesticides by someone with a commercial pesticide applicator license. Replace damaged shingles. Frightening techniques may keep woodpeckers away.
- Holes and tunnels made by carpenter bees can be sealed with cork, wooden dowels, or wood putty.
- To seal a hole in siding and stop more damage: cover with flashing (aluminum sheets) soon after damage has begun.
- Attach lightweight nylon or plastic bird netting or ¼-inch hardware cloth to the outer edge of the eaves, and then angle it down and attach it to the wood siding. The netting needs to hang out at least 3 inches from the building, or the woodpeckers will be able to reach through it.
- Drape bird netting over the canopy and trunk of a small tree that's suffering from sapsucker damage. For larger trees, loosely wrap ¼-inch hardware cloth or burlap around the trunk or limbs. If the damage is caused by one of the other species, remember that they are after insects, not sap. Downy, hairy, red-bellied and pileated woodpeckers, and the northern flicker don't generally create holes in healthy trees.
- Installing suet feeders as an alternative food source may work in some cases. However, this could attract more woodpeckers and increase the damage. This method requires further evaluation.
- A motion-sensing scare device that looks like a spider may frighten the birds away.

To stop nesting and roosting:

- Frighten the woodpeckers away from the building as soon as you hear them begin to drill the hole. Follow the tips in the section on drumming.
- If they're not nesting yet, seal all the holes (of course, make sure the birds aren't inside before you start).
- If they're already nesting, wait until the young fledge (usually, mid-summer), then seal the holes.
- Installing nest boxes on or near the structure may provide alternative nesting or roosting cavities. This technique requires further evaluation.

Preferred Killing Methods

Any lethal strategy requires a federal depredation permit from the U.S. Fish & Wildlife Service and then a state permit. Follow the conditions on your federal permit. Preferred techniques include: CO_2 chamber; shooting; and stunning followed by cervical dislocation, chest compression, or decapitation.

Control Strategies that Don't Work Particularly Well

- Ultrasonics don't work. Birds can't hear them.
- No pesticides (including repellents) are registered for use against woodpeckers on buildings or trees.

BLACK BEAR (*Ursus americanus*)

Size

When on all fours, male black bears stand about 3 feet high at the shoulder, and may be up to 6 feet long. They weigh 250–350 pounds on average, with some reported at over 600 pounds. Females are considerably smaller, commonly weighing 150–200 pounds.

Signs of Their Presence

- Tracks: Somewhat like a person's footprint, with long claw marks. Hind print is about 7 inches long by 4 inches wide.
- Scat: Can be variable, depending on what they're eating, but often roughly cylindrical. Somewhat like a dog's scat. May include seeds, plant material, and insect parts.
- Sounds: Adult females and cubs communicate with low moans and squeals. You may hear adults hiss, growl, or pop their teeth at each other—or at a person, if the bear feels threatened or is under stress. At the same time, their ears may be laid back, and the hair on their backs may be raised. If you hear those sounds and see those postures, be careful! Bears also bellow, whimper, mumble, and grunt, but people usually don't hear those noises.
- Dens: In the winter, bears den in shallow caves, under brush piles, rotten logs, or blow-downs, in depressions, hollow trees, or culverts. Leaves or grass are sometimes used as bedding. In the summer, during the day, they'll rest in shallow depressions in the forest litter, or sleep in trees.
- "Bear trees." Bears will rub against trees, and will stretch high (4½–6 feet off the ground) to claw and bite the tree. Those high scratches may be a mark of dominance (see how big I am?). The rubbing could be a form of scent marking, or maybe they're just itchy.
- Claw mark "climbing trails" on trees. These are easy to see on smooth-barked trees like beech and aspen, or on trees that were climbed repeatedly as the bears searched for food. Look for this sign on apple trees, black cherry, and mountain ash, too.

Bears will break branches as they climb; this is sometimes mistaken for ice damage.
- While foraging, bears will dig; turn over rocks and logs; and tear open logs, yellow jacket nests, and beehives.
- Trails leading to a food source.
- Flattened areas in corn or grain fields or berry patches.
- Fruit stripped off cherry and apple trees, with possible damage to the branches.
- Scattered garbage around a dumpster or garbage can.
- Bears will sometimes damage buildings or cars if they smell food within.

Diet

Opportunists. During the spring and summer, they eat mostly plants and insects, especially ants and bees. They'll feed heavily on fruits such as apples, cherries, and raspberries, and on nuts and agricultural crops, often corn. Bears will also eat amphibians, reptiles, fish, and small mammals, mostly rodents, although they can prey on fawns. In the fall, nuts are a critical food, especially acorns and beechnuts, which provide the fat that will get them through the winter. Bears will eat carrion and garbage. They sometimes kill livestock. They do love honey and will eat the bees, too.

Typical Activity Patterns

Social style: Males are generally solitary, although they'll interact at a food source. Females are often found with their cubs for about 18 months after their birth.

Daily activity: Most active at dawn and dusk, but may be active any time, depending on the season and availability of food.

Hibernator? Well… Bears are not true hibernators, but they do sleep deeply during the winter. The females give birth while they're in this sleepy state. They'll groom and nurse their young—still half-awake (perhaps you can relate)—for about four months. During a warm spell, the females will get up and go outside for a short time.

Migrates? No. They will, however, travel great distances in search of food. Males will move hundreds of miles away from their birth site, while females usually won't.

Where Found

Distribution: Black bears are found throughout North America and are the most numerous of our bear species.

Habitat: Most often found in the deep woods, in areas with large tracts of mature hardwoods or mixed forests that include some wetlands, such as swamps, rivers, streams, or lakes. They will live in second-growth forests if better habitat is limited. Black bears like areas with thick ground cover and few people, but they will venture into farm fields, and orchards. Bears are very adaptable and some frequent suburban areas looking for food. Bears have denned under decks or porches during winter.

Territory and home range: The home range of males is up to 100 square miles. Females stay close to where they were born, with a home range as small as 15 square miles.

Breeding Habits

Pair bonding style: Polygamous.

Breeding dates: May–June, sometimes extending into July, August, or even September.

Litter size: 2–3 cubs is common, up to 5 cubs is possible, if food sources are excellent. Females usually have cubs every other year.

Birthing period: Late January to early February. They give birth in their winter dens.

Weaning dates: At about 7 months old (late August to early September).

Amount of time young remain with parents beyond weaning date: They'll den with their mother when they are about 9 or 10 months old. The family stays together until early summer during the cub's second year, when the female's ready to mate again, usually in June.

Common Nuisance Situations

Time of year: From early spring (April) to December, but problems usually peak in June through September. Bears rarely cause problems while in their winter dens.

What are they doing?

Most complaints are associated with their feeding.

- Bears eat bird seed and destroy feeders, especially in the early spring and late fall. They may enter porches seeking stored bird seed.
- Raid dumpsters, garbage cans, coolers, tents, camps, and picnic tables looking for an easy meal. This is commonly seen at campgrounds and resorts throughout the summer and early fall.
- At restaurants, bears will raid dumpsters, garbage cans, and the containers that hold cooking grease. This is also most typical during the summer and fall.
- Will sometimes eat corn and grain, and may cause additional damage to the field while they're feeding. Sometimes, this damage may make it hard to harvest the crop.
- Eat fruits, such as cherries and apples.
- Occasionally attack domestic livestock or poultry. This doesn't happen often.

Debunking myths about bears:

- Bears are not true hibernators, as we mentioned. True hibernators, like bats and woodchucks, do not wake up until spring. Bears will lose about a quarter of their body weight during the winter. During their deep sleep, their body temperature drops about 7°F and their heart rate by about 50 percent.
- Bear dens are much smaller than most people think—only about 2 feet high by 5½ feet long, which is smaller than a bunk bed. Many are at ground level, beneath snags or blown-down trees.
- Bears don't often reuse their dens.
- Playing dead will not deter an attack by a black bear. This defense is meant to be used with grizzly bears.

Legal Status

Check with your state wildlife agency. Black bears may be categorized as a game species. There may be a regulated hunting season.

Best Practices

Most black bears avoid people unless they learn to associate people with food. This usually happens around dumpsters, garbage cans, campgrounds, or places where people feed bears (often, at restaurants). Bears who have developed this habit pose the greatest threat to people. The conditions need to be changed. To do that, think with your nose. This means:

Manage the garbage:

- Use bear-proof garbage containers, or place regular cans in bear-proof storage facilities. Garages (with closed doors) are safer than porches, for example.
- Use plastic bags inside garbage cans to help hide odors. Putting camphor disks (available from some drugstores), mothballs, air fresheners, disinfectant, or an ammonia-soaked rag in the can may also mask food odors.
- Remove garbage regularly, and put it at the curb in the morning, on the day it will be collected.
- Don't burn garbage because that makes it more attractive to bears. Bears especially like burnt grease, fat, bacon, and other meats.
- Clean garbage, compost, and recycling containers frequently with ammonia, bleach, or disinfectant.
- Keep the site clean.
- Clean old refrigerators and other insulated containers that are left outside.
- Do not leave dirty diapers or diaper pails outside.

Eliminate other enticing food sources:

- If people are feeding bears, persuade them to stop. It's dangerous.
- Don't feed birds during the spring and summer. Suet, bird seed, and the sweet liquid placed in hummingbird feeders all attract bears.
- Remove livestock carcasses from the site and compost them properly.
- Pick and remove all fruit from trees that are located near buildings.
- Don't feed pets outside. Even an empty dish can attract a bear.
- Keep livestock in buildings and pens, especially during the birthing seasons. If possible, those pens should be located away from woods and areas that provide good cover for bears.
- When camping, store food and organic wastes in bear-proof containers, on elevated platforms ("bear poles"), or in an airtight container that's suspended on a rope between two tall trees that are downwind of your campsite. Bear poles should be 15–20 feet above ground. The pole should be at least 6 inches wide. Wrap a 4-foot band of galvanized sheet metal around the pole at a height of 6–7 feet above ground.
- Plant vulnerable crops (corn, oats, fruit) away from woods and areas that provide good bear cover.

Don't create a scent trail to a vulnerable area:

- Turn off kitchen exhaust fans that vent to the outside when they're not in use. Make sure the vent screen is cleaned regularly.
- Don't eat or cook in your tent. Wash your hands before you handle any of your gear. Clean everything that touched food, such as dirty dishes and pots. Keep anything that might smell good to a bear far away from your tent. If you can, store food in a bear-proof container, or suspended between two trees, or in your car's trunk (although bears have been known to break into cars and tear through the back seat to get into the trunk).
- Remove the grease can from gas and charcoal grills after every use, and turn the grill on "high" for several minutes after you're done cooking.
- Cover barbeque grills with aluminum foil before cooking, then dispose of the foil properly.
- Clean barbeque pits and grills thoroughly with an ammonia-based cleaner before leaving them outside.

Protect vulnerable areas and vulnerable buildings:

- Remove brush and cover around homes, corrals, and livestock pens, creating a 50-yard barrier.
- Electric fences work well to protect a specific site, such as an apiary, cabin, or dumpster. Several factors influence the choice of material and design, including the size of the vulnerable area, the amount of bear activity, and local laws.
- If building in an area that's prone to bear damage, use strong construction materials. Solid frame construction, ¾-inch plywood, strong and tight shutters and doors, and steel plates will keep bears out.
- Locate campsites and hiking trails in areas that aren't used much by bears.
- Clear hiking trails so you can see 50 yards down the trail.

Frighten them away from a site:

First, a caution: if a bear shows any aggressive behaviors—growling, hissing, or popping its teeth, or if its ears are laid back and the hair on its back is raised—*do not attempt to harass it and do not approach the bear!* It might attack. If the bear's aggressive, back off. (See the next section for information about what to do if you're attacked by a black bear.)

If the bear is not aggressive:

- A combination of frightening techniques (noises and visual deterrents) may convince the bears to leave the area. As always, your chance of success increases if the techniques are used together and in an unpredictable fashion.
- The Critter Gitter scare device, which combines noise and flashing lights, works well.

- Visual scare devices: night lights and scarecrows may work.
- Frightening noises: propane cannons, loud music, air sirens, crackers, boat horns, banging on pots, and shouting will usually scare bears away.
- Guard dogs may be able to keep bears out of fenced areas, but few dogs are a match for a bear in a fight.
- In some states in our region, with a permit, dogs may be used to chase bears away from an area in which they've caused problems. This can be a very effective way to persuade bears to leave an area. Please note that this does not necessarily mean that dogs may be legally used to hunt bears. Contact your state wildlife agency for information.
- For WCOs with a commercial pesticide applicator license: Bear BeGone, a device that looks like a plastic barrel, dispenses pepper spray when the bear enters the can and pulls the baited trigger.

What to do if you encounter a black bear:

- In our region, black bears rarely behave aggressively toward people but there have been fatal attacks. Be alert.
- If a wild bear approaches a person in a remote area, consider that a sign of aggression. (The same behavior from a bear that is used to people and is often fed illegally may not be considered aggressive.)
- Back off slowly. Watch out for cubs.
- *Do not climb a tree or run away.*
- You can usually frighten a bear away by making a lot of noise. Shout, clap your hands, throw objects, bang on pots, use a siren or boat horn, or rev an engine. Just don't go so far that the bear feels threatened enough to attack! Bears are pretty good at telling when someone is scared. If they realize you're scared, you can't easily bluff them.

What to do if you're attacked by a black bear:

- Immediately call for help if you can: call the state wildlife agency, local law enforcement, forest rangers.
- Fight back with all that you've got. Hit the bear with rocks, sticks, your fists, or feet. Yell. Wave your arms and flap coats. If there's more than one person, stay together. If the bear is biting or mauling you, shoot it, if possible.
- *Do not climb a tree or run away.*
- *Do not play dead*—that seldom works with black bears.
- As soon as you can, slowly back away from the bear. Avoid the bear and any cubs.
- If the attack stops but the bear follows you, or you meet up with it again, wave your arms, flap your coats, yell, throw rocks and sticks and try to frighten the bear away.
- Pepper spray is effective when there's a close encounter with a black bear (its range is usually less than 30 feet). The active ingredient is capsaicin.

Trapping Strategies

Live traps:

- Check your state regulations. Permits will be needed for bear capture.
- Culvert trap: This is the most common trap used to capture bears. It's another variety of cage trap, although obviously much, much bigger than the kind you'd set to catch a woodchuck. A large culvert pipe is placed on a trailer or stand. Inside the pipe is a baited trigger, which closes the door. Culvert traps are very heavy, so they're hard to move and can't be used everywhere. They're also dangerous. Do not set one by yourself. Some have doors that weigh over 100 pounds and can crush a person. Post warning signs to keep people away from the trap.
- After the bear is trapped, the handler may use nonlethal harassment techniques, such as shooting the bear with rubber buckshot, to stress the bear (contact your state wildlife agency to find out if you'd need a permit for this harassment technique). This training is called "aversive conditioning," and is the same as shouting "*No!*" to upset a dog when it's done something wrong. If the conditioning works, it should persuade the bear to avoid the area.

- State or federal wildlife agency staff may use dart guns with drugs to help capture and handle bears. This is a complicated and risky technique best restricted to those with extensive training, experience, and skill. (Federal licenses are required, as is a protocol for the use of restricted drugs.) This method is useful in certain situations, such as when a bear's on a telephone pole or in a tree. Of course, great care is required during the entire process.
- Cable restraints: Spring-operated foot snares are sometimes used to capture nuisance bears that avoid culvert traps. The bear must be drugged before it's removed from the cable restraint, so state or federal agency staff with the appropriate permits should be involved. Bears caught in cable restraints may be transported and released. This is another aversive conditioning technique.

Preferred Killing Methods

- Follow the conditions of your permit.
- Shooting, using a 20-gauge or larger shotgun with a slug, or a center fire rifle (25-caliber or larger), or a pistol. Target the head, if no rabies testing is required, or the heart/lungs.
- Lethal injection of barbiturate

BEAVER (*Castor canadensis*)

Size

Up to 60 pounds. 25–35 inches long, excluding the 1-foot tail.

Signs of Their Presence

- Flooded streams and beaver ponds; channels (they tend to use the same routes).
- Dams and dome-shaped lodges of sticks (usually aspen, willow, birch, and maple) caulked with mud. Beavers will also build burrows in riverbanks, especially if living in a faster-moving stream or larger river, in railroad beds, and on the banks of lakes. Young beavers that recently moved out on their own are more likely to build dens in banks than they are to construct dome-shaped lodges. Many dome-shaped lodges begin as bank dens. When the water level rises, the beavers add to the den, forming the dome-shaped lodge.
- Chewed branches and trees, piles of wood chips, fallen trees, and stumps, usually found very close to water. Their tooth marks are obvious.
- Sounds: Tail slap: their alarm call. Young whine, squeal, and moan at night, but it's very hard to hear them.
- Tracks: not often seen; back foot may be nearly as large as your hand; webbed toes (Figure C-1).
- Scat: look like Ping-Pong balls made of sawdust, usually left in the water at the bottom of a dam or lodge (Figure C-2).
- Food caches: In late September, beavers begin to cache food, often within about 20 feet of the entrance to their lodge or den. Look for sticks and tree tops stuck into the bottom of the pond; the sticks are often visible above the waterline. New material added to the cache is a sign of recent activity, suggesting the beavers plan to stay for the winter.
- Beavers have distinctive and powerful scents.
- "Mud pie" scent mounds: mud heaps perfumed with the beaver's oily scent (castor) that mark the family's territory. These heaps may be up to one foot tall and three feet wide, and are found at the water's edge, or along river or pond banks.

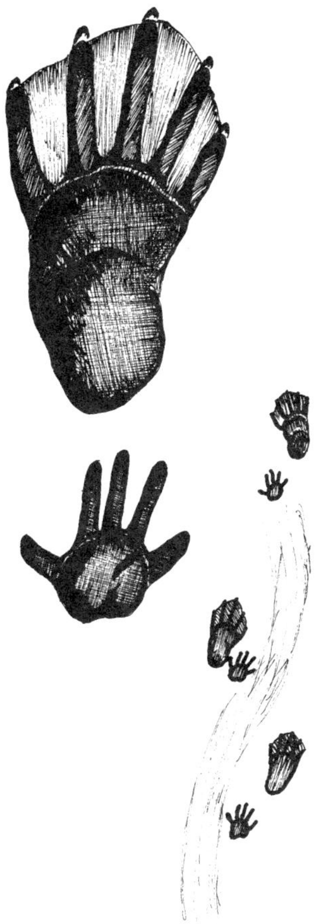

FIGURE C–1 Beaver tracks, close-up and track pattern. *(Wayne Trimm)*

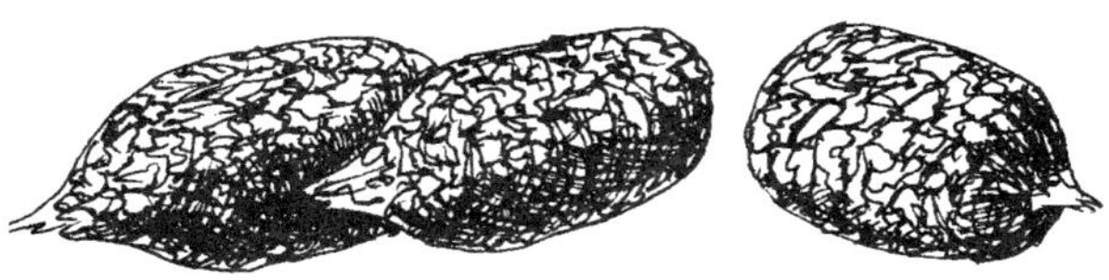

FIGURE C–2 Beaver scat. *(Wayne Trimm)*

Diet

Herbivore. They eat the leaves, twigs, and bark of trees and shrubs, especially aspen, willow, alder, birch, and maple. They also eat stems and roots of many water plants, such as cattails. Beavers will also eat acorns and corn, if the field is close to a pond or stream. In the fall they create a food cache of mostly submerged branches and poles placed near their lodge. In northern states they'll rely on this cache during the winter, after complete ice over.

Beavers gnaw around a tree, eventually cutting a deep groove. They usually fell smaller trees of 2–6 inches in diameter but are capable of cutting down much larger trees, even those 2–3 feet in diameter. And they're quick about it: beavers can cut down a 5-inch diameter willow in about three minutes. They'll trim the branches to a convenient size and then carry them in their mouths back to their impoundment. Then, as if eating a piece of corn, they'll turn the branch around to eat the bark, or, in the fall, may add it to the food cache they'll use during the winter.

Typical Activity Patterns

Social style: Sociable within families. Building and maintaining dams and lodges takes a lot of muscle. Young may remain with their parents into the spring of their third year, helping with housekeeping. The average pond is home to four to eight beavers.

Daily activity: Mainly nocturnal, with peaks at dawn and dusk. Beavers can regulate their heart rates, blood chemistry, and circulation to allow them to remain underwater for up to 15 minutes.

Hibernator? No.

Migrates? No. Once a colony has used up its food, it will move to a new area. When the young disperse, they may travel as far as 50 miles to find a territory.

Where Found

Distribution: Throughout the region.

Habitat: Ponds, lakes, marshes, rivers, and streams with good cover nearby. They prefer streams that have a gentle slope of less than three percent. Beavers prefer deep water, and will dam areas to create deep ponds.

Territory and home range: Territorial by family group. Their home range is defined by a dam or series of dams, and the nearby foraging area, which is rarely more than 100 feet from the ponds.

Breeding Habits

Pair bonding style: Monogamous. Both parents cooperate in rearing the young. Females are dominant.

Breeding dates: January–March. In more southern locales, beavers begin breeding in December. Gestation takes about 4 months.

Litter size: 1–8 kits, usually 2–4.

Birthing period: May–July, depending on when breeding occurred. This period extends from March–July in southern areas. Newborn kits are fully furred, able to walk and even swim.

Weaning dates: from 1½–3 months.

Amount of time young remain with parents beyond weaning date: 2–3 years.

Common Nuisance Situations

Time of year: Any time of year. From ice-out through June, new problems are often associated with the dispersal of the two-year-olds. From October through ice-up, beavers are preparing for winter by cutting many trees and working on their dams.

What are they doing?

- Plug culverts, drainage ditches, and drainpipes, which floods or washes out roadbeds, railroad beds, or driveways. The structural integrity of the road or railroad track is related to the stability of its bed.
- Beaver dams force water to collect against roadbeds. This may saturate the roadbed, causing potholes, settling, and instability.
- Their dams may flood upstream areas. Flooding may damage roads or homes or kill crops and trees. It may make certain areas unusable.
- Beavers cut down or girdle trees and shrubs for food. They can fell a tree that's up to 2–3 feet in diameter.
- Beaver dams transform the environment. This will be good for some species and bad for others. Whether you see this as a problem or a gift depends on your perspective.
- Flooding may contaminate water supplies or interfere with sewer systems.
- Disease risks: Rabies, tularemia, and giardiasis (an intestinal infection caused by the protozoan *Giardia lamblia*; there's still much to learn about this parasite and how the disease is transmitted).

Best Practices

This is an overview. Solutions for beaver damage problems can be far more complex than those for some of the other species in this book. There are environmental, legal, and safety issues to consider. On the practical front, repairing a hole that a squirrel chewed in a wall is much easier than replacing a culvert, for example. Additional training may be needed to help you successfully use an integrated approach to beaver damage management. It's equally important to talk to your customers to make sure you understand their goals.

Beavers create wetlands, which are highly valuable ecosystems for many species ranging from fish to waterfowl to people. Although there is no guarantee that beavers will remain in a particular pond for a long time (they often abandon ponds after a few years when their food runs out), consider the overall ecological consequences before you remove all of the beavers from a site.

Once beavers are gone and their dam deteriorates, their pond might drain and transform into a different habitat. Beavers may be abundant, but some of the species that rely on their wetlands are declining. For example, in the northern parts of our region, some hibernating amphibians depend on beaver ponds. If you drained a pond between mid-October and early May, those amphibians might freeze.

Several species of amphibians are declining in the East, and this trend worries many ecologists.

When you can, encourage your customers to think beyond their property, to consider the effects of their management decisions on other local wildlife. There are regional implications, too, as wetlands are among the more threatened ecosystems in our area.

If your management involves actions that alter wetlands, be aware that you may need additional permits, perhaps even from different agencies.

When necessary, consult with technical experts, such as your state agency's wildlife staff, highway departments, and site engineers. Breaching a dam the wrong way might flood downstream areas. Improper modifications to culverts might create dangerous road conditions. You could accidentally create a problem that's worse than the one you were hired to solve.

With careful planning, you may be able to satisfy your customers' needs—protecting roads, timberland, and agricultural sites from flooding, for example—while maintaining the beaver wetlands. Remember the idea of the "two-zone strategy" from Chapter Two—the moat around the castle analogy? At larger sites, this idea might be very helpful.

Unfortunately, in some places in our region beaver management is controversial, one factor that has impeded the sharing of information and the testing of new techniques. For example, biologist Skip Lisle has experimented with modifications to the deep-water fence, a classic water level control device (WLCD). His design is called the "Beaver Deceiver." Although not yet adequately studied, it's promising and is currently being used on over 100,000 acres in the Penobscot Nation in Maine, among other municipalities. Staff from Maine's Department of Inland Fisheries and Wildlife report that it seems to work quite well, cautioning that the quality of the manufacturing of the device is crucial to its success.

We urge you to maintain an open and skeptical mind; to search for the most current information; to consider ideas from all sources; and to tinker with techniques and tools, including WLCDs, as you would with other control devices such as traps. The effectiveness of the older style WLCDs is debatable—but some of those problems may have been solved.

Defining Success with Your Customer

Complaints about beavers usually focus on flooding, tree-cutting, or some combination of those two. These problems are caused by their feeding and construction activities (building of dams and lodges). You may need to know more details before you can solve the problem. For example, if one customer says, "I just don't want the beavers to cut down these three trees," then you might suggest a barrier to keep the beavers away from those trees. But if the customer says, "I really don't want them to cut these three trees or wash out this road," you have a very different situation.

The same problem at a different site may require a different solution. Take flooding, for example. You may need to totally exclude or remove beavers near water supplies, sewer systems, certain roadbeds, or power line right-of-ways, where flooding cannot be tolerated. But if the beavers are on a large private property, they may be tolerated or even welcome because of the benefits offered by beaver wetlands. Perhaps that landowner is happy with the pond and wants to control water levels to prevent flooding. The landowner might be able to secure a permit to allow the dam to be breached so that a WLCD can be installed. That might be the perfect solution for some people.

Here are examples of questions that may help you choose the right strategy for your customer.

- What exactly are you trying to stop or prevent?
- Can any level of beaver activity be tolerated at the site?
- Do you want the beavers to stay or go?
- Would you prefer only nonlethal techniques, only lethal techniques, or some combination of the two?

- If your site offers good habitat for beaver, it may remain attractive to them for years. In this case, would you prefer to use a long-term solution now?

So why are beavers attracted to certain sites?

- The feel and sound of flowing water are the cues that attract beavers to a "hole," whether that opening is a breach in their dam or a culvert. These cues trigger their instincts to build or repair a dam.
- Faster-moving water also sounds louder and is more likely to attract beavers than slow-moving, quiet water.
- Think of a culvert from a beaver's point of view: it looks like a smallish hole in an otherwise terrific dam. Easily fixed, eh? Plugging a culvert often creates a large pond for less effort than would be required in natural sites.
- The size of the "hole" (culvert or breach in dam) usually determines whether or not beavers will try to plug it. The smaller the hole, the more compelling the sound of water rushing through it, and the more likely they are to try to plug it.
- The shape of the culvert matters, too: beavers are less likely to plug a pipe-arch shaped culvert than a round one.

Some general tips:

- Don't crawl into a culvert or stand in the water in front of the culvert when you're trying to unblock it. While trying to breach a beaver dam, don't stand on the downstream side. Any of those locations could be dangerous.
- You'll need a good pair of waders and may want a wet suit.
- Trapping is often an important part of the solution. During the legal fur trapping season your customers can invite licensed trappers onto their property to remove the beavers.
- Some beavers are more determined to fix a breached dam than others. This can be a problem if you're trying to use a water level control device to prevent flooding. It's possible that if you removed the most determined individuals, usually adults, the other beavers would accept the device. However, another beaver is likely to move in.

How to protect vulnerable trees and shrubs from being gnawed by beavers (these tips won't stop flooding, which could kill the plants):

- Wrap hardware cloth loosely around trees and shrubs to a height of at least 3 feet to prevent beavers from gnawing on them. Match the height of the hardware cloth to the typical snow depth for the area, because deep snow will allow the beavers to reach much higher on the tree. The metal should be several inches from the plant, to give it room to grow. Attach the metal to the ground firmly to keep the beavers from chewing on the tree's roots. This barrier will have to be replaced every few years as the tree grows.
- A 3-foot-high fence (welded wire, 6 by 6 inch mesh) or foot-high electric fence (high tensile, 3 strands set 4 inches apart) can protect groups of plants. Make sure the fence can't be breached during deep snow. If the beaver burrow under the fence, add a strip of wire on the ground in front of the fence.
- The National Wildlife Research Center (USDA APHIS-WS) suggests that painting trees with a mix of 1 quart of sand in 1 gallon of exterior paint (oil or latex) may prevent beavers from chewing on the trees.

First, Two Definitions:

Deep water fence (a.k.a. "culvert fence"): A device that protects the *upstream* (intake end) of a culvert, bridge, or spillway by preventing beavers from getting *into* that structure and building their plug *within* the culvert or spillway. This fence is a barrier. It does not alter the water level.

Water level control device (a.k.a. "WLCD," "water leveling device," or "pond leveling device"): A system of pipes and cages (or pipes alone) that maintains the pond at a stable, desired level by matching the pond's outflow to its inflow.

General tips for managing water levels (at sites with a beaver dam or a culvert):

- Work with your customers and technical experts to determine the degree to which you need to drop the water level. This depends on the pond's slope and other site factors. Your "draw down" requirement determines the depth at which you'd install a WLCD, for example.
- Beavers usually build their dams on the upstream side of a culvert, so that's where you want to install the WLCD. This way, they're less likely to build a dam *within* the culvert. That's harder to deal with, so try to prevent it. For added security, protect the culvert's outlet, too. This will keep the beavers from entering the culvert.
- If you install a WLCD in a culvert, make sure it can handle as much water as the culvert did before it was modified. Otherwise, it could cause flooding.
- Accurately estimate the rate of water flow, because if the WLCD is too small, there may be persistent flooding problems.
- There have been improvements in WLCDs that may reduce the cost and maintenance requirements while making installation easier. For greater success, use modern devices and ensure that they are properly installed and maintained.
- Because you may encounter some older-style beaver control devices, we've included descriptions of some of these "classic" constructions. Some of them may still be useful in combination with modern devices. In most cases, for example, a pitchfork guard wouldn't be your best choice to stop beavers from plugging a culvert. But say you've taken care of that problem with a modern WLCD and want to ensure that the beavers can't swim into the culvert from the other end and build a dam *inside* the culvert. Adding a pitchfork guard over the culvert's outlet (at the downstream end) would work.
- WLCDs can be used to protect beaver wetlands while minimizing the risk of damage. In some cases, an entirely nonlethal approach may work. Integrated strategies might combine the use of WLCDs with the harvesting of beaver to maintain sustainable population levels.
- Regular inspection and maintenance of WLCDs is crucial, especially during the first two weeks after installation. The degree of required maintenance will depend on the device and the site. A properly built modern device may require only 1–4 inspections/year, but poorly built or improperly installed devices, and some of the older styles, may require much more attention.
- Keep WLCDs clean of debris. If you don't, it may just turn into a dam! A 6 by 6 inch mesh size will keep the beavers out but allow debris (and fish) to pass through the fence.
- Larger enclosures require less frequent cleaning.
- You may need to install a debris dam in front of your exclusion device.
- Some options, such as installing a WLCD or replacing a culvert, may be too expensive for some customers. These options may pay for themselves in the long term, if the devices work.
- If you install WLCDs or fences, consider offering maintenance services, too.
- Don't drain ponds between mid-October and early May in the northern parts of our region, because doing so might threaten amphibians that rely on those ponds as hibernating sites (many other factors currently threaten the survival of those species, so please don't tip the scale the wrong way).
- Check your insurance to make sure it covers activities such as breaching a beaver dam. Many don't.
- And of course, remember there are no magic bullets in wildlife control. Install a WLCD, for example, and the beavers might respond by building a dam directly downstream of the device, raising the water level back to the damaging level.

Special considerations for sites with culverts:

- The best WLCDs keep beavers away from the culvert's intake pipe and regulate the water level in the pond. They muffle the sound and feel of rushing water, reducing the beavers' compulsion to build dams.

- Replace smaller, round culvert pipes with oversized ones, where possible. Unfortunately, the up-front cost may be too high for your customers, although in the long run, this may be more cost-effective than yearly maintenance.
- Low-profile box culverts may be the most practical solution for roads that require a lot of fill to meet the size demands of a beaver-resistant culvert design.
- Don't create depressions at the inlet when you install a culvert, because beavers may just decide to enlarge the small pond you've created.
- In all but a few situations (see next section), no grates or guards over your culverts! Beavers will plug them, too.
- If beavers have a history of plugging a certain culvert, they may attempt to plug your replacement culvert, even if it's oversized. If this happens, trap the family group or install a deep-water fence or a WLCD such as a flex pipe system.
- If you're installing a new culvert, you may be able to prevent damage by choosing one that's the right shape and size for the site and by installing it correctly. The width of the culvert must be equal to, or greater than, the width of the stream. Use box or pipe-arch culverts with an inlet opening of at least 18 square feet. Better yet, bring the inlet size up to 38 square feet, or larger. A culvert with smooth inner walls will be quieter. New culverts should be installed level with the stream, which dampens the sound of the culvert and reduces the sensation of water flow—the two key cues that attract beavers to culverts. If the culvert slopes downward throughout its length, it will be noisier, and the beavers will sense the water flow all through its length. In addition to muffling the sound and speed of the water, the level installation helps to concentrate the sound at the downstream end of the culvert.

Making the most of water level control devices (WLCDs):

- Some water level control devices exclude beavers *and* muffle the water flow, which makes the "hole" less attractive to the beavers. If you began your control efforts by installing a barrier such as a culvert fence and it was compromised, you can try adding pipes, which act as mufflers.
- When installing a WLCD, remember to make allowances for heavy rains and winter runoff, which can cause higher fluctuations in water level. Success with WLCDs depends on your ability to hide the intake pipe or keep it from being detected by the beavers. Place the intake pipe at least 35 feet upstream of the dam, and set it in the deepest water possible.
- Make sure the style of WLCD you've chosen matches the site, because some of these devices work well only in specific situations. For example, the Clemson pond leveler was designed for use in small streams with a flow rate of less than 10 cubic feet/second. Pond drain pipes require a water depth of at least 1½ feet; in shallower water, they fill with silt. You know the drill—right tool, right place.
- When installing a WLCD, you'll need to leave enough water in the pond for the beavers to survive the winter (at least 3 feet of water at the lodge—and remember to factor in the typical depth of ice for that area; 3 feet of solid ice will not do the beavers any good). While it's true that beavers create deep channels, if you drop the water level too low, they may be tempted to foil your device, build another dam, or leave the area.
- Particularly with some of the older WLCD models, installation is trickier in shallow water. Use those in ponds that are at least 4 feet deep.
- Older WLCD designs may need more frequent maintenance than modern devices. Pitchfork guards may require weekly inspections and cleaning. Check them after the leaves drop in the fall, after storms, and soon after ice-out, because there's more debris moving through the water at

those times. If the maintenance is inadequate the device may fail.

- Where do older-style WLCDs work best?
  - Temporary flooding of the site back to the original pond level won't cause much harm.
  - Pond is at least 3–4 feet deep.
  - There's clay or gravel on the bottom of the pond.
  - People can easily reach the devices for inspection, and will do the maintenance.

Barriers that keep beavers from vulnerable areas:

- *Deep water fence*, such as the Beaver Deceiver

See http://dnr.metrokc.gov/wlr/Dss/beavers/construction-sequence.htm.

The deep water fence works best at:

- road culverts
- beaver dams on larger streams, where the installation is easier
- ponds with hard bottoms

Caution: the installation is trickier on narrow streams, but possible. You'd need a fence that's at least 20 feet long, and must be able to maintain water on both sides of the fence.

Deep water fences prevent beavers from blocking the intake of a road culvert. Properly installed and maintained, these fences should last 7–10 years. They are barriers. By themselves, deep water fences do not muffle the sound and flow of the water, so the beavers may still be attracted by those cues. If the beavers construct a dam against the fence, install pipes (a.k.a. "tubes") into the fence. The pipes will muffle the flow of the water and may prevent the beavers from damming the fence again.

To help prevent beavers from building a dam across the front of the fence, do not block the stream channel. Choose a shape for the fence that fits the site. Some WCOs claim that an equilateral trapezoid (which looks a bit like a squared-off D) is the most effective shape for a deep water fence, because it's stable, offers a longer perimeter than other shapes, and features flared sides that direct any beaver dam created against it into an unnatural direction. (The narrow part would be installed against the culvert, with the sides flaring out.) Others find great success with rectangular fence designs that are 20 feet long and 5 feet wide.

The length of the fence should be matched to the site conditions, ranging from 15–120 feet along its longest side (a total perimeter length of 60 feet is often adequate). A longer fence is recommended when dealing with a large stream or a site that's very attractive to beavers because of its good habitat or noisy culvert.

Construct your deep water fence using 6-gauge, 6 by 6 inch welded structural-steel wire mesh (a.k.a. concrete reinforcement wire). Secure the fencing with wooden or heavy-duty steel posts. The posts must be well braced. Attach the fencing to the outside of the posts, and make sure it's tight against the bottom of the pond; or better yet, add a full floor to keep the beavers from digging underneath the fence. Beavers are excellent diggers. The fence must extend 18–24 inches above the water level. Some WCOs routinely add a roof as added protection because beavers can climb that high.

- **Electric breach guard**

See Handout #11 at www.dec.state.ny.us/website/dfwmr/wildlife/beaver/handou11.htm.

The electric breach guard works best at:

- the upstream side of a breached dam

It's not suitable for

- sites prone to vandalism or theft
- remote locales, because the batteries need to be checked weekly

This is a grounded electric fence that is used to keep beavers from repairing a breach in their dam. It's installed on the upstream side of the breach. The hot wire is suspended across the water, with a few wires dangling from it, down to about an inch above the water's surface. When a beaver swims up to inspect the breach in the dam, it touches the dangling wires and receives a nasty shock. Use only deep-cycle RV batteries. *Do not use fence energizers with household current because that's dangerous.* Although it's fairly easy to carry and install, the energizer and battery are expensive, and both are often stolen. Check the condition of the battery once each week.

- **Pitchfork guard**

See Handout #3 at http://www.dec.state.ny.us/website/dfwmr/wildlife/beaver/handout.htm.

The pitchfork guard works best:

- at culverts that are only blocked occasionally, if intended as the primary protection
- as a secondary device to protect the culvert's outlet

It's not suitable for:

- sites that lack proper maintenance, because pitchfork guards need more frequent cleaning than other WLCDs

A metal grate made of ½–¾-inch heavy steel rods that are 6 inches apart, braced at the top with two horizontal rods. The grate is pushed into the bottom, and is also held in place by the current. Because there are only vertical rods, it's a little less likely to collect debris than wire mesh. This device won't stop beavers from plugging the culvert—in fact, it may actually promote dam-building—but it will keep the beavers from building the dam inside the culvert. Removing a dam that's inside the culvert is much harder than plucking out one that's in front of the culvert, especially if the highway crew has the right equipment. So, if the beavers plug a culvert only every now and then, and the crew has the right tools, this device may be enough to solve the problem. Pitchfork guards require frequent cleaning.

Water level control device designs:

- **Deep water fence combined with pipes**

See http://dnr.metrokc.gov/wlr/Dss/beavers/construction-sequence.htm and Handout #7 at http://www.dec.state.ny.us/website/dfwmr/wildlife/beaver/handout7.htm.

This combination device (deep water fence with a pipe) works best at:

- water depth of at least 3 feet
- sites where a deep water fence by itself didn't provide enough control
- road culverts
- beaver dams on larger streams
- sites where little water level fluctuation can be tolerated
- ponds with hard bottoms

Caution: the installation is trickier on narrow streams, but possible. You'd need a fence that's at least 20 feet long, and must be able to maintain water on both sides of the fence.

Follow the instructions above for the construction of the deep water fence. Then add a pipe at the desired water level. The pipe runs from the fence out into the water or through the dam. Keep the intake end (the upstream side) submerged to muffle the sound of the flowing water. Either install a 90° elbow joint facing downward on the upstream side of the pipe, or angle the pipe downwards along its whole length.

The length of the pipes should be matched to the stream's volume, with longer pipes for bigger flowages. You can add as many pipes as necessary. Some WCOs prefer to use 8- to 10-inch diameter PVC sewer pipes, which the beavers can't chew through, while others recommend corrugated polyethylene flex pipes.

PVC pipes can be supported on an H-shaped frame of 2-by-4s nailed between two paired posts (they can be level). Wrap wire around the posts to keep the beavers from chewing on them, or use steel frames.

Perforate any flex pipe over 6 inches in diameter to allow air to escape so the pipe won't float, and secure it to the bottom using fence posts or concrete blocks (two blocks for every 20 feet of pipe). Beavers can gnaw on polyethylene pipe, so wrap the pipes with galvanized welded wire mesh. Flex pipes must extend further than PVC pipes, at least several feet from the dam.

- **Clemson pond leveler**

See http://virtual.clemson.edu/groups/psapublishing/Pages/AFW/afw1.pdf and http://www.co.snohomish.wa.us/documents/Departments/Public_Works/SurfaceWater-Management/Flexleveler.pdf.

The Clemson pond leveler works best at:

- road culverts
- beaver dams on small streams (flow rate of less than 10 cubic feet/second)
- water level control structures

It's not suitable for:

- sites where the water surges violently
- flowages that exceed the capacity of one or two levelers
- large watersheds
- sites with multiple beaver dams and only a slight drop in elevation
- lake outlets where ice movement in the spring might damage device
- questionable for areas including extensive drainage ditch systems or large agricultural fields

This is a perforated PVC sewer pipe fit into a cylindrical wire cage. The entire cylinder is installed through the beaver dam, culvert, or blocked pipe. The closed end of the pond leveler should be upstream of the beaver dam, about 10–12 feet away. It must always be submerged. The downstream end of the pond leveler should stick out no more than 3 feet from the dam. On a standard Clemson leveler, the inlet is 10 inches diameter, while the outlet is a touch smaller, at 8 inches diameter. This device should be used only where the pond is fed by a small stream or spring (flow should be no more than 10 cubic feet/second). Pond levelers may be overwhelmed by unusually high rainfall, so occasional flooding must be tolerable. They exclude beavers and muffle the water flow.

There are other pond leveler designs. Some WCOs substitute flexible single- or double-wall polyethylene pipes for the PVC pipe. A circular wire cage that's 4 feet high (enclosed with a top) is installed at least 10 feet away from the upstream side of the dam. The pipes extend from the wire cage through the dam (they can be supported on an H-shaped frame of 2-by-4s secured to paired posts, as described earlier). Brace the posts and wrap them in wire. On the downstream side of the dam, the pipes follow the natural contours of the pond's bottom, to which they're secured with stakes every 6 feet to prevent movement. Drill holes in the top of the pipe to release air.

- **Pond drain pipes**

See Handout #9 at http://www.dec.state.ny.us/website/dfwmr/wildlife/beaver/handout9.htm.

Pond drain pipes work best at:

- ponds that are at least 4 feet deep after the installation of the pipes (in areas where ponds freeze)
- sites where seasonal flooding is tolerable
- areas where proper maintenance is available
- ponds with hard bottoms

Cautions: When breaching the dam, make sure you don't blow out the entire dam, which could cause flooding downstream.

These are long PVC pipes with one perforated end that are usually suspended at an incline at least 3 feet above the bottom of the pond (they can be supported on posts or an H-shaped frame). The pipes lead from a deep part of the pond through the dam, at the desired water level. The intake of the pipe will need protection and can be capped (if using perforated pipe) or it can be covered with wire mesh. These WLCDs are less expensive and weigh less than some of the previously mentioned devices, but require more frequent cleaning (use a telescoping rod, such as the kind used to clean chimneys).

Trapping

General tips:

- It takes experience and skill to trap beavers effectively. Don't wing it! Beaver easily become trap-shy, so a botched trapping attempt may leave the customer with a worse problem. There are excellent courses, videos, and books that explain how to trap beaver.
- Trapping can be used in combination with nonlethal controls or by itself. It may be a particularly important option at sites where beavers have been plugging very large culverts (over 38 square feet), or where a culvert can't be replaced with an oversized model.

- If lethal trapping is used as the sole control technique, be aware that other beavers may move into the site (depending on the conditions and the size of the beaver population), so you may need to continue trapping on a yearly basis.
- If the customer wants to minimize the number of beaver removed from the pond, trap as close to the site of the nuisance activity as your permit allows. Breach the dam (if allowed). The first beavers to investigate this hole are most likely the ones causing the most damage.
- If you want to remove the entire family, begin trapping away from the lodge so you don't spook the beavers. Removing all the beavers may take time. Don't start a job unless you can finish it properly.
- Otter frequently use beaver ponds and may be killed in traps intended for beaver. Follow the tips in the following section to avoid accidentally trapping river otter and muskrat.
- Trapping in front of a breach in a dam is difficult. The beaver may have a big stick in its mouth as it approaches the dam (because it wants to plug that hole). The trap may catch the stick instead of the beaver, allowing the beaver to escape. You may have greater success if you try to catch the beaver where it's looking for material to fix the hole. If you'd like to set near the dam (and your permit allows this) try placing your trap perpendicular to the dam, or about 10 feet away.

How to avoid accidentally trapping river otter:

- Look for otter and their sign.
- Get in, catch the beavers you can, and get out. Don't spend weeks trying to catch the last beaver or a trap-shy individual. Instead, give them a few weeks to calm down and then return with specialized traps.
- If using body-gripping traps, choose a #330 with an offset trigger. This trap provides enough room for otters to pass through without triggering the trap. Smaller traps, such as #220s and #280s, do not. Replace 4-way triggers with 2-way triggers. Crimp the trigger wires together and place them as far to the side as possible. This trigger modification doesn't affect the capture rate for beaver but reduces the accidental capture of otter by half.
- Using a tension-adjustable trigger may reduce the otter catch even further.
- Avoid channel sets, especially in main channels.
- Stay away from dams and other crossover locations.
- Use baited sets, where possible.
- Use castor mound sets with trap set 8–10 inches deep, for hind foot catches.
- Avoid setting traps near beaver lodges or burrows made in banks, even if those areas are old and seem to be unused.
- Carry a catchpole to carefully release any otter caught in a foothold trap.
- Use snares (a.ka. "cable restraints") if they're legal in your state.
- For more information, see:
    - http://www.dec.state.ny.us/website/dfwmr/wildlife/wildgame/330modi.pdf
    - http://www.dec.state.ny.us/website/dfwmr/seasons/wortf.html
    - http://www.dec.state.ny.us/website/dfwmr/wildlife/wildgame/Trappers21.pdf

Live traps:

- Snares (a.ka. "cable restraints"). Use a #4 beaver or coyote snare, which is a 40-inch-long steel cable that's 3/32 inch thick, with a swivel and slide lock. Maximum loop size of 9–10 inches. Suspend the loop about 4 inches above the runway. Ideal sites include feeding and travel routes on land or near slides over dams or banks. Anchor to a large log, tree, or root within 10 feet of the snare, or use rerod stakes. (Anchor the stakes to the ground with 14-gauge wire that's braided for strength. Ensure that the anchors swivel freely.)
- There are several other different types of live traps specially designed for beaver (Bailey, Hancock, Breathe Easy, Koro, Ram Ezee). Some are modified cage traps, some are funnel traps, and the Bailey

and Hancock traps resemble suitcases. Cost and effectiveness vary. Some WCOs and trappers report a significantly greater success rate with the Hancock design compared to the Bailey. (See the description of the suitcase trap in the "live traps" section of Chapter Five.)

- When setting a suitcase style beaver trap, ensure that the beaver will be able to surface to breathe after it's trapped. If possible, set the trap in a small channel in 10–12 inches of water. With care, this trap can be set in deeper water (use a deep stake so the top of the trap remains above the water's surface), hung off a dock, or staked so it flips the beaver onto land. Secure the trap so the beaver can't drag it into deep water and drown.
- Beavers might break their teeth while trying to escape from a suitcase trap.
- Suitcase traps are large, heavy, and strong. Take care while setting them. If the trap fired prematurely and hit you in the head, it could knock you unconscious. Manufacturers suggest that users wear a helmet.

Lethal traps:

- Body-gripping trap, #330, with a modified trigger to minimize the risk of catching otters.
- Trap placement is critical when using foothold traps. Beaver have short front legs. Consider where they will put down their feet. Pay attention to the water depth and distance from the shoreline. You might catch only belly fur.
- Foothold traps, #3, 4, or 5, equipped with submersion wires or cables that have sliding locks. Set the traps in at least 3 feet of water to keep the beaver from surfacing. This technique requires more skill than the use of the body-gripping trap. Set the device to drown the animal in the shortest time possible.
- Specialized foothold traps for beaver (CDR 7.5, MB 750, Bridger #5) are expensive but well worth the cost, because they increase the chance of a good catch. They're especially useful for hind foot catches.
- Snares can also be used as lethal traps by adding a submersion set (a slide wire, anchor, and slide lock that holds the beaver underwater). Set the device to drown the animal in the shortest time possible.
- Bait with castoreum (the secretion from the beaver's glands) or twigs or branches from aspen, alder, willow, or maple.
- Under ice, use channel, slanted-pole, and bait-pen (an underwater cubby) sets.

Preferred Killing Methods

- Lethal trap
- Lethal injection of barbiturate, if possible
- Gunshot (.22 caliber bullet) to the head, if no rabies testing is required, or to the heart/lungs

Acceptable Killing Methods

- Cable restraint (snare) with a submersion set
- Foothold trap with a submersion set

Control Strategies that Don't Work Well

- Installing several smallish culvert pipes instead of a single large one is a foolish choice. Doesn't matter how many pipes there are, if they're too small, chances are good they'll be plugged.
- Some differences in culvert pipe materials (smooth vs. corrugated pipe) and installation design (flush vs. projecting pipe inlet) do not matter to beavers. They'll plug any of them if the culvert is too small.
- Large-scale removal of plants along roadsides may discourage beavers in certain situations but can lead to erosion and could hurt other wildlife. This approach also requires regular maintenance.
- Although you may be able to secure a permit to destroy a beaver dam or lodge, this won't solve the problem by itself. Breaching a dam quickly can flood roads or downstream areas, which could be dangerous or cause a lot of damage for your neighbors. That's why you have to consult with your state wildlife agency first.

- Harassment doesn't work that well. Repeatedly destroying secondary dams might convince the beavers to abandon the site, if that's the goal.

For more information, see also:

- http://icwdm.org/
- http://www.beaversww.org/solutions.html
- http://el.erdc.usace.army.mil/wrap/
- Lisle, S. (1999). "Wildlife Programs at the Penobscot Nation," *Transactions of the 64th North American Wildlife and Natural Resource Conference,* Wildlife Management Institute, Washington, D.C.
- Strong, Paul. 1997. *Beavers: Where Waters Run.* NorthWord Press, Inc. Minocqua, WI. 143 pp.

WHITE-TAILED DEER (*Odocoileus virginianus*)

Size

Bucks (males) typically stand 3–3½ feet tall at the shoulder and weigh 125–200 pounds. Males grow antlers each spring and summer and shed them in the winter. Does (females) are smaller, lighter, and lack antlers.

Signs of Their Presence

- Deer are commonly seen.
- Tracks: Look like a pair of bottom-heavy crescent moons.
- Scat: Jellybean-shaped pellets, usually seen in piles.
- Evidence of their feeding: browsed branches have a ragged, broken end. Rabbits or rodents clip off branches neatly. Also, deer can browse to a height of 6 feet above ground. Unless there's deep snow, a rabbit or rodent would need a ladder to do that.
- Trails. Narrow, well-worn paths.
- Beds. An oval depression in grass, where the deer slept or rested. In snow, this may be a deep indentation.
- Scrapes: Bucks will tear up soil and plants using their hooves and antlers. May see a scrape that's 1–5 feet wide, often with nearby rubs.
- Buck rubs. When the antlers first grow, they're covered with "velvet," a layer of hair and blood vessels. This dries and is shed or rubbed off; bucks usually rub against saplings, shrubs, or rocks, probably to mark their territories.
- Shed antlers may be found in the woods, although this is rare because they're usually quickly eaten by mice, squirrels, and porcupines. Antlers are rich in minerals such as calcium.

Diet

Herbivores. Deer eat mostly grasses, herbs, crops, fruits, leaves, twigs, and buds in the spring and summer. Their fall diet includes beechnuts, acorns, and wild cherry seeds. By winter they're surviving on twigs and buds. Deer prefer to dine on these ornamentals and crops: white cedar, yew, rhododendrons, azaleas, fruit trees, sumac, oaks, ash, various maples and dogwoods, corn, alfalfa, soybeans, buckwheat, clover, apples, lettuce, celery, potatoes, turnips, sugar beets, and strawberries. When malnourished, deer will eat spruces, alder, tamarack, most pines, junipers, hawthorn, and sweet fern. Deer eat about 3% of their body weight each day. For an average buck, that's about 6–8 pounds of plants, twigs, and buds.

Typical Activity Patterns

Social style: Males keep to themselves for much of the year but during the spring and summer they'll form bachelor groups. Females care for their fawns by themselves. Groups of does and their young often feed together. In northern climates, during the coldest part of the winter, deer may gather for a few months in a sheltered part of the woods, called a "deer yard," that protects them from deep snow and wind. Typical deer yards include swampy areas with spruce, balsam, cedar, and pines with some hardwood trees for food.

Daily activity: Most active at dawn and dusk, but may feed throughout the night or day.

Hibernator? No.

Migrates? No. However, trips to and from winter deer yards may exceed 20 miles.

Where found

Distribution: Widespread and abundant throughout our region.

Habitat: Deer are more likely to be found in the forest edge, not in the deep woods (except for a few months in the winter when those living in northern areas retreat to their sheltered yard). Ideal deer habitat is a mix of woods and agricultural fields with brushy undergrowth, but deer are highly adaptable and can live in suburban areas with a mix of lawns, gardens, and patches of forest.

Territory and home range: During the breeding season, or "rut," males are territorial. Some will defend large areas, while others will be shut out. Males range over a larger area than females; in urban areas that could be 200 acres, in rural areas as much as 600 acres. The core home range of females is usually about 25 acres in suburban areas.

Breeding Habits

Pair bonding style: Polygamous.

Breeding dates: Mid-September through late February, with peak in mid-November.

Number of young: 1–2 fawns, occasionally triplets if food sources are excellent.

Birthing period: Late May to mid-July in northern sections of our area. In the southern locales, the birthing period lasts from late May through August.

Weaning dates: 3–4 months old.

Amount of time young remain with parents beyond weaning date: Up to a year, sometimes longer for females. Most males leave their birth territory between 12–16 months after birth. They may travel 8 miles to establish their new home range.

Common Nuisance Situations

Time of year: Any time of year. Damage to landscape plants and ornamentals is usually most severe in the late winter through early spring, when other foods are scarce.

What are they doing?

- In the East, deer have essentially become our forest and wildlife managers. Many people are unhappy with the results. Deer populations in the Northeast are so high that they over-browse many areas, severely reducing or eliminating certain plant species. Where once there was a thick understory composed of many different plants, the deer may leave behind an open, virtually park-like forest floor that's dominated by grasses and ferns. Trees such as oak and sugar maple, which deer love to eat, are being replaced by beech and other species they don't prefer. Wildflowers, such as Canada mayflower and several species of trillium, have been eaten out of existence in some places. This dramatic change in our forests will hurt some wildlife, such as ground-nesting birds and amphibians, which need the cover of the shrub layer.

- Deer prefer to eat many economically valuable trees (maple, oak, ash, birch). The species they leave behind are less marketable. This is already hurting the forest industry to the tune of $750 million/year, according to government estimates.
- Eat garden crops, field crops, fruits, and ornamental plants. The losses are estimated at more than $250 million for damage to landscape plants nationwide.
- Their browsing has caused long-term damage that reduces the yield of fruit trees or permanently disfigures ornamental trees and shrubs.
- Collide with vehicles causing accidents (estimated cost of $1 billion/year, nationally), and 150–200 deaths annually.
- The combined losses amount to more than $2 billion/year, nationwide.
- Disease risks: Lyme disease, leptospirosis, and chronic wasting disease (to date, no person has ever contracted CWD from deer or elk). Deer are host to a variety of parasites, such as botflies, liver flukes, ticks, lice, and worms.

Legal Status

Game species with regulated hunting seasons. Any control technique that involves the handling of deer would almost certainly require a state permit.

Best Practices

- Deer populations can double in size every 2–3 years, if unchecked. Solutions that address a community-wide problem will probably include a combination of options, such as harvesting female deer, fencing, replanting with less palatable species, and repellents. ("Buck-only" harvests will not reduce or stabilize the population.)
- In rural areas, deer populations can be reduced through regulated hunting seasons. On privately owned land, a biologist can advise the landowners about how many deer should be harvested to minimize damage; the landowners can negotiate with individual hunters or a hunting club to ensure that adequate numbers are killed. In suburban areas, some states are allowing sharpshooters to remove deer from communities under special permits. Qualified WCOs may be hired to assist with these programs.
- Site-specific problems can often be solved with fencing, replanting, regulated harvests, and the application of repellents every 4–5 weeks. There are some relatively effective deer repellents that contain putrescent egg. They work best in small areas. Check label directions, as most repellents are approved for woody ornamentals (nonfood crops). For WCOs with a commercial pesticide applicator license, see the deer fact sheet http://wildlifecontrol.info/ccewdmp/Publications/Deer_factsheet.pdf for detailed information about repellents.

Reduce food sources:

- If people are feeding the deer, persuade them to stop. Even off-site feeding to try to lure the deer away from the vulnerable area may only increase problems. Because of the potential for disease transmission, several states have laws that prohibit deer feeding.
- Switch to plants that deer find less tasty. There's no plant that's completely deer-proof—if they're hungry enough, they'll even eat plants with little nutritional value—but under normal conditions, they'll avoid certain plants. For a list of shrubs that deer tend to avoid, see http://wildlifecontrol.info/ccewdmp/publications.htm.

Protect vulnerable gardens, crops, and landscapes:

- Fences work well to protect a specific site, such as a home garden, agricultural field, airport, or schoolyard. There are many choices for the material and design of the fence. Several factors influence the choice, including the size of the vulnerable area, the intensity of deer feeding, the amount of damage that can be tolerated, and local laws, which may restrict the use of some fence designs.

- Individual plants, or small clumps of flowers, can be protected with plastic netting or wire cages. Wrap the plastic netting around the plants, or place the wire cage over them. The deer will be able to eat any part of the plant that sticks out of the net, but this reduces damage substantially.
- Vegetable and flower gardens can also be protected with a temporary electric fence used during the growing season. Install the fence before there's damage to prevent a problem from developing, or at the very first sign of damage. Never directly connect a fence to household current—it should run through a UL-approved fence charger. Adding an attractant, such as peanut butter, to an electric fence will entice the deer to touch the fence with their noses, so they'll get shocked. Wrap a large clump of peanut butter in aluminum foil and attach it to the fence. This may teach the deer to avoid the area.
- Small groups of trees or shrubs (within a circle of about 20 yards) can be enclosed with snow fencing during the winter and early spring. This will provide decent but not perfect protection. Deer do not like to jump into small enclosed areas.
- For year-round protection, the best barrier is a permanent woven-wire fence that's at least 8 feet high. In some cases, a high tensile electric fence with a 7-wire design works well, too (best if the spacing between the wires is no more than 9 inches). Entrances must be protected with gates or cattle guards.
- Combination fences work well. To reduce the overall cost, you can use woven wire on the bottom and then switch to high-tensile electric wire for the top 4 feet of the fence. This is cheaper than a fence constructed of only woven wire.
- Dogs may be able to keep deer out of fenced areas. Some growers have enclosed their orchards with a hidden barrier such as Invisible Fence, allowing dogs to freely patrol the area. One study showed that two dogs could patrol about 60 acres in the summer, 10 acres in the winter. This technique is most effective with large dogs that patrol aggressively and are kept outdoors all the time.

WCOs with a commercial pesticide applicator license:

- There are several effective commercial repellents registered for deer. Most of them contain an active ingredient that either tastes or smells bad to deer, such as rotten eggs, soap, and hot pepper (active ingredient is capsaicin). Refer to *Managing White-Tailed Deer in Suburban Environments: A Technical Guide* (http://wildlifecontrol.info/NEWDMC/PDFs/Deer_management_mechs.pdf) for details about repellents.
- For a small vegetable garden or flower bed, string a rope fence 30 inches above the ground, with the posts set 3–4 feet apart, and spray the rope with an odor-based repellent. Or you can spray the repellent onto cloth strips and attach them to the rope every 3–4 feet.
- Adding repellents to an electric fence may also teach the deer to avoid the area. Repellents seem to be more effective than attractants, but you must have the pesticide applicator license to use them.

If you're asked for advice about how communities can prevent or reduce the number of traffic accidents caused by collisions with deer, here's what you can say:

- Contact the Department of Transportation and the state wildlife agency for information.
- Reducing the size of the local deer herd may reduce the number of collisions. Although that seems like common sense, there's only limited research on this point.
- Keep roadside areas well-mowed and keep shrubs well-groomed, so drivers have a better view of deer approaching from the edges.

- Encourage drivers to slow down during dawn and dusk, especially during May and June, and again from October through December.
- Deer tend to travel in family groups, so if you see one, look for more. Slow down.
- More people have been hurt when a driver tried to avoid hitting a deer and swerved into the oncoming lane or another object. It is better to stay in your lane and collide with a deer rather than hit something else. There's really no way to predict the direction the deer will choose to run.
- If the problem is concentrated along a short section of road, fencing may be the best option.

Preferred Killing Methods

- Follow the conditions of your permit
- Shooting (using a shotgun with 20-gauge or larger slugs, a center-fire rifle, or other implement specified in the permit)
- Lethal injection of barbiturate, if possible
- Stunning using a penetrating captive bolt pistol, followed by exsanguination

Control Strategies that Don't Work Particularly Well

- No scare devices are recommended because they don't work well and many of them can annoy the neighbors. Deer generally get used to scare devices within a few days. So don't waste your time with lights, loud noises, scarecrows, fireworks, gunshot, crackers, bangers, propane cannons, deer whistles, ultrasonic devices, strobes, water sprays, or sirens. Deer can hear ultrasonic sound but it usually doesn't frighten them because they don't associate the sound with a predator or other dangerous situation.
- Human hair and other home remedies usually don't consistently repel deer.
- Bars of tallow-based soaps hung in trees may reduce damage, if the deer aren't feeding intensively in the area, but if the soap melts and runs down the tree it may attract voles.
- Roadside reflectors have been used with varying success to reduce the number of collisions between deer and cars. Deer may get used to them. It may work better in rural areas because deer in suburbs are more used to human activity and lights.
- Deer whistles and ultrasonic devices will not help prevent car collisions with deer.

MUSKRAT (*Ondatra zibethicus*)

Size

18–26 inches long, including the 10- to 12-inch-long tail. Weigh 1½–4 pounds; most average about 2½ pounds. They look like a large rat with a long, narrow tail (the tail is flattened vertically).

Signs of Their Presence

- Tracks: Hind print about twice as long as front, may overlap front track. May see foot drag and tail drag.
- Scat: Oval pellets, often seen in clusters on rocks, logs, or any object that sticks out above the waterline. Usually, the ½-inch-long pellets are stuck together, but may flatten with age.
- Dens (also called "bank burrows"). Muskrats may burrow into the banks of streams or ponds. The entrance, usually 5–6 inches in diameter, is often found about 6 inches below the water's surface. This tunnel, which connects to the den, could be up to 45 feet long. Muskrat dens are found above the high-water line (this may still be 4–5 feet below ground level). The den is usually only a bit bigger than the tunnel that leads to it, often 6–8 inches wide. A small air shaft at the top connects it to the surface; it's often "screened" with loose twigs or plants. Dens generally have several underwater entrances. In fast-moving water, muskrats are more likely to burrow into the bank than to construct a lodge.
- Muskrats may build a den in the outer wall of a beaver lodge, even when the beavers are using their lodge.
- Lodges. If the water is shallow and plants are abundant, muskrats may build a lodge that looks a bit like a smaller version of a beaver lodge. Muskrat lodges are usually a dome-shaped hut of cattails, sticks, and leaves piled on a platform of mud and rotted debris. They're often slightly lopsided, 3–6 feet wide, and about 1½–4 feet above the high-water line. The walls of the lodge are often a foot thick, and the inside is hollowed into several chambers. There are several underwater entrances, called "plunge holes." Muskrats will add to the lodge as long as they're using it. (Same is true for feeding platforms.)
- "Runs." As the muskrats enter and leave their dens, their hind feet scour out a path in the muck in the bottom of the pond. You may be able to see the run in clear water, or feel the smoothed trail with your hands or feet. Muskrats will also travel over land, especially in the fall and right after ice-out. Trails, both in the water and on land, are kept open by frequent use and pruning, and are noticeable. You may see tracks and scat in their trails.
- Feeding stations. They may look like a small lodge. Muskrats will tow food out to this platform. They'll also push mud and aquatic plants up through a hole in the ice (called a "push-up"). When the mud freezes, it keeps the hole open and creates a shelter for the muskrats, protecting them from both predators and the cold. They'll rest in the push-up instead of returning to the lodge, which may be further away. You may see floating blades of cattails, sedges, or other plants near their feeding platforms, or piles of clamshells on the platform.
- Scent posts. Male muskrats will secrete an oily, pungent liquid on a scent post, which is often a twist of grass at the water's edge, to mark their territories. The musky odor of muskrats (both genders) is especially noticeable during the breeding season.

Diet

Mostly aquatic plants and some field crops, such as corn, soybeans, wheat, oats, grain sorghum, sugarcane, and rice grown as a flooded crop. Muskrats prefer cattails, pickerelweed, bulrushes, sedges, arrowheads, reeds, pondweeds, water lilies, and young willow. They can survive entirely on upland plants such as Bermuda grass, clover, and Johnsongrass, if grown near a farm pond. They will eat freshwater clams, crayfish, mussels, snails, crustaceans, small fish, turtles, and frogs when aquatic plants are scarce.

Typical Activity Patterns

Social style: In the winter, 3–4 muskrats may share a lodge temporarily to stay warm, but they are not colonial, like beavers. As the breeding season approaches they will become territorial. Dens are occupied only by a female and her young while she's raising them. When populations are high, males are more likely to fight.

Daily activity: Most active at dawn and dusk, but may be active throughout the night or day. Spends most of its time in the water and can remain submerged for a long time.

Hibernator? No.

Migrates? No. During the spring (usually late February–early March), males will leave in search of mates. In the fall, young muskrats may seek new territories. These dispersals are more likely to happen when their populations peak. The muskrats may travel over a half-mile from their home.

Where Found

Distribution: Widespread and abundant throughout the region.

Habitat: Wetlands (both fresh and slightly salty water), especially with still or slow-moving water and dense cattail stands. They're found in marshes, beaver ponds, lakes, swamps, streams, drainage ditches, canals, reservoirs, and mine pits. Expect to find muskrats in beaver ponds, farm ponds, and any semipermanent waterway.

Territory and home range: Both sexes are aggressively territorial just before and during the breeding season and may kill other muskrats, even their own young, during a squabble. Home range is small, usually within 200 yards of den.

Breeding Habits

Pair bonding style: Polygamous. A pair may remain together for a breeding season. Both male and female may build and maintain the den, but the female mostly cares for the young by herself.

Breeding dates: Mid-March–September (they may begin breeding in early March in southern areas). Females may breed while still nursing.

Litter size: Usually 1–8, typically 5–6 young. Average 2–3 litters/year.

Birthing period: 1st litter: April–May. 2nd litter: June–July (about half of New York's muskrats have a second litter). 3rd litter, less common: August. The young are born in a grass-lined nest in the lodge or den.

Weaning dates: About 1 month old.

Amount of time young remain with parents beyond weaning date: Young are driven off after weaning.

Special note: Muskrat populations boom and crash every 10 years, in a regular cycle. Droughts and floods may cause population crashes, too.

Common Nuisance Situations

Time of year: Spring.

What are they doing?

- Most muskrat damage is caused by their burrowing.
- They burrow into earthen dams, dikes, levees, and railway embankments, weakening their structures. They also burrow into the banks of ponds, canals, and irrigation and drainage ditches. Their tunnels may drain a small farm pond. They may damage floating docks, marinas, and boathouses.
- Muskrats will cross yards. Some people are frightened of them, or mistake them for Norway rats.
- Occasionally eat field crops. May cause substantial financial losses in states with major rice and aquaculture operations, because they eat rice, cut it down to use as building material for their lodges, and damage the field by burrowing through levees.
- Damage aquaculture sites by burrowing into levees or pond banks.
- Damage ornamental aquatic gardens by eating water lilies or cattails or other plants.

- When their populations grow too high, they may "eat out" all of the aquatic plants in the area, reducing the quality of the habitat for other species, such as waterfowl.
- Disease risks: tularemia, hemorrhagic septicemia, leptospirosis, salmonellosis, ringworm, pseudotuberculosis. They are hosts for many ticks, mites, fleas, and various worms.

Legal Status

Check with your state wildlife agency. Muskrats are often categorized as a furbearer. There may be a regulated trapping season. Actions that affect wetlands may require additional state permits.

Best Practices

To prevent damage, "overbuild" a dam or farm pond:

- Normal water level should be at least 3 feet below the top of the dam.
- Spillway should be wide enough to prevent the water level from rising more than about 6 inches during heavy rainfall.
- Inside face of dam (towards water) should be built at a 3:1 slope (3 feet out for every foot of height).
- Outer face of dam should be built at 2:1 slope.
- Top of dam should be at least 8 feet wide, best if it's 10–12 feet wide.

To solve an existing conflict, reduce their food sources (this may be necessary if there's extensive damage):

- Around a pond, mow grassy areas frequently.
- Remove the aquatic plants that muskrats eat. They prefer starchy foods including cattails, pickerelweed, bulrush, smartweed, duck potato, horsetail, water lily, sedges, young willow trees, and rice, when grown as a flooded crop.
- Remove the following upland plants from around the pond (and replant with other species): Bermuda grass, clover, Johnsongrass, and orchard grass. Muskrats can survive entirely on those plants.
- If muskrats are traveling from the pond to feed in fields or gardens, fence those areas. Muskrats will eat corn, soybeans, wheat, oats, grain sorghum, sugarcane, and ornamental flowers.
- If you don't want to remove plants from your pond, try to attract muskrat predators instead. Increase cover in and along the pond's edge to attract mink, one of the main predators of muskrat.

Protect vulnerable ponds and dams:

- Before attempting to modify a wetland environment, speak to staff in your state wildlife agency. In some cases, permits would be needed.
- Riprap the inside of the pond's dam with rocks or cover it with vinyl-coated welded wire. Rocks should be flat and closely-fitted. The rock layer must be at least 8 inches thick, extending 3 feet below and 1 foot above the normal water line.
- Draw down ponds during the winter (requires a permit). Remove the muskrats, and make sure they're gone. Reduce the water level to expose their burrows (this often means lowering the water from 1½–3 feet). Then fill in any burrows and tamp down the soil. Cover them with rocks.

Trapping Strategies

Trapping is often a critical part of a strategy to control muskrat damage. When possible, invite licensed trappers to remove muskrat during the legal season. If required, follow the conditions of your permit.

To avoid catching beaver, otter, mink, and birds:

- Avoid trapping if there's sign of otter nearby, if possible.
- Use natural baits. In the spring, try muskrat musk. In the fall and winter, use sweet-smelling oils such as oil of sweetflag, spearmint, or anise; apples, parsnips, or carrots also work well. Do not use baits made from beaver or otter glands.
- Use only small foothold or body-gripping traps, and anchor all traps so they'll hold any live animal that may be accidentally caught. Put crossed hoops over a floating log set to discourage birds from landing on the log.

- Use catchpoles to release any otters or beavers that are accidentally caught in the trap.
- Don't set traps in spillways, channels, large bank holes, or other natural funnels.

Live traps:

- Cage traps, 8 × 8 × 24 inches, baited with apples or carrots

Lethal traps:

- Body-gripping trap, #110 or #120, set in the water. An underwater set is more selective and will reduce nontarget captures.
- Foothold trap, #1 or #1½, set in 1–2 inches of water and anchored in at least 18 inches of water.
- Set traps in active runs as close to the den's entrance as possible. Traps must be set 5 feet or more away from a muskrat lodge, unless allowed by permit. Pole sets, under-ice sets, and float sets also work.
- If the water is less than 18 inches deep, use a body-gripping trap (preferable) or a guarded foothold trap.

Preferred Killing Methods

- Lethal trap
- Foothold trap in a submersion set
- Shooting (practical if removing a few muskrat from a small pond)

Acceptable Killing Methods

- CO_2 chamber
- Stunning and chest compression
- Stunning and cervical dislocation

Control Strategies that Don't Work Particularly Well

Conventional frightening techniques aren't effective against muskrats.

EASTERN WILD TURKEY (*Meleagris gallopavo silvestris*)

Size

Males (toms) weigh 11–25 pounds; females (hens) weigh 8–12 pounds. Young are called "poults."

Signs of Their Presence

- Sightings of the bird. Often seen feeding in fields or dusting along dirt roads.
- Feathers and droppings.
- Scratchings: In the spring and fall, can see scratchings on the ground under nut and fruit trees. When searching for food in the woods, turkeys usually scratch, kicking the leaves back and to the side as they feed. Then they switch feet. This leaves an inverted-V mark.
- Tracks: Good bet it's a turkey if the distance from the tip of the middle toe to the back of the heel is greater than 4 inches. Near water, you might confuse their tracks for those of the Great blue heron.
- Nest: A simple depression, usually 8–10 inches in diameter, made when the hen rubs her breast into the leaves. The nest is usually found in brushy areas, such as near raspberry bushes, fallen branches, along the edge of a weedy field, and in a brushy field that's changing back to forest.
- Sounds: Males gobble, usually as part of their courtship display (but may gobble while roosting, too). Hens cluck, purr, and cackle in response. Gobbling is most commonly heard in the early morning, but during the breeding season the toms may gobble at any time of day. The sound may be heard up to a mile away. Both toms and hens make a variety of other noises, including yelps, purrs, and cackles.

Diet

Omnivores. Turkeys will eat insects, salamanders, small frogs, and plants. Their diet changes seasonally.

Spring: They'll eat nearly any available plant, such as grass shoots; sedges; buds, flowers, and leaves of shrubs and trees; the roots, tubers, and bulbs of perennials; dried fruits; and nuts.

Spring and summer: High-protein foods, such as insects, spiders, centipedes, millipedes, snails, and slugs are critical to the poults for feather development. Adults eat those foods as well as grasses, seeds, salamanders, and frogs.

Late summer: Grasshoppers, beetles, crickets, and the fruits and seeds of nearly any plant.

Fall: Just about any nut or fruit crop, including acorns, beechnuts, hazelnuts, wild cherries, and grapes; waste grain; insects; salamanders and frogs.

Winter: Nuts and seeds; grains, including corn; insects; and snails. Turkeys will often feed where deer have pawed through deep snow and in corn fields, especially if manure was spread on the field, and in spring seeps.

Typical Activity Patterns

Social style: Gregarious. Several hens and their broods may flock together during mid-summer into fall. From the end of the breeding season through the summer, bachelor flocks are common, and usually stay together through late winter. But in late fall, a few of the toms may join the flocks of hens and poults. During the winter, several flocks may congregate in areas with good food supplies.

Daily activity: Diurnal. Turkeys are active throughout the day, depending on the weather. While feeding in a part of their range, they tend to roost in the same area each night. During periods of deep snow, they may stay on the roost for several days.

Hibernator? No.

Migrates? No.

Where Found

Distribution: Stable or increasing throughout the region.

Habitat: Turkeys are highly adaptable but do best in areas that offer mature woods, clearings, and fields. They roost in large trees and feed under the hardwoods, in openings, and in agricultural fields. In the winter, they seek south-facing slopes with hardwoods (for nuts) and springs and seeps (for insects and salamanders). Turkeys are also found near corn fields, orchards, dairy farms (especially those with available silage and fields spread with manure), and pastures. In suburban areas turkeys will frequent birdfeeders and home gardens.

Territory and home range: Turkeys move around a lot, seeking nesting sites in the early spring, brood areas in the summer, and woodland cover in the fall. In the early winter, they often move from forested or hilly country to agricultural areas, where they may depend heavily upon grain left in the fields after harvest and seeds they can pick out of manure. These seasonal movements can be significant in some regions—a flock may range over many square miles. In the winter and during nesting, their home range is often limited to 100–200 acres.

Breeding Habits

Pair bonding style: Polygamous.

Breeding dates: Late March–June.

Egg-laying dates: April–July. They lay one egg each day, for about 13 days (but may skip a day). Hens lay one clutch but will nest again if their eggs are destroyed.

Clutch size: 8–15, usually 10–12 (they take about 2 weeks to lay all of their eggs).

Eggs hatch: About 26–28 days after the female begins incubating the eggs. The poults leave the nest right after hatching, to forage for insects with the hen, usually in a clearing or field. They can fly when they're about 10 days old. Poults chill easily until they're about 5 weeks old, so the female will brood them when it's damp or cool.

Amount of time young remain with hen: Until the flock breaks up the next spring. At that time, the young seek their own home ranges. Young males may disperse earlier than young female turkeys.

Common Nuisance Situations

Time of year: Any time of year.

Turkeys are often blamed for crop damage that was caused by another species. These large birds are easily seen in the fields during the day, but they may be eating insect pests—not the crop! Remember to investigate carefully, looking for tracks, tooth marks, and other animal signs that will help you find the real culprit.

What are they doing?

- Eat crops, such as grapes, berries, corn, and small grains. They'll eat mature corn in the field and harvested corn from outside storage cribs, and may damage newly planted seeds and seedlings.
- They eat dairy silage and may peck holes in the plastic covers protecting the silage, which can cause it to spoil.
- Tear up turf on golf courses (the greens) and newly established lawns.

Although uncommon, WCOs may receive more calls about these behaviors:

- Dig up flower beds and home gardens. They've been reported nesting in a flower bed next to a home.
- They may peck and scratch at cars and other reflective surfaces, such as windows.
- Their droppings may foul cars, decks, porches, and other areas.
- They'll sometimes stop traffic as they pause while crossing roads. They can pose a hazard to planes on runways (not often in the air, as do other birds).
- Turkeys may act aggressively toward people or pets, usually during their mating season in the spring.

Legal Status

Check with your state wildlife agency. Turkeys are often categorized as a game species. There may be a regulated hunting season, and you may need a special permit to control nuisance turkeys.

Best Practices

In rural areas, the regulated harvest of males in spring, or birds of either sex in the fall, may be an important contribution to strategies intended to reduce damage.

Reduce food sources and roost sites:

- If anyone is feeding the turkeys, persuade them to stop.
- Keep the area underneath bird feeders clean.
- Rake up and remove nuts and fruits, if possible.
- Netting could be practical for protecting high-value wine grapes, or a few valuable fruit trees or shrubs in a suburban area.
- Plant a variety of corn with a tighter husk, so the kernels are better protected.
- Replace flowering bulbs and other vulnerable plants with varieties turkeys find less tasty.
- Trim branches or remove trees from favored roosts.

Protect people and vulnerable objects:

- If a turkey is behaving aggressively, try to scare it away. Establish yourself as the "top turkey." Wave your arms, make noise, threaten the bird with a broom, spray it with water from a hose. Just don't harm it—that's illegal, in this situation.
- Park the car in a garage, or as far away from the wooded side of the parking lot as practical, or keep it covered. Cover the rear-view mirror with a paper bag to reduce its reflectivity. (For those who don't like to wash their cars, this can be an excuse. A dirty car won't be as reflective, either.)
- Cover windows or other reflective surface that the birds are pecking or scratching with tarps or opaque plastic sheets.

Protect vulnerable sites:

- Fencing may keep turkeys away from a small area that they aren't using much, such as a deck, porch, or flower garden. A 2-foot fence should be adequate, because turkeys usually try to walk around fences rather than fly over them.
- If it's possible, totally enclose the area with the barrier, perhaps adding netting as a cover for the fence (they can fly).
- Frightening noises from propane cannons or bangers may work for protecting crops or stored silage.
- Remote-activated sprinklers, windmills, Mylar tape, balloons, scarecrows, and predator models may also scare off the turkeys.
- Patrol dogs (larger breeds that run and chase, such as border collies or terriers) may be able to chase turkeys out of vineyards, orchards, and other enclosed areas. Dogs must be trained to chase but not capture the birds. Some growers place platforms in the enclosure to give the dogs a better view of the area.
- You can suggest that your customers allow hunting on their land. This may help to reduce local populations and alleviate some of the pressure on the site.

Trapping Strategies

It's unlikely that a WCO will trap turkeys to solve a nuisance problem, because of several practical issues. You would probably need a state permit. Turkeys are wary and not easy to capture. Also, it takes specialized equipment and a lot of time and effort. No lethal traps are approved or even available for turkeys. The nonlethal methods described above are a much more practical approach to dealing with the problem, especially in urban areas.

Preferred Killing Methods

- Shooting, using a shotgun with # 4, 5, or 6 shot, or center-fire or rimfire rifle
- CO_2 chamber
- Stunning and decapitation
- Lethal injection of barbiturate, if possible

Acceptable Killing Methods

- Stunning and cervical dislocation

Control Strategies that Don't Work Particularly Well

- There are no pesticides (including repellents) registered for use against turkeys.
- Any nighttime efforts, because turkeys roost in trees at night.
- Conventional fencing is unlikely to work in large areas, or areas that are used heavily, because the turkeys will probably just fly over the fence. If the attractant is strong—such as a crop—the fence isn't as likely to work.
- It may be impractical, impossible, or too costly to protect most crops with netting.

Is it a pet or a pest? An overview of the issues related to the control of feral cats and dogs

As a WCO, you may not be allowed to handle domestic cats (*Felis catus*) or dogs (*Canis f. familiaris*). In some states in our region, they are not regulated by the state wildlife agency. Yet you may receive calls. This is an overview of the issues related to the control of feral cats and dogs.

House Cats (*Felis catus*)

Many cat situations are quite straightforward for WCOs. An elderly mother passes away and the daughter wants the frightened cat captured so she can take it to her home. This type of situation holds the same type of risks as any normal wildlife call. That's not true once you're confronted with a stray cat, a feral cat (a "house" cat that lives outdoors and acts like a wild animal), or a cat of mysterious origins.

There are several reasons why some WCOs are tempted to handle feral cats. Cat control may not be mandated. Some shelters have a no-kill policy, which limits their ability to accept more animals. WCOs may sympathize with people who are experiencing problems and reason that they have the appropriate skills and tools to capture and handle nuisance cats in a safe and compassionate manner, while the average person probably doesn't.

The problems are real. Farmers and restaurant owners may find themselves overrun with cats, because the cats have been abandoned there, or they're attracted by a food source, such as a dumpster. The noise, especially during the breeding season, can be extremely annoying. Feral and free-roaming cats may carry as many as a half-dozen diseases. People are more likely to approach and pet a cat, especially a kitten, than a wild animal, so their risk of being exposed is higher. Cats are predators and kill birds and other wildlife. Some people who feed birds want to protect them from cats, which they consider unnatural predators. Others feel guilty or sad when they see a malnourished cat outdoors, and feel motivated to address its suffering.

Legal Risks

Cats are not wildlife, even if they were born in the wild and act like wild animals. The domestic cat, *Felis catus*, is an exotic species. It may not be covered by the regulations governing your state wildlife agency. This means that your WCO license may not give you the authority to handle cats.

Also, from a legal standpoint, cats are property. Private citizens have no legal authority to dispose of "found property" (and "finding" a stray cat without a collar is like finding a wallet without ID on the street). This means that in some states in our region, if someone disputed your activity, you could be charged with both a criminal lawsuit (under animal cruelty statutes) and a civil lawsuit (under property laws). If that doesn't make you queasy, here's the kicker: in some parts of our region, the laws describing cat control are vague.

Here's an example of how that could affect someone who's handling feral cats. In New York, there's no law requiring the licensing of cats. How do you decide if a particular cat is a free-roaming and beloved pet, a stray, or a feral cat, and whether it's "owned" or "un-owned"? Do you canvas the entire neighborhood before you set a trap? What if someone lies to you? Imagine someone hired you to capture and kill a colony of feral cats living in their barn. You do everything right, but then someone sues you, claiming that you killed their pet. How do you defend yourself? Without tags or licenses, there's no way to prove whether or not a cat is a pet. Even cats that appear to be wild or feral may be owned (such cats are often ear-tipped, but not always).

It's much easier to make a mistake with cats than with wildlife. A truly feral cat usually avoids human contact, but distinguishing between a free-roaming pet and a stray can be hard. It may not be easy to tell from their behavior or their appearance. What if a free-roaming cat wandered into an area at the wrong time, and was trapped along with the feral cats?

If you become known as the neighborhood "cat guy," new problems might arise. You wake up one morning and discover a box of kittens on your doorstep. They're so young they'd require hand-feeding, and they're sick. You don't have the facilities or time to raise them and seek adoptive homes. You make your usual phone calls, but no luck. So you decide that the merciful and practical thing to do is euthanize them. The "owner" hears about this later, and is spitting mad. The public may be more upset with you for euthanizing them than with the owner for abandoning them. Now are you nervous?

Financial Risks

With all this legal liability, people who handle cats must make a lot of money, right? Guess again. Many people don't want to pay for these services. Some may think that the government takes care of it, which may be true for dogs, but not always for cats. Some people may appreciate your skills and your right to make a living but they're just tired of having to deal with—and pay for—someone else's mistake.

Don't think that all you're offering is a bit of your time and expertise, and that once you've caught the cats you can drop them off at an animal shelter. Private shelters need to keep their euthanasia rates low because that's what appeals to donors. They survive on donations. Even if they agree that euthanasia is the right choice, they might be unwilling or unable to do it. You may do better with a municipal shelter because protecting public safety is part of their mission, so they're concerned about reducing the disease risks associated with feral cats.

Veterinarians sometimes provide discounts, most commonly to nonprofits but also to customers with whom they have a well-established professional relationship. One person with a great deal of cat experience warns that if you become involved in actually arranging for the neutering of the cats or the adoption of kittens, the venture will not be profitable. It will turn into a community service rather than a business.

Risks to Your Personal Reputation and WCO Business

Even though you may be acting as a private citizen when you work with cats, this could affect your WCO business. Imagine that someone writes a nasty letter to the newspaper. Maybe the paper covers the story, runs your picture, and mentions your business' name. That's not the kind of advertising you probably want.

Your feral cat work might jeopardize your WCO business, but your WCO work could also pose problems for your cat sideline. Some cat activists are against trapping and the WCO industry. They may not trust you, yet they're an important source of information and support. It would be best to keep both efforts separate. For example, don't use equipment with your WCO identification tags to trap cats.

Ethical Questions

You're used to considering many factors as you make decisions in your WCO work. Here are a few new wrinkles.

If someone intends to maintain and feed a feral cat colony, and hires you to provide services that will help them do that, are you violating the ethical code you apply to your WCO work? You know that by feeding the cats, the person will also be feeding wildlife, such as raccoons. This could create problems for the neighbors. Feeding cats also attracts other cats to the site, so the size of the colony might increase. Would you be helping to create or worsen a nuisance animal situation? And what effects might this feral cat colony have on local wildlife populations and nearby natural areas?

And yet, some people form an even stronger bond with cats than they do with wildlife. Some people go to a great deal of trouble to help take care of feral cat colonies. They may receive strong psychological benefits from the companionship of the cats, from the chance to do something they consider meaningful, and from being outdoors. It may reduce their isolation and prevent depression, and may be a source of joy. Some wildlife biologists believe that in certain areas, feral cat colonies don't pose much of a threat to wildlife. How do you balance the potential harm and good caused by a particular colony?

Let's face it, most of us feel a little differently about cats than we do about voles. The thought of a cat being killed may be a little harder to accept. People who have problems with nuisance wildlife can seek professional help, but that may not be true with cats. So some desperate people may try to do it on their own, with horrifying results. You could provide badly needed expertise. How would you feel if you do (or if you don't) offer your skills?

Ecological Questions

What effects do free-roaming cats have on wildlife? You will hear many strong opinions, especially from birders, but it's actually not been well-studied. It's a hard question to answer scientifically. Usually several factors cause a population to decline. How much is due to habitat destruction and how much to predation? That can be hard to determine. Then you'd need to isolate the effects of cats from other predators, which is also not an easy task.

So far, we can say that free-roaming cats can definitely hurt wildlife populations on "islands." From a conservationist's point of view, the term "island" refers to three scenarios: a piece of land surrounded by water; an urban park, in which a habitat is surrounded by development; and a pocket of habitat for an endangered or threatened species. In these islands, predation by cats may be a serious problem. For example, on some islands, populations of prey species have not been exposed to predators, so the introductions of any predator can be devastating. If the prey is an endangered species, losing even a few of them to a predator—any predator—could matter a great deal. In such fragile situations any additional stress could tip things the wrong way. (Yet on other islands, cats may help some vulnerable species by killing other more dangerous predators, such as rats.) Another complication to this island scenario is that when cats are fed, so are other predators such as raccoons, skunks, foxes, and opossums, further increasing predation pressure.

Most free-roaming cats don't live on islands or near endangered species. Unfortunately, some people have established feral cat colonies near natural areas that do support threatened wildlife, sometimes on public land.

Hard Science about Feral Cats is Hard to Find

If you're interested in this subject and want to learn more, evaluate your information sources carefully. It's hard to find unbiased and reliable information about feral cats. When you consider the ratio of bad information to good information, it's probably worse for cats than it is for most of the wildlife species you handle. There haven't been many studies of free-roaming cats, and some of them were seriously flawed. There are a handful of really important questions about free-roaming cats that no one can answer in a fair, scientific fashion, such as: how many free-roaming cats are there in the United States? What do they eat?

What Should You Tell People Who Call with Cat Problems?

It's best to avoid any situation involving cats of unknown origin. Suggest the callers contact a local shelter or their municipality.

For More Information

- If you're interested in working with cats, read the applicable laws carefully.
- For an excellent overview of the subject, read *Community Approaches to Feral Cats: Problems, Alternatives, and Recommendations* by Dr. Margaret R. Slater. This 2001 handbook is available from the Humane Society Press in both print and online versions (see http://www.hsus.org/ace/14631). It's one of the more comprehensive, unbiased, and well-researched reports on feral cats.
- "Free-roaming and feral cats—their impact on wildlife and human beings" by Dr. Gary J. Patronek, a special report published in the January 15, 1998 issue of the *Journal of American Veterinary Medicine Assn.*, vol. 212, no. 2, pgs. 218–226.

- "Domestic cat colonies in natural areas: A growing exotic species threat" by Alice L. Clarke and Teresa Pacin. 2002. *Natural Areas Journal* 22: 154–159.
- Alley Cat Allies offers many useful links and answers to FAQs, "frequently asked questions."
  1801 Belmont Road NW, Suite 201
  Washington DC 20009
  (202) 667-3630
  http://www.alleycat.org/faqs.html#11
- Feral Cat Coalition also offers useful articles.
  9528 Miramar Road, PMB 160
  San Diego CA 92126
  http://www.feralcat.com
- The San Francisco SPCA offers several fact sheets, including one for addressing conflicts between neighbors.
  2500 16th Street
  San Francisco CA 94103
  (415) 554-3000
  http://www.sfspca. org/feral/factsheets.shtml

Domestic Dogs (*Canis f. familiaris*)

What if someone calls you with a complaint about a dog? This is often a much simpler situation. In most states, local municipalities are required to have a dog control officer. In most cases, you can refer people to their local dog control officer or the local police, who'll deal with the problem.

There may be times when a dangerous situation develops with a nearby dog when you're on a job. If it's not a life-threatening emergency, capture and restrain the dog, then call the dog control officer and the police, and let them take over. Even though the laws about dog control may be specific, situations can still become emotional. Some of the same risks involved in cat control apply to dogs. If you're interested in becoming a dog control officer, contact your local municipality.

Appendix D
Contacts and Equipment Suppliers

APPENDIX D:
Contacts and Equipment Suppliers

Your state agency's wildlife office (region __________)
contact
phone
fax
address
Email
website

Local health authority
contact
phone
fax
address
Email
website

Police
contact
phone
fax
address
Email
website

Animal shelter
contact
phone
fax
address
Email
website

Cooperative Ext. Master Gardener hotline

Veterinarian
contact
phone
fax
address
Email
website

Emergency veterinary care (after hours)
contact
phone
fax
address
Email
website

Animal control officer
contact
phone
fax
address
Email
website

Town clerk
contact
phone
fax
address
Email
website

EQUIPMENT SUPPLIERS

This is not an exhaustive list. Inclusion does not indicate an endorsement nor does omission indicate disapproval by the authors or publisher. This list was last updated in November 2006.

Baits and Lures

Jameson's Ultra Blend
Box 579
Brownsville PA 15417
(742) 938-2002
Email: wcs@dp.net

Marsyada Scents
722 North St.
W. Hazleton PA 18202
(570) 459-1403

Bird Control (Noisemakers, Exclusion)

Bird Barrier America, Inc.
20925 Shico Rd.
Carson CA 90746
(800) 503-5444
www.birdbarrier.com

Bird-B-Gone, Inc.
23918 Skyline Rd.
Mission Viejo CA 92692
(800) 392-6915
www.birdbgone.com

Birdbusters
300 Calvert Ave.
Alexandria VA 22301
(800) NO-BIRDS
www.birdbusters.com

Bird Scare Predator Eye, Inc.
132 E. Demont Ave., #122
St. Paul MN 55117
(651) 482-1241
www.pestbirdscare.com

Bird-X, Inc.
300 N. Elizabeth St.
Chicago IL 60607
(800) 662-5021
www.Bird-X.com

Cat Claw
PO Box 5250
Johnstown PA 15904
(800) 832-2473
www.catclaw.com

Critter Gitter
manufactured by Amtek:
(800) 762-7618 (for distributors)
www.amtekpet.com/products.html#crittergitter

D&S Specialty Products
13611 N.E. 126th Place, #200
Kirkland WA 98034
(800) 820-1980
www.flybye.com

InterNet, Inc.
7300 49th Ave.
Minneapolis MN 55428
(800) 328-8456
www.Internetmesh.net

Margo Supplies, Ltd.
PO Box 5400
High River AB T1V 1M5
Canada
(403) 652-1932
www.margosupplies.com

Nixalite of America, Inc.
1025 16th Ave.
East Moline IL 61244
(888) 624-1189
www.nixalite.com

Reed-Joseph International Co.
PO Box 894
Greenville MS 38702
(800) 647-5554
www.reedjoseph.com

SEA Technologies, Inc. (manufacturer of the Avian Dissuader)
PO Box 195
382 E. Oak
St. Louisville KY 40150
(888) 732-2246
Fax: (775) 703-2074
Email: info@aviandissuader.com
www.aviandissuader.com

Sutton Ag. Enterprises
746 Vertin Ave.
Salinas CA 93901
(831) 422-9693
suttonag@earthlink.net

The Tanglefoot Co.
314 Straight Ave. S.W.
Grand Rapids MI 49504
(616) 459-4139
www.tanglefoot.com

Wildlife Control Supplies (Alan Huot)
PO Box 653
Simsbury CT 06070
(877) 684-7262
www.wildlifecontrolsupplies.com

Wildlife Control Technology, Inc.
2501 N. Sunnyside Ave.
Fresno CA 93727
(800) 235-0262
www.wildlife-control.com

Bird Distress Calls

The Critter Control
9435 E. Cherry Bend
Traverse City MI 49684
(800) 451-6544
www.crittercontrol.com

Johnny Stewart Wildlife Calls
5100 Fort Ave.
Waco TX 76710
(254) 772-3261
www.johnnystewart.com

JWB Marketing LLC (manufacturer of the Electronic Bird Gard, a programmable device that plays bird distress calls)
101 Hurlbut St.
Westwood NJ 07675
(800) 555-9634
birdgard@ix.netcom.com
www.birddamage.com or
www.birdcontrolsupplies.com

Macaulay Library of Natural Sounds
Cornell Laboratory of Ornithology
159 Sapsucker Woods Rd.
Ithaca NY 14850
(607) 254-2404
www.birds.cornell.edu

Wildlife Damage Control
340 Cooley St.
Springfield MA 01128
(413) 796-9916
Fax: (413) 782-6719
www.wildlifedamagecontrol.com

For crow distress calls, order: Controlling Urban Crow Roosts with Taped Distress Calls (Publication 21561) from:
University of California
ANR-Communication Services
6701 San Pablo Ave., 2nd Fl.
Oakland CA 94608
(510) 642-2431
Email: jerry.hendrex@ucop.edu
http://anrcatalog.ucdavis.edu

Chimney Covers

Chim-A-lator Co.
5025 208th St.
Farmington MN 550240
(651) 460-1080 or (800) 729-9505
www.reedfirstsource.com/members/result~msa~10013389.asp

Copperfield Chimney Supply, Inc.
304 S. 20th St.
Fairfield IA 52556
(800) 247-3305
www.copperfield.com

DoItYourself.com, Inc.
PO Box 157
Raritan NJ 08869
(866) 835-5643
Fax: (908) 541-1154
Email: sales@doityourself.com
www.doityourself.com

HomeSaver
(866) HOMESAVER
www.homesaver.com

HY-C Company, Inc.
2107 N. 14th St.
St. Louis MI 63106
(800) 325-7076
Fax: (314) 241-2277
www.hyccompany.com

Northline Express
9229 Murphy Rd.
Roscommon MI 48653
(866) 667-8454
Fax: (989) 275-5967
sales@NorthlineExpress.com
www.northlineexpress.com

Vestal Mfg.
PO Box 420
Sweetwater TN 37874
(615) 337-6125
www.rumford.com/Vestal.html

Volko Supply Co., Inc.
205 Herricks Road
Garden City Park NY 11040
(800) 685-8263
Fax: (516) 741-4470
Email: volkosupply@yahoo.com
www.volko.com

Whitecaps
1920 Tower Industrial Dr.
Monroe NC 28110
(800) 335-2534
Fax: (704) 226-9797
Email: customchimney@aol.com
www.whitecaps.com

Wildlife Control Supplies (Alan Huot)
PO Box 653
Simsbury CT 06070
(877) 684-7262
www.wildlifecontrolsupplies.com

Direct Capture Equipment

Animal Care Equipment and Services, Inc.
PO Box 3275
613 Bert Way
Crestline CA 92325
(800) 338-ACES
www.animal-care.com

The Critter Control
9435 E. Cherry Bend
Traverse City MI 49684
(800) 451-6544
www.crittercontrol.com

Fuhrman Diversified, Inc.
2912 Bayport Blvd.
Seabrook TX 77586
(281) 474-1388
www.fieldcam.com

Ketch-All Company
4149 Santa Fe Rd., #2
San Luis Obispo CA 93401
(805) 543-7223
www.ketch-all.com

Tomahawk Live Trap Co.
PO Box 323
Tomahawk WI 54487
(800) 272-8727
www.livetrap.com

Wildlife Control Supplies (Alan Huot)
PO Box 653
Simsbury CT 06070
(877) 684-7262
www.wildlifecontrolsupplies.com

Dryer Vents and Vent Pipe Shields

Atlanta Supply Co.
643 Tenth St.
Atlanta GA 30318
(800) 972-5391
Fax: (404) 876-7582
Email: sales@atlantasupply.com

The Critter Control
9435 E. Cherry Bend
Traverse City MI 49684
(800) 451-6544
www.crittercontrol.com

Do it Best Corp.
PO Box 868
Fort Wayne IN 46801
(260) 748-5300
Fax: (260) 493-1245 (for corp.)
(260) 748-7175
Fax: (260) 748-5664
(doitbest.com)

Email: mail@doitbest.com
www.doitbest.com

DoItYourself.com, Inc.
PO Box 157
Raritan NJ 08869
(866) 835-5643
Fax: (908) 541-1154
Email: sales@doityourself.com
www.doityourself.com

Hart's Hearth
Uplink Companies
2523 Middle Rd.
Trenton NC 28585
(800) 924-0850
Fax: (561) 258-8030
www.hartshearth.com

RepairClinic.com, Inc.
47440 Michigan Ave., Suite 100
Canton MI 48188
(800) 269-2609
(734) 495-3150
www.repairclinic.com

Wildlife Control Supplies, Inc.
PO Box 125
Cortland IL 60112
(815) 286-9251

General Exclusion Equipment

The Critter Control
9435 E. Cherry Bend
Traverse City MI 49684
(800) 451-6544
www.crittercontrol.com

Critter Gitter
manufactured by Amtek:
www.amtekpet.com/crittergitter.htm
(800) 762-7618 (for distributors)

Critter Ridders (Suppliers of Guest Control Weep-Hole Screen)
Simmons Pest Management, Inc.
2938 Ridgeway Rd.
Memphis TN 38115
(901) 365-BUGS
www.critterridders.com

Pocatello Supply Depot (carries the Electronic Guard, a scare device)
USDA APHIS
238 E. Dillon St.
Pocatello ID 83201
(208) 236-6920

PVE Wildlife (Todol foam gun)
PO Box 84
Pennsburg PA 18073
(800) 724-9468
www.pvewildlife.com/index.htm

Wildlife Control Supplies (Alan Huot)
PO Box 653
Simsbury CT 06070
(877) 684-7262
www.wildlifecontrolsupplies.com

Wildlife Control Supplies, Inc.
PO Box 125
Cortland IL 60112
(815) 286-9251

Inspection Supplies

Ben Meadows
PO Box 5277
Janesville WI 53547
(800) 241-6401
www.benmeadows.com

Forestry Suppliers, Inc.
PO Box 8397
Jackson MS 39284
(800) 647-5368
www.forestry-suppliers.com

Gempler's
100 Countryside Dr.
PO Box 270
Belleville WI 53508
(800) 382-8473
www.gemplers.com

Lab Safety Supply, Inc.
PO Box 1368
Janesville WI 53547
(800) 356-0783
www.labsafety.com

Professional Equipment
30 Oser Ave., Suite 500
Hauppauge NY 11788
(800) 334-9291
www.professionalequipment.com

Stephenson Equipment
952 Main Street
Nashville TN 37206
(800) 251-4931

Wildlife Control Supplies (Alan Huot)
PO Box 653
Simsbury CT 06070
(877) 684-7262
www.wildlifecontrolsupplies.com

Louvers

Am Conservation Group, Inc.
430 Sand Shore Rd, Suite 7
Hackettstown NJ 07840
(800) GOOD-BUY
Fax: (908) 852-6444 or -7614
Email: amcg@nac.net
www.amconservationgroup.com

DoItYourself.com, Inc.
PO Box 157
Raritan NJ 08869
(866) 835-5643

Fax: (908) 541-1154
Email: sales@doityourself.com
www.doityourself.com

The Home Depot
(800) 430-3376
www.homedepot.com

Lomanco
PO Box 519
Jacksonville AR 72076
(800) 643-5596
www.lomanco.com

Lowe's
(800) 44-LOWES
www.lowes.com

Name Tags

Funke Trap Tags & Supplies
2151 Eastman Ave.
State Center IA 50247
(641) 483-2597
Fax: (641) 483-9903
Email:
funketag@funketraptags.com
www.funketraptags.com

Larry A. Finch
PO Box 435
West Unity OH 43570
(419) 924-2234

Odor Neutralizers

Neutroleum Alpha (for skunk scent)
Pocatello Supply Depot
USDA APHIS
238 E. Dillon St.
Pocatello ID 83201
(208) 236-6920

Wildlife Control Supplies (Alan Huot) (Citrus Neutrox, Bioshield, Atom Mist atomizer)
PO Box 653
Simsbury CT 06070
(877) 684-7262
www.wildlifecontrolsupplies.com

One-way Doors

Tomahawk Live Trap Co.
PO Box 323
Tomahawk WI 54487
(800) 272-8727
www.livetrap.com

Penetrating Captive Bolt Pistols

Koch Supplies, Inc.
528 East 19th Ave.
North Kansas City MO 64116
(800) 456-5624
www.kochsupplies.com

Rat Guard (One-way Toilet Flap Valves)

J. T. Eaton & Co., Inc.
1393 E. Highland Rd.
Twinsburg OH 44087
(800) 321-3421
(216) 425-8353
www.jteaton.com

Levenson's, Inc.
1407 Harney St.
Omaha NE 68102
(402) 341-0088

Rubber Buck Shot

Cabela's
(800) 237-4444
www.cabelas.com

Natchez Shooters Supplies
PO Box 182212
Chattanooga TN 37422
(800) 251-7839
www.natchezss.com

Wildlife Traps

Adirondack Outdoor Company
PO Box 86T
Elizabethtown NY 12932
(518) 873-6806
Email:
info@adirondackoutdoor.com
www.adirondackoutdoor.com

Animal Care Equipment and Services, Inc.
PO Box 3275
613 Bert Way
Crestline CA 92325
(800) 338-ACES
www.animal-care.com

Bird Barrier America, Inc.
20925 Shico Rd.
Carson CA 90746
(800) 503-5444
www.birdbarrier.com

Butera Manufacturing Industries
PO Box 349
Wickliffe OH 44092
(216) 761-8800
Fax: (216) 716-8809

CDR Trap Co., L.L.C.
240 W. Muskingam St.
Freeport OH 43973
(740) 658-4469
www.CDRTraps.com

The Critter Control
9435 E. Cherry Bend
Traverse City MI 49684
(800) 451-6544
www.crittercontrol.com

Cumberland's Northwest Trappers Supply, Inc.
PO Box 408
Owatonna MN 55060
(507) 451-7607
Fax: (507) 451-5869
Email: trapper@clear.lake.com
www.northwesttrappers.com

Dobbins Products
208 Earl Dr.
Goldsboro NC 27530
(919) 580-0621
www.trapperman.com

Duffer's
PO Box 9
Bern KS 66408
(785) 336-3901
duffers@mewlan.com

Duke Company
PO Box 555
West Point MS 39773
(662) 494-6767
Fax: (518) 882-5212
www.duketraps.com/8contact/contact.html

DWL Brand Mfg.
PO Box 174
Galway NY 12074
(518) 882-9145
www.dwlbrand.com

The Egg Trap Company (raccoon traps)
PO Box 334
Butte ND 58723
(701) 626-7150
Email:
theeggtrapcompany@srt.com
www.theeggtrapcompany.com

Gremar, Inc.
PO Box 65003
West Des Moines IA 50265
(800) 597-3867
www.gremarinc.com

JC Conner Supply
7522 Mt. Zion Cemetery Rd.
Newcomerstown OH 43843
(740) 498-6822

Kness Mfg. Co., Inc.
2053 Hwy. 5 South
PO Box 70
Albia IA 52530
(800) 247-5062
www.kness.com

Koro Traps
(204) 422-8636 (to find nearest dealer)

M & M Fur Co.
PO Box 15
Bridgewater SD 57319
(800) 658-5554
www.sd-exports.org/m_and_m_fur_company/index.htm

Memphis Net & Twine Co., Inc.
PO Box 80331
Memphis TN 38108
(888) 674-7638
Fax: (901) 458-1601
http://memphisnet.net/10/FishCatalog/AnimalTraps.html

Minnesota Trapline Products (beaver speciality traps)
6699T 156th Ave., NW
Pennock MN 56279
(320) 599-4176
Fax: (320) 599-4314
Email: info@minntrapprod.com
www.minntrapprod.com

Montgomery Fur Co. (Bridger traps)
1539 W. 3375 S.
Ogden UT 84401
(801) 394-4686
Fax: (801) 394-9828
Email: montgomeryfur@hotmail.com

Northern Sport Co.
9191 Leavitt Rd.
Elyria OH 44035
(440) 986-3366
Fax: (440) 986-2072
www.molnaroutdoor.com

Oneida Victor Inc., Ltd. (manufacturer of Conibears)
PO Box 32398
Euclid OH 44132
(216) 761-9010
Fax: (216) 761-8809

Quad Performance Products Co. (Lil' Grizz Get'rz raccoon trap)
Route 1
PO Box 114
Bonnots Mill MO 65016
(573) 897-2097
Email: getrz@prodigy.net

R.C. Best Traps
PO Box 103
Knox PA 16262
(814) 797-1546

R-P Outdoors
PO Box 1170
Mansfield LA 71052
(800) 762-2706
(318) 872-8824

Safeguard Products
PO Box 8
New Holland PA 17557
(800) 433-1819
www.safeguardproducts.com

Schmitt Enterprises, Inc.
PO Box 44
New Ulm MN 56073
(507) 359-4149
Fax: (507) 359-2472

The Snare Shop
13191 Phoenix Ave.
Carroll IA 51401
(712) 822-5318
Fax: (712) 822-5319
www.snareshop.com

Southeastern Outdoor Supplies, Inc.
537 Old Quarry Rd.
Bassett VA 24055
(276) 638-4698

Sterling Fur Company
11268C Frick Rd.
Sterling OH 44276
(330) 939-3763
Fax: (330) 939-5135

Sullivan's
429C Upper Twin
Blue Creek OH 45616
www.sullivansline.com

Supplies For Wildlife
PO Box 84
Pennsburg PA 18073
(800) 724-9468

The Tanglefoot Co. (glue boards for rats and mice)
314 Straight Ave. SW
Grand Rapids MI 49504
(616) 459-4130
www.tanglefoot.com

Tomahawk Live Trap Co.
PO Box 323
Tomahawk WI 54487
(800) 272-8727
www.livetrap.com

Tru-Catch Traps
PO Box 816
Belle Fourche SD 57717
(800) 247-6132
www.trucatchtraps.com

Whitmore Micro-Gen Corp.
10700 Sentinel Dr.
San Antonio TX 78217
(800) 777-8570

Wickenkamp Live Trap MFG
129 Buckeye
Hedrick IA 52563
(641) 661-2700

Wilco Distributors, Inc.
1215 W. Laurel Ave.
PO Box 291
Lompoc CA 93436
(805) 735-2476
www.wilcodistributors.com

Wildlife Control Supplies (Alan Huot)
PO Box 653
Simsbury CT 06070
(877) 684-7262
www.wildlifecontrolsupplies.com

Woodstream Corp. (Havahart live traps, body-gripping traps, Victor snap traps, glue boards and mole traps)
69 N. Locust
Lititz PA 17543
(717) 626-2125
www.woodstreamcorp.com

Appendix E
Resources

APPENDIX E: Resources

The following list is not comprehensive but should give you a sense of what's readily available. Inclusion does not indicate an endorsement nor does omission indicate disapproval by the authors or publisher. This list was last updated in May 2007.

ORGANIZATIONS

The following list focuses on those organizations whose mission includes nuisance wildlife control activities.

Federal Agencies

United States Department of Agriculture, Animal and Plant Health Inspection Service, Wildlife Services Unit (USDA APHIS-WS)
Charles Brown, Director
Eastern Regional Office
920 Main Campus Drive, Suite 200
Raleigh NC 27600
(919) 855-7200
Fax: (919) 855-7215
www.aphis.usda.gov/wildlife_damage

Formerly known as the "Animal Damage Control Unit." Provides advice, particularly about Canada geese and roosting crows and blackbirds. The website includes fact sheets on specific tools, such as low-powered lasers and the electronic guard, among other topics.

United States Fish and Wildlife Service (USFWS)
Federal Permit Office
United States Department of Interior
Contact: Nadine Fleming
PO Box 799
Hadley MA 01035
(413) 253-8643
www.fws.gov

International Organizations

Association of Fish & Wildlife Agencies (AFWA), Furbearer Resources Technical Work Group
444 N. Capitol St. NW, Suite 544
Washington DC 20001
(202) 624-7890
www.fishwildlife.org

Conducts research on best management practices for trapping furbearers, which includes some nuisance species.

National Organizations

The Fund for Animals
200 W. 57th Street
New York NY 10019
(888) 405-3863
http://www.fundforanimals.org

The Humane Society of the United States
2100 L Street NW
Washington DC 20037
(301) 258-8276
Fax: (301) 258-3078
http://www.hsus.org/wildlife/urban_wildlife_our_wild_neighbors

National Animal Control Association (NACA)
PO Box 480851
Kansas City MO 64148
(800) 828-6474
Email: naca@interserv.com
www.nacanet.org

Publishes NACA News and NACA Training Guide; operates the NACA Training Academy.

National Trappers Association (NTA)
524 5th Street
Bedford IN 47421
(812) 277-9670
Fax: (812) 277-9672
Email: ntaheadquarters@nationaltrappers.com
www.nationaltrappers.com

Offers liability insurance to members; publishes American Trapper.

National Wildlife Control Operators Association (NWCOA)
1832 North Brazil Ave.
Indianapolis IN 46219

(317) 895-9060
www.nwcoa.com

The Wildlife Society
5410 Grosvenor Lane, Suite 200
Bethesda MD 20814
(301) 897-9770
Fax: (301) 530-2471
Email: TWS@wildlife.org
www.wildlife.org

Sponsors the Wildlife Damage Management Working Group, which publishes an electronic newsletter, Interactions.

Regional Organizations

Northeast Wildlife Damage Management Cooperative
Contact: Paul Curtis, pdc1@cornell.edu, or Gary San Julian, jgs9@psu.edu

A partnership of universities and federal and state agencies from CT, DE, MD, MA, NH, NJ, NY, PA, RI, VT, VA, and WV. Jointly run by Cornell University and Penn State University.

Southeastern Cooperative Wildlife Disease Study
Department of Population Health
589 D.W. Brooks Drive
Wildlife Health Building
College of Veterinary Medicine
University of Georgia
Athens GA 30602
(706) 542-1741
Fax: (706) 542-5865
www.uga.edu/scwds/index2.htm

A cooperative effort of the wildlife agencies of 15 southeastern states and Puerto Rico, the Biological Resources Divisions of the U.S. Dept. of the Interior, and the Veterinary Services of USDA APHIS to provide diagnostic, research, educational, and consultation services about wildlife diseases. Publishes the Field Manual of Wildlife Diseases in the Southeastern United States *and a quarterly newsletter,* SCWDS Briefs.

BOOKS/MANUALS/ GUIDELINES

Available from Libraries and Bookstores

Field guides are an indispensable tool for WCOs. Select a few for whichever species you handle, such as field guides for mammals, birds, reptiles, amphibians, or animal tracks and signs. There are several excellent, well-known series such as those published by Peterson, Audubon, Stokes, and National Geographic. Pick your favorites. Here are a few lesser-known guides that may also interest you:

- *Amphibians and Reptiles of the Great Lakes Region.* J. H. Harding. 1997. University of Michigan Press.
- *Bird Tracks and Sign: A Guide to North American Species.* M. Elbroch, E. Marks, and D. C. Boretos. 2001. Stackpole Books.
- *Bull's Birds of New York State.* E. Levine, ed. Comstock.
- *Tracking and the Art of Seeing: How to Read Animal Tracks and Sign,* 2nd ed. P. Rezendes. 1999. Harper Resource.

Other Notable Books

The following should be available through bookstores and online sources.

- *America's Neighborhood Bats: Understanding and Learning to Live in Harmony with Them.* M. D. Tuttle. 1988. University of Texas Press.
- *Being Kind to Animal Pests: A No-Nonsense Guide to Humane Animal Control with Cage Traps.* S. Meyer.
- *Bird Hazards to Aircraft.* H. Blokpoel. 1976. Books Canada.
- *City Critters: How to Live with Urban Wildlife.* D. Bird. 1986. Eden Press.
- *Common-Sense Pest Control: Least-toxic Solutions for Your Home, Garden, Pets and Community.* W. Olkowski, S. Daar, and H. Olkowski. 1991. The Taunton Press.
- *Resolving Human-Wildlife Conflicts: The Science of Wildlife Damage Management.* M. Conover. 2002. Lewis Publishers, Inc.

From Cornell University

You can order the following from the Department of Natural Resources, Room 108, Fernow Hall, Ithaca NY 14850; (607) 255-2115; Email: dlt5@cornell.edu.

- *Human-Wildlife Conflict Management: A Practitioner's Guide.* D. Decker, T. B. Lauber, W. F. Siemer. 2002. Northeast

Wildlife Damage Management Research and Outreach Cooperative.

- *Managing Canada Geese in Urban Environments: A Technical Guide.* A. E. Smith, S. R. Craven, and P. D. Curtis. 1999. Jack H. Berryman Institute Publication 16, and Cornell University Cooperative Extension.
- *Managing Nuisance Beavers Along Roadsides: A Guide for Highway Departments.* P. G. Jensen, P. D. Curtis, and D. L. Hamelin. 1999. Cornell Cooperative Extension.
- *Managing White-Tailed Deer in Suburban Environments: A Technical Guide.* A. J. DeNicola, K. C. VerCauteren, P. D. Curtis, and S. E. Hygnstrom. 2000. Cornell Cooperative Extension, the Wildlife Society—Wildlife Damage Management Working Group, and the Northeast Wildlife Damage Research and Outreach Cooperative.
- *Community-Based Deer Management: A Practitioner's Guide.* D. Decker, D. B. Raik, W. S. Siemer. 2004. Northeast Wildlife Damage Management Research and Outreach Cooperative.

From The Wildlife Society

See contact information under "National Organizations."

- *A Manual of Wildlife Conservation.* R. D. Teague, ed. 1971. The Wildlife Society.
- *Research and Management Techniques for Wildlife and Habitats,* 5th ed. T. A. Bookhout, ed. 1994. The Wildlife Society.

From Other State Agencies and Organizations

- *Beaver Damage Control Techniques Manual.* 1997. New York State Department of Environmental Conservation. Ordering info:
  Division of Fish, Wildlife, and Marine Resources, Bureau of Wildlife
  625 Broadway
  Albany NY 12233-4750
  (518) 402-8919
  Fax: (518) 402-9027
  Email:
  fwwildlf@gw.dec.state.ny.us
  www.dec.state.ny.us/website.dfwmr/wildlife
- *Controlling Deer Damage.* F. R. Henderson and C. Lee. 1992. Kansas State University Cooperative Extension Service.
- *Field Guide to Wildlife Diseases: General Field Procedures and Diseases of Birds.* M. Friend, J. C. Franson, and E. A. Ciganovich, eds. 1999. U.S. Department of the Interior, U.S. Geological Survey, Biological Resources Division Information and Technology Report; 1999-001.
- *Field Manual of Wildlife Diseases in the Southeastern United States.* W. R. Davidson and V. F. Nettles. 1988. Southeastern Cooperative Wildlife Diseases Study. [To order, call the AAVIM at (800) 228-4689, fax (706) 742-7005, or write to AAVIM, 220 Smithonia Road, Winterville GA 30683]
- *A Homeowner's Guide to Northeastern Bats and Bat Problems.* L. Williams-Whitmer and M. Brittingham. 1995. State College: Penn State University Cooperative Extension.
- *NPCA Field Guide to Structural Pests.* To order, contact NPMA at (800) 678-6722.
- *NTA Trapping Handbook: A Guide for Better Trapping.* T. Krause. 1984. National Trappers Association, Bloomington, Illinois.
- *Pest Control Technician Safety Manual.* L. Pinto and S. Kraft. 2000. Pinto & Associates, Maryland.
- *Pocket Guide to the Humane Control of Wildlife in Cities and Towns.* G. R. Hodge, ed. 1991. Humane Society of United States.
- *Prevention and Control of Wildlife Damage.* S. E. Hygnstrom, R. M. Timm, G. E. Larson, eds. 1994. Cooperative Extension Division, Institute of Agriculture and Natural Resources, University of Nebraska. Available online at the Internet Center for Wildlife

Damage Management (http://icwdm.org) as a two-volume book, on CD, and free download; revised edition scheduled for publication in 2008.

- *Skunk Ecology and Damage Management Techniques for Homeowners.* R. Olson and A. M. Lewis. 1999. University of Wyoming Cooperative Extension.
- *Truman's Scientific Guide to Pest Control Operations,* 5th ed. G. W. Bennett, J. M. Owens, and R. M. Corrigan. 1997. Purdue University/Advanstar Communications.
- *Vertebrate Pest Handbook.* Austin M. Frishman. 1999. To order, contact Advanstar Communications, Customer Service, 131 West 1st Street, Duluth, MN 55802 or call (218) 740-7200.
- *Wild Neighbors: The Humane Approach to Living with Wildlife.* J. Hadidian, G. Hodge, and J. Grandy, eds. 1997. The Humane Society of the United States.
- *Wildlife Damage Inspection Handbook,* rev. ed. S. Vantassel and T. Olander. Wildlife Damage Control. Available in CD and spiral-bound formats from www.wildlifedamagecontrol.net/wdih.php.
- *Wildlife Pest Control Around Gardens and Homes.* T. P. Salmon and R. E. Lickliter. 1984. University of California Division of Agriculture and Natural Resources Publications. Contact: 6701 San Pablo Ave., Oakland, CA 94608; (800) 994-8849 or (510) 642-2431.
- *Wildlife Removal Handbook: A Guide for the Control and Capture of Wild Urban Animals.* S. Vantassel. 1994. Wildlife Damage Control. To order, contact http://www.wildlifecontrolsupplies.com/Merchant2/merchant.mvc?Screen=CTGY&Store_Code=WDC001&Category_Code=WDC00B3.

FACT SHEETS/BROCHURES

From Cornell University

Wildlife Damage Management Fact Sheet Series (primary authors, Paul Curtis and Kristi Sullivan) online at http://wildlifecontrol.info/ccewdmp/publications.htm.

- Bats
- Black Bears
- Crows
- Homes, Grounds, Gardens
- Moles
- Raccoons
- Reducing Deer-Vehicle Crashes
- Snakes
- Striped Skunks
- Tree Squirrels
- Voles
- White-tailed Deer
- Woodchucks
- Woodpeckers

The following brochures are online at http://www.nysipm.cornell.edu.

- *Evict and Exile Mice from Your Home.* J. Shultz, C. Koplinka-Loehr, K. Sharpe. 2000. New York State IPM Program.
- *IPM for Homes.* C. Koplinka-Loehr. 2000. New York State IPM Program.

From Pennsylvania State University

Wildlife Damage Control series (primary authors L. M. Williams-Whitmer and M. C. Brittingham):

- Black Bear
- Chipmunks
- Controlling Birds Around Farm Buildings
- Controlling Birds at Aquaculture Facilities
- Controlling Birds on Fruit Crops
- Cottontail Rabbits
- Geese, Ducks, and Swans
- Moles
- Rabies
- Voles

See also Integrated Pest Management for the Deer Tick. 1994.

From USDA APHIS-WS

The following fact sheets are available from http://www.aphis.usda.gov/ws/pubs.html.

- Urban and Suburban Coyotes
- Low-Powered Lasers: Another Nonlethal Tool for Resolving Wildlife Damage

- Livestock Guarding Dogs
- The Electronic Guard: A Tool in Predation Control
- WS Assistance with Waterfowl
- WS Assistance at Airports

From Other Agencies and Organizations

- Rabies Fact Sheet. 2000. New York State Department of Health. Publication 3010.
- "Solving Problems with Your Wild Neighbors" fact sheet series, Fund for Animals. See www.hsus.org/wildlife/urban_wildlife_our_wild_neighbors/solving_problems_with_your_wild_neighbors.

TECHNICAL ARTICLES/ REPORTS/BIBLIOGRAPHIES

- "2000 Report of the AVMA Panel on Euthanasia." B. V. Beaver et al. 2001. *Journal of the American Veterinary Medical Association* 218 (5): 669–696.
- Bat Rabies in New York State. 1999. New York State Department of Health.
- Bibliography of Cooperative Extension Service Literature on Wildlife, Fish, and Forest Resources. R. L. Ruff et al. 1993. University of Wisconsin.
- Controlling Animals that Cause Damage or Create a Nuisance. L. W. Adams. Urban Wildlife Resources.
- "Integrated Rat Management." L. Simon and W. Quarles. *Common Sense Pest Control Quarterly* vol. 12, no. 1, winter 1996.
- EPA study of secondary poisoning studies with eagles, owls, coyotes, dogs, etc. http://www.epa.gov/oppsrrd1/REDs/2100red.pdf.
- "Responses of woodchucks to potential garden crop repellents." R. K. Swihart and M. R. Conover. 1991. *Journal of Wildlife Management* 55: 177–181.
- "Stopping gophers and moles." W. Quarles. 2001. *Common-sense Pest Control Quarterly* 18 (4).
- "Toward a professional position on the translocation of problem wildlife." S. Craven, T. Barnes, and G. Kania. 1998. *Wildlife Society Bulletin* 26(1): 171–177.
- University of Connecticut Occupational Health and Safety Program for Animal Handlers. 4th revision, 2007. University of Connecticut. http://www.ehs.uconn.edu/Word%20Docs/Animal%20Handlers%20Program%20Rev%204.pdf
- Wildlife and Diseases: Biohazards Associated with Exposure to Animals. R. A. French. 1997. Northeastern Research Center for Wildlife Diseases, University of Connecticut Department of Pathobiology.
- Wildlife fertility control. K. A. Fagerstone, M. A. Coffey, P. D. Curtis, R. A. Dolbeer, G. J. Killian, L. A. Miller, and L. M. Wilmot. 2002. *Wildlife Society Technical Review* 02-2, 29 pp.

MAGAZINES/JOURNALS/ NEWSLETTERS

American Trapper
National Trappers Association (NTA)
524 5th Street
Bedford IN 47421
(812) 277-9670
Fax: (812) 277-9672
Email: ntaheadquarters@nationaltrappers.com
www.nationaltrappers.com

Fur-Fish-Game
2878 E. Main Street
Columbus OH 43209
(614) 231-9585
Fax: (614) 231-5735
ffgservice@ameritech.net

Fur Taker Magazine
Fur Takers of America
PO Box 18248
Louisville KY 40261
www.furtakersofamerica.com

Journal of Applied Ecology
Blackwell Science Ltd
Osney Mead
OXFORD
OX2 0EL
United Kingdom
Phone: +44 1865 206206
Fax: +44 1865 721205
Email: jnl.orders@blacksci.co.uk
www.blackwellpublishing.com/journal.asp?ref=0021-8901&site=1

Journal of Mammalogy (published by The American Society of Mammalogists)
Allen Marketing & Management
810 East 10th Street
PO Box 1897
Lawrence KS 66044

(785) 843-1235
Fax: (785) 843-1274
Email: asm@allenpress.com
www.mammalsociety.org/pubjom

Journal of Wildlife Management
(published by The Wildlife Society)
www.wildlife.org/publications/index.cfm?tname=journal

Pest Control Technology
4012 Bridge Avenue, Cleveland, OH 44113
(800) 456-0707
www.pctonline.com/magazine

Pest Control magazine
(888) 527-7008
www.pestcontrolmag.com

The Trapper and Predator Caller magazine
Circulation Dept. ABAU33
700 E. State Street
Iola WI 54990
(800) 258-0929
www.krause.com/static/outdoors.htm

Trapper's World
PO Box 96
Galloway OH 43119
(614) 878-6011

Wildlife Control Technology (W. C. T.) magazine
PO Box 480
Cortland IL 60112
www.wctech.com

WEBSITES

Animal Diversity Web, University of Michigan Museum of Zoology (an online encyclopedia and museum of animal natural history, distribution, classification, and more): http://animaldiversity.ummz.umich.edu

Bat Conservation International: www.batcon.org

The Berryman Institute for Wildlife Damage Management: www.berrymaninstitute.org

Best Management Practices for Trapping Coyotes in the Eastern United States: http://www.pgc.state.pa.us/pgc/lib/pgc/trapping/eastern_coyote.pdf

Centers for Disease Control and Prevention: www.cdc.gov

- Hantavirus: www.cdc.gov/ncidod/diseases/hanta/hps/index.htm
- Histoplasmosis: www.cdc.gov/ncidod/dbmd/diseaseinfo/histoplasmosis_g.htm
- LCM virus: www.cdc.gov/ncidod/dvrd/spb/mnpages/dispages/arena.htm
- Lyme disease: www.cdc.gov/ncidod/dvbid/lyme/index.htm
- Rabies: www.cdc.gov/ncidod/dvrd/rabies
- Raccoon roundworm: http://www.cdc.gov/ncidod/dpd/parasites/baylisascaris/factsht_baylisascaris.htm
- Toxoplasmosis: www.cdc.gov/ncidod/dpd/parasites/toxoplasmosis/factsht_toxoplasmosis.htm
- West Nile Virus: www.cdc. gov/ncidod/dvbid/westnile/index.htm

Dobbins' Products (includes many videos, publications, trapping tips, and a way to find trapping mentors): www.trapperman.com

International Hunter Education Association: www.ihea.com

Internet Center for Wildlife Damage Management: http://icwdm.org

Michigan Wildlife Diseases: http://www.michigan.gov/dnr/0,1607,7-153-10370_12150_12220—-,00.html

National Wildlife Health Center: www.nwhc.usgs.gov

- Emerging Diseases in Wildlife: http://www.nwhc.usgs.gov/our_research/invasive_pathogens_and_emerging_diseases_of_wildlife_in_the_united_states.jsp
- Field Manual of Wildlife Diseases: http://www.nwhc.usgs.gov/publications/field_manual/

National Wildlife Research Center (USDA APHIS-Wildlife Services): www.aphis.usda.gov/ws/nwrc

Nebraska Cooperative Extension: www.ianr.unl.edu/pubs/wildlife/index.htm

NYS Integrated Pest Management (IPM) Program: http://nysipm.cornell.edu

Penn State University Cooperative Extension: www.extension.psu.edu

USDA APHIS-Wildlife Services: www.aphis.usda.gov/wildlife_damage

University of Kentucky Cooperative Extension Service: www.ca.uky.edu/ces

Utah State University Publications: http://extension.usu.edu/

htm/publications/by=category/category=63

Wadsworth Center Rabies Laboratory: www.wadsworth.org/rabies

Wildlife Damage Management Program (Cornell): www.dnr.cornell.edu/ext/wildlifedamage and http://wildlifecontrol.info/

LISTSERVES

National Wildlife Control Operators Association listserves:

- NWCOAlistserve@yahoogroups.com, open to the public. Intended as a networking opportunity for wildlife control operators.
- NWCOA.NET, members-only listserve. For information, see their website at www.nwcoa.net.

Sullivan's Trap Line
A trapper's forum. Go to their website (www.sullivansline.com/tline/tlhome.htm) and follow the directions to subscribe.

Wildlife Health listserve
Run by the National Wildlife Health Center (NWHC), a Science Center of the Biological Resources Division of the U.S. Geological Survey, this list focuses on wildlife health issues. To subscribe, send an email message to listserver@relay.doit.wisc.edu with the following in the body of the message: SUBSCRIBE WildlifeHealth firstname lastname. Substitute your own first and last names in the spaces indicated. For more information about the list, contact Joshua Dein, Joshua_Dein@nbs.gov.

Wildlife Society listserves (several). Go to their website and follow the directions to subscribe. http://www.wildlife.org/professional/index.cfm?tname=listserv

VIDEOS

- *Bat-free Belfries: A Guide to Bat-Proofing.* L. M. Williams-Whitmer and M. C. Brittingham. 1995. Pennsylvania State University.
- *The Facts About Bats: Exploring Conflicts and Designing Solutions* (29 min.). 2002. Cornell Cooperative Extension.
- *Suburban Deer Management: Voices, Views, Visions* (28 min.). 1993. Cornell Cooperative Extension. 147VSDM.
- *Suburban Goose Management: Searching for Balance* (28 min.). 1998. Cornell Cooperative Extension. 147VSGM. See complementary manual, Managing Canada Geese in Urban Environments.
- *Summertime Beaver Control* (1 hr. 37 min.). Charles and Paul Dobbins. Dobbins' Products.
- *Whitetails at the Crossroads* (30 min.). 1996. Cornell Cooperative Extension. 147VWC.

COURSES/WORKSHOPS/CONFERENCES

See the "Professional's Corner" of the Internet Center for Wildlife Damage Management for recent listings: http://icwdm.org

Wildlife Damage Management Conference
Offered in the spring of years ending in an odd number, such as 2007, 2009. Contact the Wildlife Society Wildlife Damage Management Working Group for more information.
http://icwdm.org

Vertebrate Pest Conference
Offered in the spring of years ending in an even number (2008, 2010), this conference is hosted by a different university each time. To receive info, fill out the mailing list form on this website: www.vpconference.org

Northeast Fish and Wildlife Conference
Rotates around the 13 northeastern states, usually hosted by that state's wildlife agency.

WCT/NWCOA National Conference
An annual conference specifically for the private wildlife control industry.

National Goose Control Academy
An annual two-day workshop that includes classroom and field training. Contact Kirk La Pierre, (201) 933-9700 or (877) 667-6674, kirk@goosecontrol.com or Tim Julien, (317) 895-9069, tjulien@iquest.com.

National Wildlife Damage Management Academy
An annual course offered by NWCOA.

NACA Training Academy
Provides three levels of certification for animal control professionals. See www.nacanet.org/training.htm.

Trapper's College and Furbearer Management Short Course
Offered by Fur Takers of America (www.furtakersofamerica.com/college.htm).
Utah State University
WILD 4810, Directed Reading: Wildlife Damage management, 2 credits.
http://distance.usu.edu/htm/howto
USU Distance Education Office
(800) 233-2137
Email: distance.info@usu.edu

GLOSSARY

B

banger also known as a bird bomb; an explosive cartridge that makes a loud bang to repel birds. The shell is often launched from a handheld pistol launcher, similar to a modified starter pistol. This is one of many "pyrotechnic" or fireworks-based repellents. Bangers can be used at medium range (50–100 feet).

barbiturate a group of drugs (for example, sodium pentobarbital) that sedate and can kill animals by interfering with the central nervous system. Their use is heavily restricted at federal and state levels.

best practice an effective method for solving a nuisance wildlife problem that also minimizes risks to the environment and our health and well-being. Best practices support the problem-solving strategy known as "integrated wildlife damage management," which balances concerns about safety; the humane treatment of wildlife; practicality; landowner rights; the protection of wildlife populations and habitats; and ethical, legal, financial, and aesthetic issues.

bird bomb see *banger*.

body-gripping trap usually a square, spring-loaded, lethal trap available in many sizes (some specialized models are round). The animal must pass through the jaws of the trap to be captured. Ideally, this trap catches the animal directly behind the head, snapping the part of the spine that's in the upper third of the neck (the cervical spine area). This provides a quick death. Body-gripping traps are often called "Conibears," which is the name of a popular model manufactured by Woodstream, but there are many manufacturers of other body-gripping traps; the traditional mousetrap is a body-gripping trap, for example.

C

cable restraint a.k.a. a relaxing snare; a looped cable designed to capture and restrain wild animals, including coyotes, foxes, bears, and beavers. Similar to a catchpole without the pole, cable restraints are made of galvanized metal cable. They generally feature a lock, multiple swivels, loop stops, springs, and a breakaway device. This live capture device is suspended along the animal's travel path. When foxes and coyotes walk into the loop, they'll continue forward and try to push their way through it, as they would with a natural obstacle such as brush. Their movement causes the loop to tighten around the neck. The fox or coyote will then try to back out again, as if it has encountered a natural obstacle, but will stop as soon as it reaches the end of the cable. The lock relaxes slightly to reduce pressure on the animal's neck without giving it the chance to escape.

carbon dioxide (CO_2) chamber an enclosed space into which carbon dioxide gas is added at a controlled rate to kill an animal. Carbon dioxide gas causes rapid unconsciousness followed by death, usually within five minutes, and is one of the most practical humane killing techniques available to WCOs.

carnivore meat-eater. Identified by its sharp canine teeth, the pointed, conical teeth located between the incisors and the first bicuspids.

carrying capacity Biological carrying capacity is the maximum population of a given species that can be supported by a given habitat. This ever-fluctuating quantity is affected by many factors such as the food supply; the species' ability to tolerate neighbors, both wild and human; and the season. Some areas can support more animals during the summer, when food and shelter are readily available, than they can during the winter, when those resources are scarce. Social carrying capacity is the maximum number of animals people can tolerate in a given habitat or location.

catchpole a.k.a. a snare pole; a device used to capture and restrain an animal. Basically, this is a long stick with a noose (cabled loop) on one end. You place the loop over the animal's head, and then tighten the cable to hold the animal. When capturing some species, such as bobcat, you'd place the loop over the animal's head and over one front leg.

cervical dislocation a killing method commonly referred to as "breaking the neck" that would more accurately be described as "snapping the spine." The goal is to quickly separate the spinal cord from the brain to provide a fast and painless death. The separation must take place at the base of the brain or within the upper third of the neck (the cervical spine area). It's used primarily for small- to medium-sized birds (duck-sized or smaller) and small mammals, such as mice and rabbits.

cracker see *shell-cracker.*

crepuscular animals that are more active at dawn or twilight than at other times.

D

decapitation a killing method in which the head is quickly cut from the body. Like cervical dislocation, the goal is to quickly separate the spinal cord from the brain to provide a fast and painless death. It's used primarily for birds that are too large for cervical dislocation, such as geese, and sometimes for snakes.

depredation permit a special permit issued by state and federal government wildlife agencies that allows the use of certain wildlife control techniques on federally- or state-protected wildlife or game species. Without the permits, these techniques (usually taking or killing an animal or its eggs and nest) would be illegal because of the special protection given to those species.

diurnal animals that are more active during the daytime.

E

edge the zone between two or more adjacent habitats. Edges often have relatively abundant or widely varying sources of food and cover.

euthanasia the word means "good death" and it describes a humane killing method that provides as painless a death as possible, specifically, by killing the animal quickly, or causing rapid unconsciousness, then rapid death.

exclusion techniques and products that prevent wildlife from entering an area, such as protective barrier fences. Barriers do not change the nature of the site; they work by making the target inaccessible. There are many materials and designs for fences, including the popular wire mesh and electric fences. Birds are sometimes repelled by a grid of wires, called "post-and-wire grids."

exsanguination known commonly as "bleeding out," this describes the cutting of the major blood vessels to rapidly drain blood from the body. This technique ensures death after an animal has been stunned into unconsciousness.

F

feral describes a member of a domesticated species, such as a cat or dog, that lives and behaves like a wild animal.

firearm a device that fires a projectile, using an explosive charge, force of spring, air, or other gas as a propellant. This includes pistols, shotguns, rifles, handguns, and air rifles (pellet guns).

fledge when a young bird leaves the nest. These young birds, called fledglings, can fly but are often still dependent on their parents for some, or all, of their food.

foothold trap refers to a variety of traps that restrain animals by holding the foot. They may be used as live or lethal traps. Examples of foothold traps include the coilspring trap and the Lil' Grizz Get'rz cylindrical trap, designed to selectively capture raccoons.

G

gestation the length of a pregnancy—the amount of time it takes from conception until birth.

H

herbivore plant-eater. Identified by its flat molars, which are used to grind plants. Molars are found in the back of the jaw.

home range the area in which an animal (or a mated pair) lives, hunts, and mates throughout its life. Generally, this area is not defended. The animal doesn't necessarily use its whole home range every day. The home ranges of males and females of the same species may overlap, and a male's home range may include that of several females. Many factors influence the size, shape, and use of a home range, such as the gender and age of the animal, the season, the amount of food available, the nearness of mates, and the degree of fighting among its kind. The boundaries of a home range are often marked by scats, urine, or other signs.

host an animal or plant that provides nourishment or a protected home for another animal, such as a parasite, or a disease organism.

humane a practice or product that causes no unnecessary pain or stress for the animal to whom it's applied.

I

integrated wildlife damage management (IWDM) a strategy for solving conflicts between humans and wildlife while reducing risks to people, wildlife, and the environment. IWDM focuses on reducing wildlife damage—not necessarily wildlife populations. It's a holistic approach that helps the operator decide if control is needed, when and where it should be applied, and which control methods would likely work best (usually a combination of best practices). IWDM balances concerns about human safety, the humane treatment of wildlife, the rights of landowners, public health, the protection of the environment, financial and aesthetic losses, societal values, and legal and practical limitations. Although there's a lot to it, IWDM can be remembered with three words: *effective, selective, humane.*

L

limiting factor anything that puts limits on the growth of a species' population; for example, bluebird populations are limited by the availability of nest cavities.

listserv, listserve a computerized "bulletin board." People subscribe to a listserve, which usually focuses on a subject, such as wildlife control. There are two types: open listserves and moderated ones. In the open ones, when someone posts a message to the "bulletin board" (the listserv) it is immediately sent to everyone, often as an email message. Others can then respond to the message. If there's a moderator, messages are sent to that person, who then decides whether or not to post them for the whole group and if they need editing.

live trap a trap that's meant to capture the animal without killing it. Includes a variety of traps, such as cage (box) traps, foothold traps, nets, multiple capture traps, cable restraints, and many bird traps. These traps are often made of wire or plastic and come in many sizes. Cage traps are sometimes called "Havaharts" after a popular model, but there are many types and many manufacturers. See Chapter Five for details.

louver a framed ventilation opening that's covered with horizontal slats, usually located in an attic wall.

N

nestling a young bird that's in the nest, under the care of adults, usually its parents. Birds are nestlings after they hatch, for as long as they remain in the nest. Once they can fly and leave the nest, they're called fledglings for as long as their parents care for them.

niche this describes both the animal's job and the part of the environment that it uses (the area and the time when the animal is active). So, for example, bats and chimney swifts have the same specialized job: insect-eaters that feed on the wing. They may even sleep in the same chimney, but bats are active at night, while chimney swifts are active during the day. They occupy similar, but not exactly the same, niches.

nocturnal animals that are generally more active at night.

O

omnivore an animal that eats both meat and plants. Omnivores have both canines and molars, but neither tooth is as specialized as those found in carnivores or herbivores. An omnivore can eat meat but it can't rip flesh the way a carnivore can, nor can it grind plants with an herbivore's ease. Some animals classified as carnivores or herbivores by their teeth are actually omnivorous. For example, foxes (a carnivore) readily eat large amounts of berries, while the herbivorous squirrels eat insects, baby birds, birds' eggs, and even each other—and not only when the food supply is tight.

opportunist an animal that will eat whatever's available, within reason. Animals that are opportunists are more likely to become nuisances because they adapt better to human conditions.

P

penetrating captive bolt also known as a captive bolt pistol; a killing device that forces a bolt into the brain of the animal, causing rapid unconsciousness and death. Captive bolt guns use either gunpowder or compressed air to propel the bolt through the skull. The animal must be adequately restrained to allow proper placement of the bolt.

pesticide any substance designed to prevent, destroy, repel, or mitigate any pest. (Nuisance wildlife are considered "pests.") Repellents, fumigants, rodenticides, and poison baits are all examples of pesticides. Pesticides are regulated by the U.S. Environmental Protection Agency and the state, following the Federal Insecticide, Fungicide, and Rodenticide Act (FIFRA) and state laws. All pesticides are registered as either "general use" or "restricted use" products. If a product has an EPA registration number on its label, it's a restricted use pesticide. Restricted use pesticides may be used only by, or under the direct supervision of, a certified pesticide applicator. Contact your state's pesticide regulatory agency for details.

polygamous mating with more than one individual. Commonly used to describe a situation in which one male mates with several females.

porcupine wire a mechanical device intended to repel birds from landing on surfaces such as building ledges. It's made of sharp, stainless steel prongs that stick out in many angles. The prongs are attached to a base for easy installation on ledges, roof peaks, windowsills, and ornamental architectural features.

propane cannon a mechanical device that makes a loud explosion to repel birds. This is one of many "pyrotechnic" or fireworks-based repellents. Unlike many other noisemakers, propane cannons can be used with a timer. Most of the other devices must be fired by a person.

pyrotechnics fireworks. This term is used loosely to refer to repellents that use explosive charges that are like fireworks, such as screamers. Most of these devices are used in bird control.

R

repellent an object, substance, or technique that repels, or drives an animal away. There are several kinds of repellents: scare devices; chemical repellents; and guard animals. Most repellents frighten an animal away from a site or make the desired object (such as a crop or nesting site) intolerable by making it smell or taste nasty, or feel bad to the touch. Animals may be "hazed," or driven away from a site when chased by predators (dogs, falcons) or vehicles (radio-controlled planes, boats). Scare devices include Mylar tape, strobe lights, models of predators, and "scare-eye" balloons; and noisemakers, such as propane cannons, bangers, clappers, crackers, and distress calls.

S

screamer an explosive cartridge that makes a long, drawn out whistle to repel birds. They can be launched from a handheld pistol launcher or from a 12-gauge shotgun. The shell flies out about 100 feet, screaming and whistling all the way. This is one of many "pyrotechnic" or fireworks-based repellents.

shell-cracker also known as a cracker; an explosive cartridge that makes a loud bang that sounds like an M-80 firecracker to repel birds. These shells are fired from a 12-gauge shotgun, exploding about 75 yards away. This is one of many "pyrotechnic" or fireworks-based repellents.

soffit the section of a building underneath the eaves.

stunning a.k.a. a "blow to the head" or "lethal blow"; may be used for two different purposes: it may be intended to make the animal unconscious so another killing method may be used safely (called "stunning"), or it may be intended as a primary killing method, in which case it's usually referred to as a "lethal blow." A lethal blow is a quick, very forceful blow to the back of the head that is meant to kill the animal. It's appropriate for small animals, such as birds, rabbits, and small mammals (mice to squirrel size). Stunning is a less forceful blow delivered to the center of the head.

T

taking and take Check your state laws for the precise definition of this term. In general, to "take" means to kill wildlife. In some states, "taking" also refers to pursuing, capturing, or placing a device intended to capture the animal. It may even cover actions that disturb or harass wildlife.

territory an exclusive area that is often vigorously defended for a certain time. May be inhabited by an individual, a mated pair, or a group of animals. Territories may be defined by the way they're used; for example, a breeding territory, feeding territory, or nesting territory. An animal defends its entire territory, which is usually smaller than its home range. The defense of the territory may involve ritualistic behaviors that drive off rivals, or actual fighting.

trapping Check your state laws for the precise definition of this term. In general, to "trap" means taking, killing, and capturing wildlife with traps, cable restraints, and other devices commonly used to take wildlife. It may include the acts of setting these devices and the killing of wildlife that have been trapped.

V

vector also known as a carrier; an animal that can carry and transmit a disease agent from one animal to another. For example, mammals are vectors, or carriers, of rabies.

Z

zoonotic, zoonoses pronounced "zu-not-tick" and "zu-nos-ees"; diseases that people can catch from wildlife.

INDEX